CRIMINOLOGICAL THEORY

Titles of Related Interest

Tibbetts and Hemmens: *Criminological Theory: A Text/Reader*
Lawrence and Hemmens: *Juvenile Justice: A Text/Reader*
Stohr, Walsh, and Hemmens: *Corrections: A Text/Reader*
Spohn and Hemmens: *Courts: A Text/Reader*
Lilly, Cullen, and Ball: *Criminological Theory* (4th edition)
Wright, Tibbetts, and Daigle: *Criminals in the Making*
Walsh and Ellis: *Criminology: An Interdisciplinary Perspective*
Hagan: *Introduction to Criminology* (6th edition)
Crutchfield, Kubrin, Weiss, and Bridges: *Crime: Readings* (3rd edition)
Van Dijk: *The World of Crime*
Felson: *Crime and Everyday Life* (4th edition)
Hagan: *Crime Types and Criminals*
Hanser: *Community Corrections*
Mosher and Akins: *Drugs and Drug Policy*
Bachman and Schutt: *The Practice of Research in Criminology and Criminal Justice* (3rd edition)
Bachman and Schutt: *Fundamentals of Research in Criminology and Criminal Justice*
Loggio and Dowdall: *Adventures in Criminal Justice Research* (4th edition)
Gabbidon and Greene: *Race and Crime* (2nd edition)
Helfgott: *Criminal Behavior*
Bartol and Bartol: *Forensic Psychology* (2nd edition)
Banks: *Criminal Justice Ethics* (2nd edition)
Lippman: *Contemporary Criminal Law* (2nd edition)
Lippman: *Criminal Procedure*
Hemmens, Brody, and Spohn: *Criminal Courts*
Holmes and Holmes: *Profiling Violent Crime* (4th edition)
Holmes and Holmes: *Sex Crimes* (3rd edition)
Alvarez and Bachman: *Violence*
Howell: *Preventing and Reducing Juvenile Delinquency* (2nd edition)
Cox, Allan, Hanser, and Conrad: *Juvenile Justice* (6th edition)
Mahan and Griset: *Terrorism in Perspective* (2nd edition)
Martin: *Understanding Terrorism* (3rd edition)
Martin: *Essentials of Terrorism*

CRIMINOLOGICAL THEORY

STEPHEN G. TIBBETTS
California State University, San Bernardino

Los Angeles | London | New Delhi
Singapore | Washington DC

Los Angeles | London | New Delhi
Singapore | Washington DC

FOR INFORMATION:

SAGE Publications, Inc.
2455 Teller Road
Thousand Oaks, California 91320
E-mail: order@sagepub.com

SAGE Publications Ltd.
1 Oliver's Yard
55 City Road
London EC1Y 1SP
United Kingdom

SAGE Publications India Pvt. Ltd.
B 1/I 1 Mohan Cooperative Industrial Area
Mathura Road, New Delhi 110 044
India

SAGE Publications Asia-Pacific Pte. Ltd.
33 Pekin Street #02-01
Far East Square
Singapore 048763

Acquisitions Editor: Jerry Westby
Editorial Assistant: Erim Sarbuland
Production Editor: Catherine M. Chilton
Copy Editor: Pam Schroeder
Typesetter: C&M Digitals (P) Ltd.
Proofreader: Annette Van Deusen
Indexer: Molly Hall
Cover Designer: Gail Buschman
Marketing Manager: Erica DeLuca
Permissions: Karen Ehrmann

Printed in the United States of America

Library of Congress Cataloging-in-Publication Data

Tibbetts, Stephen G.

Criminological theory : the essentials/Stephen G. Tibbetts.

p. cm.
Includes bibliographical references and index.

ISBN 978-1-4129-9234-3 (pbk.)

1. Criminology. I. Title.

HV6025.T513 2012 364.01—dc22 2010044629

This book is printed on acid-free paper.

12 13 14 15 10 9 8 7 6 5 4 3 2

Brief Contents

Detailed Contents

Preface

This textbook is intended to cover the essential topics of criminological theory in a more brief and efficient manner than other larger, more comprehensive texts on this topic. *Criminological Theories: The Essentials* presents a comprehensive overview of the major concepts and perspectives of virtually all major theories in the evolution of criminological theory, reviewing some of the most recent empirical research on each theory that is currently available. Furthermore, in each chapter, as well as an entire concluding chapter, this book examines the various policy implications that can be derived from each type of criminological theory, as well as what can possibly be done but has not yet been tested.

A number of excellent criminology theory textbooks are available to students and professors, so why this one? This book can serve as the primary text for an introductory undergraduate course in criminological theory or as the primary text for a graduate course, given the depth and comprehensive nature of the discussion of virtually all theories in the historic and modern criminological literature. It is important to note that the book provides a comprehensive, yet concise, survey of the current state of existing scientific literature in virtually all areas of criminological theory, as well as giving a history of how we got to this point regarding each theoretical model and topic area. A key feature of this text is an added section for each chapter that examines various policy implications that have resulted from most of the dominant theories in the discipline, as well as results from empirical evaluation studies of programs based on theories presented in each section.

Structure of the Book

This book uses a rather typical outline for a criminological theory textbook, beginning with an introduction of the definitions of crime and criminology and measuring crime, as well as what such measures of crime reveal regarding the various characteristics that are most associated with higher offending rates. This is a very important aspect of the book because each theory or model must be judged by how well it explains the distribution of crime rates among these various characteristics. In the Introduction, the criteria that are required for determining causality are also discussed, including an examination of how extremely difficult (often impossible) this is to do in criminological research because we can't randomly assign individuals to bad parenting, unemployment, low IQ, and so on.

This book presents 12 chapters that chronologically trace the history and development of criminological theory with an emphasis on when such perspectives became popular among theorists and mainstream society. Thus, we will start with the earliest models (Preclassical and Classical School) of criminal theorizing in the 18th century. Then, we will examine the evolution of the Positive School perspective of the 19th century, which began with biological theories of crime. Next, the book will present the various other positive theories that were proposed in the early 20th century, which include social structure models and social process theories that were presented in the early or mid-1900s. Then, we will explore theoretical models that were presented in the latter 20th century, such as social conflict and Marxist and feminist models of criminality. Chapter 10 then presents the more contemporary theoretical explanations of criminality, which include developmental and life-course models and integrated theories of crime. Finally, although policy implications are discussed at the end of each chapter introduction, we finish with a summary chapter that specifies the types of policies suggested by the various paradigms and recent scientific findings regarding each of the major theoretical models presented throughout the book.

This book is divided into 12 chapters that mirror the sections in a typical criminology textbook, each dealing with a particular type or category of theories in criminology. Thus, each of the chapters concludes with an evaluation of the empirical support for the theories and policy implications derivable. These sections are as follows:

1. Introduction to the Book: An Overview of Issues in Criminological Theory

First, this book provides an introductory chapter dealing with what criminological theory is, as well as examining the concepts of crime and the criteria used to determine whether a theory is adequate for explaining behavior. This section introduces the facts and risk factors by which all of the theoretical models presented in the following sections will be evaluated. Also included is a discussion of the criteria involved in determining whether a given factor or variable actually causes criminal behavior.

2. Preclassical and Classical Theories of Crime

In this section, we will examine the types of theories that were dominant before logical theories of crime were presented, namely supernatural or demonic theories of crime. Then, we will examine how the Age of Enlightenment led to more rational approaches for explaining criminal behavior, such as that of the Classical School and neoclassical theory. We will also explore the major model that evolved from the Classical School, deterrence theory, as well as policy implications and applications that stem from this theory.

3. Modern Applications of the Classical Perspective: Deterrence, Rational Choice, and Routine Activities or Lifestyle Theories of Crime

In this section, we will review more contemporary theoretical models and empirical findings regarding explanations of crime, which focus on deterrence, and other recent perspectives—such as rational choice theory, routine activities theory, and the lifestyle perspective—which are based on the assumption of individuals choosing their behavior or targets based on rational decisions. Some of these perspectives focus more on the individual's choice of the perceived costs or benefits of a given act, whereas other models focus on the type of location chosen to commit crime or daily activities or lifestyles that predispose a person to certain criminal behavior. This section concludes with a discussion of policy implications that relate to this theory.

4. Early Positive School Perspectives of Criminality

This section will examine the early development of theoretical models that proposed that certain individuals or groups were predisposed to criminal offending. The earliest theories in the 19th century proposed that certain physical traits were associated with criminal behavior, whereas perspectives in the early 20th century proposed that such criminality was due to intelligence. This section also examines body type theory, which proposes that the physical body type of an individual has an effect on criminality. This section will also examine modern applications of this perspective and review the empirical support such theoretical models have received in modern times.

5. Modern Biosocial Perspectives of Criminal Behavior

In this section, we will review modern studies regarding the link between physiology and criminality. Such theoretical models include family studies, twin and adoption studies, cytogenetic studies, and studies on hormones and neurotransmitters. We will examine some of the primary methods used to examine this link as well as more rational and recent empirical studies, which show a relatively consistent link between physiological factors and criminal behavior.

6. Early Social Structure and Strain Theories of Crime

This section reviews the development of the social structure perspective starting in the 19th century and culminating with Merton's theory of strain in the early 20th century. A variety of perspectives based on Merton's strain theory will be examined, but all of these models have a primary emphasis on how social structure produces criminal behavior. We will examine some empirical studies that have tested the validity of these early social structure theories, as well as discussing policy implications that these models suggest.

7. The Chicago School and Cultural and Subcultural Theories of Crime

In this section, we examine the evolution and propositions of the scholars at the University of Chicago, who produced the most advanced form of criminological theorizing in the early 20th century. In addition to the evolution of the Chicago School and its application of ecological theory to criminal behavior, we will also examine the more modern applications of this theoretical framework for explaining criminal behavior among residents of certain neighborhoods. We will discuss several theoretical models that examine cultural or subcultural groups that differ drastically from conventional norms. Finally, we will examine the various policy implications and policies that can be derived from the theoretical perspectives presented in this section.

8. Social Process and Learning Theories of Crime

This section examines the many perspectives that have proposed that criminal behavior is the result of being taught by significant others to commit crime. When it was first presented, it was considered quite novel. We will examine the evolution of various theories of social learning, starting with the earliest, which were based on somewhat outdated forms of learning theory, and then, we will progress to more modern theories that incorporate contemporary learning models. We will also examine the most recent versions of this type of theoretical perspective, which incorporate all forms of social learning in explaining criminal behavior. Finally, we will discuss the various policies that have been based on the social learning models presented in this section.

9. Social Reaction, Critical, and Feminist Models of Crime

In this section, we examine a large range of theories with the common assumption that the reasons for criminal behavior lie outside the traditional criminal justice system. Many social reaction theories, for example, are based on labeling theory, which proposes that it is not the individual offender who is to blame but rather the societal reaction to early antisocial behavior. Furthermore, this section will examine the critical perspective, which blames the existing legal and economical structure for most of the assigned "criminal" labels that are used against most offenders. Also, we will discuss the major perspectives of criminal offending regarding females as compared to males, as well as their differential treatment by the formal criminal justice system. Finally, we look at how explaining low levels of female offending might be important for policies regarding males, as well as policy implications regarding the other theoretical perspectives presented in this section.

10. Life-Course Perspectives of Criminality

This section will examine the various theoretical perspectives that emphasize the predisposition and influences that are present among individuals who begin committing crime at early ages versus later ages. We also examine the various stages of life that tend to have a high influence on an individual's state of criminality (e.g., marriage), as well as the empirical studies that have examined these types of transitions in life. Finally, we examine the various types of offenders and the types of transitions and trajectories that tend to influence their future behavior, along with various policy implications that can be suggested by such models of criminality.

11. Integrated Theoretical Models and New Perspectives of Crime

In this section, we present the general theoretical framework for integrated models. Then, we introduce some criticisms of this integration of traditional theoretical models. In addition, we present several integrated models of criminality, some of which are based on microlevel factors and others that are based on macrolevel factors. Finally, we examine the weaknesses and strengths of these various models based on empirical studies that have tested their validity, as well as some policy implications that can be derived from such integrated theories of criminality.

12. Applying Criminological Theory to Policy

This final section will review the most recent empirical evidence regarding how various theoretical models and findings reviewed in this book can be used (or have been used) to inform policies to reduce criminal behavior among offenders. These studies show that many theoretical perspectives suggest some effective policy recommendations, whereas other theoretical frameworks have been shown to be less effective. It should be noted that there are many more modern policy applications than the theories presented in this book, but this section emphasizes the most notable. Furthermore, there are many implications that have not yet been attempted but likely will be based on the policies attributed to theories we cover in this book. Whether they will be successful is an issue to be covered by empirical evaluation studies and will warrant another book entirely.

Instructor Teaching Website. A variety of instructor's materials are available. For each chapter, this includes summaries, PowerPoint slides, chapter activities, web resources, and a complete set of test questions.

Student Study Site. This comprehensive student study site features chapter outlines students can print for class; flash cards; self-quizzes; web exercises; links to journal articles, online videos, and NPR program archives; and more.

⬚ Acknowledgments

I would first of all like to thank Executive Editor Jerry Westby. Jerry's faith in and commitment to this and previous projects are greatly appreciated. I would also like to acknowledge the efforts of his staff, including Editorial Assistant Erim Sarbuland. They kept up a most useful dialogue among author, publisher, and a parade of excellent reviewers, making this text the best that it could possibly be. Our copy editor at SAGE, Pam Schroeder, also did a truly fantastic job; it was a pleasure working with her. Thank you one and all.

I also would like to thank the various individuals who helped me complete this book. First and foremost, I would like to thank Craig Hemmens at Boise State University for the immense amount of work and effort he put into an earlier draft of this book; without his contributions, this book would not exist. Also, I would like to thank the professors I had as an undergraduate at the University of Florida, who first exposed me to criminological theory. These professors include Ronald Akers and Lonn Lanza-Kaduce, with a special thanks to Donna Bishop, who was the instructor in my first criminological theory course. I would also like to thank the influential professors I had at the University of Maryland, including Denise Gottfredson, Colin Loftin, David McDowell, Lawrence Sherman, and Charles Wellford. I would also like to give a very special acknowledgement to Raymond Paternoster, who was my primary mentor and adviser and exerted an influence words can't describe. Paternoster introduced me to a passion for criminological theory that I hope is reflected in this book.

I would like to thank Alex Piquero, with whom I had the great luck of sharing a graduate student office in the mid-1990s and who is now a professor at Florida State University. Without the many collaborations and discussions about theory that I had with Alex, this book would be quite different. Alex and his wife, Nicole Leeper Piquero, have been consistent key influences on my perspective and understanding of theories of crimes, especially contemporary perspectives (and both of the Piqueros' works are extensively cited in this book).

In addition, I would like to thank several colleagues who have helped me subsequent to my education. First, I would like to thank John Paul Wright at University of Cincinnati, Nicole H. Rafter at Northeastern University, Kevin Beaver at Florida State University, and Matt DeLisi at Iowa State University for inspiring me to further explore more biosocial and developmental areas of criminality. Also, I would like to thank Bryan McKeon and Logan Brown at California State University, San Bernardino (CSUSB), who provided much help in finding some grammatical and content errors in earlier drafts of this book. In addition, I would like to especially thank Pamela Schram and Larry Gaines, fellow professors at CSUSB, who have provided the highest possible level of support and guidance during my career. Furthermore, it should be noted that Pamela Schram provided key insights and materials that aided in the writing of several sections of this book.

On a more personal note, I would like to thank Juanita Maria Orona Tucker and Uriel Garcia Ramos, in Redlands, California, for all of the moral support they have given me and my family over the last decade. Their immense amount of assistance and friendship was invaluable in enabling me to be able to write this book, as well as engage in many other professional activities that I could have never done without them being part of my life.

Finally, I owe the most gratitude to my wife, Kim, who patiently put up with me typing away for the past few years while working on this book. Her constant support and companionship are what keep me going and inspire me on a day-to-day basis.

I am also very grateful to the many reviewers who spent considerable time reading early drafts of this work and who provided helpful suggestions for improving the content and presentation of the material. Trying to please so many individuals is a challenge but one that is ultimately satisfying and undoubtedly made the book better than it would otherwise have been. I offer heartfelt thanks to the following experts: Shannon Barton-Bellessa, Indiana State University; Michael L. Benson, University of Cincinnati; Robert Brame, University of South Carolina; Tammy Castle, University of West Florida; James Chriss, Cleveland State University; Toni DuPont-Morales, Pennsylvania State University; Joshua D. Freilich, John Jay College; Randy Gainey, Old Dominion University; Robert Hanser, Kaplan University; Heath Hoffman, College of Charleston; Thomas Holt, University of North Carolina, Charlotte; Rebecca Katz, Morehead State University; Dennis Longmire, Sam Houston State University; Gina Luby, DePaul University; Michael J. Lynch, University of South Florida; Michelle Hughes Miller, Southern Illinois University, Carbondale; J. Mitchell Miller, University of Texas, San Antonio; Travis Pratt, Arizona State University; Lois Presser, University of Tennessee; Robert Sarver, University of Texas, Arlington; Joseph Scimecca, George Mason University; Martin S. Schwartz, Ohio University; Ira Sommers, California State University, Los Angeles; Amy Thistlethwaite, Northern Kentucky University; Kimberly Tobin, Westfield State College; Michael Turner, University of North Carolina, Charlotte; Scott Vollum, James Madison University; Courtney Waid, North Dakota State University; and Barbara Warner, Georgia State University.

Stephen G. Tibbetts
CSUSB
August 2010

▧ Dedication

I dedicate this book to my parents, Jane and Steve, who have provided the kind of support and love throughout my life that enabled me to follow my dreams and actually acquire them. They are the best parents anyone could ever ask for, and I will forever appreciate their inspiration in both my work and life overall. I can't thank you two enough for everything you have done for me.

CHAPTER

1

Introduction to the Book

An Overview of Issues in Criminological Theory

Welcome to the world of criminological theory! It is an exciting and complex endeavor that explains why certain individuals and groups commit crimes and why other people do not. This book will explore the conceptual history of this endeavor as well as current theories. Most of us can relate directly to many of these theories; we may know friends or family members who fit dominant models of criminal behavior.

This introduction begins by describing what **criminology** is; what distinguishes it from other perspectives of crime, such as religion, journalism, or philosophy; and how definitions of crime vary across time and place. Then, it examines some of the major issues used to classify different theories of criminology. After exploring the various **paradigms** and categories of criminological **theory,** we discuss what characteristics help to make a theory a good one—in criminology or in any scientific field. In addition, we review the specific criteria for proving causality—for showing what predictors or variables actually cause criminal behavior. We also explain why—for logistic and ethical reasons—few theories in criminology will ever meet the strict criteria required to prove that key factors actually cause criminal behavior. Finally, we look at the strengths and weaknesses of the various measures of crime, which are used to test the validity of all criminological theories, and what those measures reveal about how crime is distributed across various individuals and groups. Although the discussion of crime distribution, as shown by various measures of criminality, may seem removed from our primary discussion regarding theories of why certain individuals and groups commit more crime than others, nothing could be further from the truth. Ultimately, all theories of criminal behavior will be judged based on how much each theory can explain the observed rates of crime shown by the measures of criminality among individuals and groups.

✄ What Is Criminology, and How Does It Differ From Other Examinations of Crime?

Criminology is the scientific study of crime, especially why people commit crime. Although many textbooks have more complex definitions of crime, the word *scientific* separates our definition from other perspectives and examinations of crime.[1] Philosophical and legal examinations of crime are based on logic and deductive reasoning, for example, by developing propositions for what makes logical sense. Journalists play a vital role in examinations of crime by exploring what is happening in criminal justice and revealing injustices and new forms of crime; however, they tend to examine anecdotes or examples of crime as opposed to examining objective measures of criminality.

Taken together, philosophical, legal, and journalistic perspectives of crime are not scientific because they do not involve the use of the **scientific method.** Specifically, they do not develop specific predictions, known scientifically as **hypotheses,** which are based on prior knowledge and studies, and then go out and test such predictions through observation. Criminology is based on this scientific method, whereas other examinations of crime are not.

Instead, philosophers and journalists tend to examine a specific case, make conclusions based on that one example of a crime incident, and then leave it at that. Experts in these nonscientific disciplines do not typically examine a multitude of stories similar to the one they are considering, nor do they apply the elements of their story to an existing theoretical framework that offers specific predictions or hypotheses. Further, they do not test those predictions by observation. The method of testing predictions through observation and then applying the findings to a larger body of knowledge, as established by theoretical models, is solely the domain of criminologists, and it separates criminology from other fields. The use of the scientific method is a distinguishing criterion for many studies of human behavior, such as psychology, economics, sociology, and anthropology, which is why these disciplines are generally classified as **social sciences;** criminology is one.

To look at another perspective on crime, religious accounts are almost entirely based on dogmatic, authoritarian, or reasoning principles, meaning that they are typically based on what some authority (e.g., the Pope, the Bible, the Torah, or the Koran) has to say about the primary causes of crime and the best ways to deal with such violations. These ideas are not based on observations. A science like criminology is based not on authority or anecdotes but on empirical research, even if that research is conducted by a 15-year-old who performs a methodologically sound study. In other words, the authority of the scientist performing the study does not matter; rather, the observed evidence and the soundness of the methodology of how the study was performed are of utmost importance. Criminology is based on science, and its work is accomplished through direct observation and testing of hypotheses, even if those findings do not fit neatly into logical principles or the general feelings of the public.

✄ What Is Theory?

Theory can be defined as a set of concepts linked together by a series of statements to explain why an event or phenomenon occurs. A simple way of thinking about theories is that they provide explanations of why the world works the way it does. In other words, a theory is a model of the phenomenon that is being

---

[1]Stephen Brown, Finn Esbensen, and Gilbert Geis, *Criminology,* 6th ed. (Cincinnati: LexisNexis, 2007).

discussed, which in this case is criminal behavior. Sometimes, perhaps quite often, theories are simply wrong, even if the predictions they give are highly accurate.

For example, in the early Middle Ages, most people, including expert scientists, believed the Earth was the center of the universe because everything seemed to rotate and revolve around our home planet. If we wake up day after day and see the sun (or moon) rise and set in close to the same place, it appears that these celestial bodies are revolving around the Earth, especially considering the fact that we don't feel the world around us moving. Furthermore, calendars predicting the change of seasons, as well as the location and phases of these celestial bodies (such as the moon), were quite accurate. However, although the experts were able to predict the movements of celestial objects quite well and develop extremely accurate calendars, they had absolutely no understanding of what was actually happening. Later, when some individuals tried to convince the majority that they were wrong, specifically that the Earth was not the center of the universe, they were condemned as heretics and persecuted, even though their theoretical models were correct.

The same type of argument could be made about the Earth being flat; at one time, observations and all existing models seemed to claim it as proven and true. Some disagreed and decided to test their own predictions, which is how America was discovered by European explorers. Still, many who believed the Earth was round were persecuted or outcast from mainstream society in Europe at the time.

Two things should be clear: Theories can be erroneous, and accurate predictions can be made (e.g., early calendars and moon and star charts) using them, even though there is no true understanding of what is actually happening. One way to address both of these issues is to base knowledge and theories on scientific observation and testing. All respected theories of crime in the modern era are based on science; thus, we try to avoid buying into and applying theories that are inaccurate, and we continuously refine and

▲ **Image 1.1**   Theories of the Earth as the center of the universe were dominant for many centuries, and scientists who proposed that the Earth was not the center of the universe were often persecuted. Over time, the theory was proved false.

improve our theories (based on findings from scientific testing) to gain a better understanding of what causes people to commit crime. Criminology, as a science, always allows and even welcomes criticism to its existing theoretical models. There is no emphasis on authority but rather on the scientific method and the quality of the observations that take place in testing the predictions. All scientific theories can be improved, and they are improved only through observation and empirical testing.

⊠ What Is Crime?

Definitions of crime vary drastically. For example, some take a **legalistic approach** toward defining crime, including only acts that are specifically prohibited in the legal codes of a given jurisdiction. The problem with such a definition is that what is a crime in one jurisdiction is not necessarily a crime in other jurisdictions. To clarify, some acts, such as murder and armed robbery, are against the law in virtually all countries and all regions of the United States, across time and culture. These are known as acts of **mala in se,** literally meaning *evil in itself.*[2] Typically, these crimes involve serious violence and shock the society in which they occur.

Other crimes are known as acts of **mala prohibita**, which has the literal meaning of *evil because prohibited.* This acknowledges that these are not inherently evil acts; they are bad only because the law says so.[3] A good example is prostitution, which is illegal in most of the United States but is quite legal and even licensed in most counties of Nevada. The same can be said about gambling and drug possession or use. These are just some examples of acts that are criminal in certain places or at certain times and thus are not agreed upon by most members of a given community.

This book examines both *mala in se* and *mala prohibita* types of offenses, as well as other acts of **deviance**, which are not against the law in many places but are statistically atypical and may be considered more immoral than illegal. For example, in Nevada in the 1990s, a young man watched his friend (who was later criminally prosecuted) kill a young girl in the bathroom at a casino, but he told no one. Although most people would claim that this was highly immoral, at that time, the Nevada state laws did not require people who witnessed a killing to report it to authorities. (Note: As a result of this event, Nevada made withholding such information a criminal act.) Therefore, this act was deviant because most people would find it immoral, but it was not criminal because it was not technically against the laws in the jurisdiction at that time.

Other acts of deviance are not necessarily immoral but are certainly statistically unusual and violate social norms, such as purposely farting at a formal dinner. Such activities are relevant for our discussion, even if they are not defined as criminal by the law, because they show a disposition toward antisocial behavior often found in individuals who are likely to become criminal offenders. Furthermore, some acts are moving from deviant to illegal all the time, such as the use of cell phones while driving or smoking cigarettes in public; many jurisdictions are moving to have these behaviors made illegal and have been quite successful to date, especially in New York and California.

Most *mala in se* activities (e.g., murder) are highly deviant, too, meaning they are not typically found in society, but many, if not most, *mala prohibita* acts—say, speeding on a highway—are not deviant because they are committed by most people at some point. This is a good example of a *mala prohibita* act that is illegal but not deviant. This book will examine theories for all of these types of activities, even those that do not violate the law at the present time in a given jurisdiction.

◪ How Are Criminological Theories Classified? The Major Theoretical Paradigms

Scientific theories of crime can be categorized based on several important concepts, assumptions, and characteristics. To begin, most criminological theories are classified by the paradigm they emphasize. Paradigms are distinctive theoretical models or perspectives; in the case of crime, they vary based largely on opposing assumptions of human behavior. There are four major paradigms.[4]

The first of these, deterrence or rational choice theories, commonly referred to as the **Classical School** perspective, will be discussed at length later in this book. It assumes that individuals have free will and

---

[2]Ibid.

[3]Ibid.

[4]George Vold, Thomas Bernard, and Jeffrey Snipes. *Theoretical Criminology,* 5th ed. (Oxford: Oxford University Press, 2002).

choose to commit crimes based on rational, hedonistic decisions; they weigh out the potential costs and benefits of offending and then choose what will maximize their pleasure and minimize their pain. The distinguishing characteristic of these theories is that they emphasize the free choice individuals have in committing crime. The other paradigms are based on the influence of factors other than free will or rational decision making—for example, biology, culture, parenting, and economics.

Another category of theories is positivism, which is somewhat the opposite of rational choice theories. These theories argue that individuals do not have free will or rationality in making decisions to commit crime. Rather, the **Positive School** perspective assumes that individuals are passive subjects of determinism, which means that people do not freely choose their behavior. Instead, their behavior is determined by factors outside of their free will, such as genetics, IQ, education, employment, peer influences, parenting, and economics.[5] Most of the highly respected and scientifically validated criminological theories of the modern era fall into this category.[6]

Another group of criminological theories belong to the conflict or critical perspective, which emphasizes the use of law as a reaction or tool to enforce restraint on others by those in power or authority; it also involves how society reacts when a person (often a juvenile) is caught doing something wrong. These theories emphasize group behavior over individual behavior: Groups that are in power use the criminal codes as a tool in keeping people who have limited power restrained or confined.

Finally, over the last few decades, a new category has emerged, namely the integrated theoretical models, which attempt to combine the best aspects of explanatory models into a single, better theoretical framework for understanding crime. These models tend to suffer from the logical inconsistencies of integrating theoretical models that have opposing assumptions or propositions. All of these categories will become clearer as we progress through this book.

✳ Additional Ways to Classify Criminological Theories

Although the major paradigms are the primary way that criminological theories are classified, there are several other ways that they can be categorized. Specifically, theoretical models can be classified based on whether they focus on individuals or groups as their primary units of examination. For instance, some theories emphasize why certain individuals do or do not commit crime. This level of investigation, in which the focus is on the individual, is often referred to as the **microlevel of analysis,** much as microeconomics is the study of economics on the individual (person) level. When your instructors score each student on an exam, this is a microlevel analysis.

On the other hand, many theories emphasize primarily the group or **macrolevel of analysis,** much as macroeconomics is the study of economic principles at the aggregate or group level. In this book, some sections are separated by whether the individual or group level of analysis is emphasized. For example, social process theories tend to be more microlevel oriented, whereas social structure theories are more macrolevel oriented. Here's a good example. If instructors compare the mean score (or average) of one class to the mean score in another, this is a comparison of group rates, regardless of the performance of any individual in either class. Ultimately, a great theory would explain both the micro- and macrolevels of analysis, but we will see that very few attempt to explain or account for both levels.

---

[5]Ibid.

[6]Lee Ellis and Anthony Walsh, "Criminologists' Opinions about Causes and Theories of Crime and Delinquency," *The Criminologist 24* (1999):1–4.

Criminological theories can also be classified by the way they view the general perspective of how laws are made. Some theories assume that laws are made to define acts as criminal to the extent that they violate rights of individuals, and thus, virtually everyone agrees that such acts are immoral. This type of perspective is considered a **consensual perspective** (or nonconflict model). On the other hand, many modern forms of criminological theories fall into an opposite type of theory, commonly known as the **conflict perspective**, which assumes that different groups disagree about the fairness of laws and that laws are used as a tool by those in power to keep down other lower-power groups. There are many forms of both of these types of consensual and conflict theoretical models, and both will be specifically noted as we progress through the book.

A final, but perhaps most important, way to classify theories is in terms of their assumptions regarding human nature. Some theories assume that people are born good (e.g., giving, benevolent, etc.) and are corrupted by social or other developmental influences that lead them to crime. A good example is **strain theory,** which claims that people are born innocent and with good intentions but that society causes them to commit crime. On the other hand, many of the most popular current theories claim that virtually all individuals are born with a disposition toward being bad (e.g., selfish, greedy, etc.) and must be socialized or restrained from following their inherent propensities for engaging in crime.[7] A good example of this is **control theory,** which assumes that all individuals have a predisposition to be greedy, selfish, violent, and so on (i.e., they are criminally disposed), and therefore, people need to be controlled or prevented from acting on their natural, inherent disposition toward selfish and aggressive behaviors.

Another variation on this issue involves theories that are often referred to as ***tabula rasa,*** literally translated as *blank slate.* This assumes that people are born with no leaning toward good or bad but are simply influenced by the balance of positive or negative influences that are introduced socially during their development. A good example is **differential association** or reinforcement theory, which assumes that all individuals are born with a blank slate and that they learn whether to be good or bad based on what they experience. Although the dominant assumption tends to vary across these three models from time to time, the most popular theories today (which are self- and social-control theories) seem to imply the second option, specifically that people are born selfish and greedy and must be socialized and trained to be good and conforming.[8] There are other ways that criminological theories can be classified, but the various characteristics that we have discussed in this section summarize the most important factors.

✉ Characteristics of Good Theories

Respected scientific theories in all fields, whether it be chemistry, physics, or criminology, tend to have the same characteristics. After all, the same scientific review process (i.e., blind peer review by experts) is used in all sciences to determine which studies and theoretical works are of high quality. The criteria that characterize a good theory in chemistry are the same ones used to judge what makes a good criminological theory. Such characteristics include: parsimony, scope, logical consistency, testability, empirical validity, and policy implications.[9] Each of these characteristics is examined here. (It should be noted that our discussion and many of the examples provided for the characteristics are taken from Akers & Sellers, 2004).[10]

---

[7]Ibid.

[8]Ibid.

[9]Ronald Akers and Christine Sellers, *Criminological Theories,* 4th ed. (Los Angeles: Roxbury, 2004).

[10] Ibid., 5–12.

Parsimony is achieved by explaining a given phenomenon, in our case criminal activity, in the simplest way possible. Other characteristics being equal, the simpler a theory, the better. The problem with criminal behavior is that it is highly complex. However, that has not stopped some criminologists from attempting to explain this convoluted phenomenon in highly simple ways. For example, one of the most recent and popular theories (at least regarding the amount of related research and what theories the experts believe are the most important) is the **theory of low self-control** (which we discuss later in this book). This very simple model holds that one personality factor—low self-control—is responsible for all criminal activity. The originators of this theory, Michael Gottfredson and Travis Hirschi, assert that every single act of crime and deviance is caused by this same factor: low self-control[11]—everything from speeding, smoking tobacco, not wearing a seat belt while driving, and having numerous sex partners to serious crimes, such as murder and armed robbery, are caused by low self-control.

Although this theory has been disputed by much of the subsequent research on this model, it remains one of the most popular and accepted models of the modern era.[12] Furthermore, despite the criticisms of this theory, many notable criminologists still believe that this is the best single model of offending that has been presented to date. In addition, there is little doubt that this model has become the most researched theoretical model over the last two decades.[13]

Perhaps the most important reason why so much attention has been given to this theory is its simplicity, putting all of the focus on a single factor. Virtually all other theoretical models have specified multiple factors that are proposed to play major parts in determining processes that explain why individuals commit crime. After all, how can low self-control explain white-collar crime? Some self-control is required to obtain a white-collar position of employment. Regardless, it is true that a simple theory is better than a more complex one, as long as other characteristics are equivalent. However, given a complex behavior like criminal behavior, it is likely that a simple explanation, such as naming one factor to explain everything, is unlikely to be adequate.

Scope is the characteristic that indicates how much of a given phenomenon the theory seeks to explain. Other characteristics being equal, the larger the scope, the better the theory. This is somewhat related to parsimony in the sense that some theories, like the theory of low self-control, seek to explain all crimes and all deviant acts as well. So, the theory of low self-control has a very wide scope. Other theories of crime may seek to explain only property crime, such as some versions of strain theory or drug usage. However, the wider the scope of what a theory can explain, the better the theory, assuming other characteristics are equal.

Logical consistency is the extent to which a theory makes sense in terms of its concepts and propositions. It is easier to see what is meant by logical consistency by showing examples of what does not fit this criterion. Some theories simply don't make sense because of the face value of its propositions. For example, Cesare Lombroso, called the father of criminology, claimed that the most serious offenders are "born criminals," biological throwbacks to an earlier stage of evolutionary development who can be identified by their physical features.[14] Lombroso, who is discussed at more length later in this book, claimed that tattoos were one of the physical features that identified these born criminals. This doesn't make sense, however, because tattoos are not biological physical features—no baby has ever been born with a tattoo. This criticism will make even more sense when we discuss the criteria for determining causality later in this chapter.

---

[11]Michael Gottfredson and Travis Hirschi, *A General Theory of Crime* (Palo Alto: Stanford University Press, 1990).

[12]Ellis and Walsh, "Criminologists' Opinions," 3–4.

[13]Anthony Walsh and Lee Ellis, "Political Ideology and American Criminologists' Explanations for Criminal Behavior," *The Criminologist* 24 (1999): 1, 14.

[14]Cesare Lombroso, *The Criminal Man* (Milan: Hoepli, 1876).

Another prominent example of theories that lack logical consistency is the work of early feminist theorists, such as Freda Adler, who argued that, as females gain educational and employment opportunities, they will be more likely to converge with males on crime rates.[15] Such hypotheses were logically inconsistent with the data available at the time they were presented and even more today; the facts show that females who are given the most opportunities commit the fewest crimes. On the contrary, females who have not been given these benefits commit the most crimes. These are just two examples of how past theories were not logically consistent with the data at the time they were created, not to mention future research findings, which have completely dismissed their hypotheses.

Testability is the extent to which a theory can be put to empirical, scientific testing. Some theories simply cannot be tested. A good example is Freud's theory of the psyche. Freud described three domains of the psyche—the conscious ego, the subconscious id, and the superego—but none of these domains can be observed or tested.[16] Although some theories can be quite influential without being testable (as was Freud's theory), other things being equal, it is a considerable disadvantage for a theoretical model to be untestable and unobservable. Fortunately, most established criminological theories can be examined through empirical testing.

Empirical validity is the extent to which a theoretical model is supported by scientific research. Obviously, this is highly related to the previous characteristic of being testable. Virtually all accepted modern criminological theories are testable, but that does not mean they are equal in terms of empirical validity. Although some **integrated models** (meaning two or more traditional theories being merged together, which will be examined later in this book) have gained a large amount of empirical validity, these models sort of cheat because they merge the best of two or more models, even when the assumptions of these models are not compatible. Therefore, the best empirical validity from an independent theoretical model, by itself, has been found for **differential reinforcement theory,** which has been strongly supported for various crime types (ranging from tobacco usage to violence) among a wide variety of populations (ranging from young children to elderly subjects).[17]

Ultimately, assuming other characteristics being equal, empirical validity is perhaps one of the most important characteristics used in determining how good a theory is at explaining a given phenomenon or behavior. If a theory has good empirical validity, it is an accurate explanation of behavior; if it does not have good empirical validity, it should be revised or dismissed because it is simply not true.

Policy implications is the extent to which a theory can create realistic and useful guidance for changing the way that society deals with a given phenomena. In our case, this means providing a useful model for informing authorities of how to deal with crime. An example is the broken windows theory, which says that, to reduce serious crimes, authorities should focus on the minor incivilities that occur in a given area. This theory has been used successfully by many police agencies (most notably by New York City police, who

---

[15]Freda Adler, *Sisters in Crime* (New York: McGraw Hill, 1975).

[16]See Stephen G. Tibbetts and Craig Hemmens, *Criminological Theory: A Text Reader* (Thousand Oaks: Sage, 2010), 9.

[17]See studies including Ronald Akers and Gang Lee, "A Longitudinal Test of Social Learning Theory: Adolescent Smoking," *Journal of Drug Issues 26* (1996): 317–343; Ronald Akers and Gang Lee, "Age, Social Learning, and Social Bonding in Adolescent Substance Abuse," *Deviant Behavior 19* (1999): 1–25; Ronald Akers and Anthony J. La Greca, "Alcohol Use Among the Elderly: Social Learning, Community Context, and Life Events," in *Society, Culture, and Drinking Patterns Re-examined,* ed. David J. Pittman and Helene Raskin White (New Brunswick: Rutgers Center of Alcohol Studies, 1991), 242–62; Sunghyun Hwang, *Substance Use in a Sample of South Korean Adolescents: A Test of Alternative Theories* (Ann Arbor: University Microfilms, 2000); Sunghyun Hwang and Ronald Akers, "Adolescent Substance Use in South Korea: A Cross-Cultural Test of Three Theories," in *Social Learning Theory and the Explanation of Crime: A Guide for the New Century,* ed. Ronald Akers and Gary F. Jensen (New Brunswick: Transaction Publishers, 2003).

reduced their homicide rate by more than 75% in the last decade). Other theories may not be as useful in terms of reducing crime because they are too abstract or propose changes that are far too costly or impossible to implement, such as theories that emphasize changing family structure or the chromosomal makeup of individuals. So, other things being equal, a theory that has readily available policy implications would be advantageous as compared to theories that do not.

▨ Criteria for Determining Causality

There are several criteria for determining whether a certain variable causes another variable to change—in other words, causality. For this discussion, we will be using the commonly used scientific notation of a predictor variable—called X—as causing an explanatory variable—called Y; such variables are also commonly referred to as an independent or predictor variable (X) and a dependent or explanatory (Y) variable. Such criteria are used for all scientific disciplines, whether chemistry, physics, biology, or criminology. In this book, we are discussing crime, so we will concentrate on examples that relate to this goal, but some examples will be given that are not crime related. Unfortunately, we will also see that, given the nature of our field, there are some important problems with determining causality, largely because we are dealing with human beings as opposed to a chemical element or biological molecule.

The three criteria that are needed to show causality are: temporal ordering, covariation or correlation, and accounting for spuriousness.

Temporal ordering requires that the predictor variable (X) must precede the explanatory variable (Y) if we are to determine that X causes Y. Although this seems like a no-brainer, it is sometimes violated in criminological theories. For example, you'll remember that Lombroso claimed born criminals could be identified by tattoos, which obviously goes against this principle.

A more recent scientific debate has focused on whether delinquency is an outcome variable (Y) due to associations with delinquent peers and associates (X) or whether delinquency (X) causes associations with delinquent peers and associates (Y), which then leads to even more delinquency. This can be seen as the argument of which came first, the chicken or the egg. Studies show that

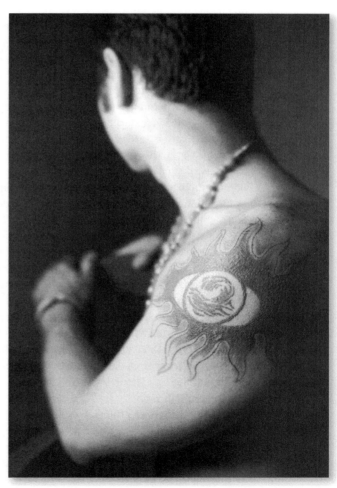

▲ **Image 1.2**   Early theories identified criminals by whether they had tattoos; at that time, this might have been true. In contemporary times, many individuals have tattoos, so this would not apply.

both processes are often taking place, meaning that delinquency and associations with delinquent peers are likely to be both predictor and explanatory variables in most cases, and this forms a reciprocal or feedback loop that encourages both causal paths.[18] Thus, temporal ordering is an important question, and often, it is quite complex and must be examined to fully understand causal order.

Correlation, or **covariation,** is the extent to which a change in the predictor (X) is associated with a change (either higher or lower) in the explanatory variable (Y). In other words, a change in X leads to a change in Y. For example, a rise in unemployment (X) someplace is likely to lead to a rise in crime rates (Y) in the same area; this would be a positive association because both increased. Similarly, an increase in employment (X) is likely to lead to a decrease in crime rates (Y) in that area; this would be a negative, or inverse, association, because one decreased and the other increased. The criterion of covariance is not met when a change in X does not produce any change in Y. Thus, if a significant change in X does not lead to a significant change in Y, then this criterion is not met.

However, correlation alone does not mean that X causes Y. For example, ice cream sales (X) tend to be highly associated with crime rates (Y). However, this does not mean that ice cream sales cause higher crime rates. Rather, other factors, in this case, warm weather, lead to increases in both sales of ice cream and the number of people who are outdoors in public areas and interacting, which naturally leads to greater opportunities and tendencies to engage in criminal activity. This leads to the final criterion for determining causality.

Accounting for **spuriousness** is a complicated way of saying that, to determine that X causes Y, other factors (typically called Z factors) that could be causing the observed association must be accounted for before one is sure that it is actually X that is causing Y. In other words, these other Z factors may account for the observed association between X and Y. What often happens is that a third factor (Z) causes two events to occur together in time and place.

A good example is the observation that a greater number of fire officers showing up at a fire is correlated with more damage. If only the first two criteria of causality were followed, this would lead to the conclusion that the increased number of fire officers (X) causes the heavier fire damage (Y). This conclusion meets the temporal ordering and covariance criteria. However, a third Z variable or factor is causing both X and Y to appear together. This Z variable is the size of the fire, which is causing more officers to show up and also causing more damage. Once this Z factor is accounted for, the effect of X on Y becomes nonexistent.

Using the Lombroso example, tattoos may have predicted criminality at the time he wrote (although criminals weren't born with them). However, Lombroso did not account for an important Z factor, namely associates or friends who also had tattoos. This Z factor caused the simultaneous occurrence of both other factors. To clarify, individuals who had friends or associates with tattoos tended to get tattoos, and (especially at that time in the 1800s) friends or associates who had tattoos also tended to commit more crime. In that era, pirates and incarcerated individuals were most likely to get tattoos. Therefore, had Lombroso controlled for the number of tattooed associates of the criminals he studied, he likely would have found no causal effect on crime from body art.

Ultimately, researchers in criminology are fairly good at determining the first two criteria of causality, temporal ordering and covariance or correlation. Most scientists can perform classical experiments that randomly assign subjects either to receive or not to receive the experimental manipulation to examine the effect on outcomes. However, the dilemma for criminologists is that the factors that appear to be important (according to police officers, parole agents, or corrections officers) are family variables, personality traits, employment variables, intelligence, and other similar characteristics that cannot be experimentally manipulated to control

---

[18]Terence Thornberry, "Toward an Interactional Theory of Delinquency," *Criminology* 25 (1987): 863–887.

for possible Z factors. After all, how can we randomly assign certain people or groups to bad parents or bad educations, no jobs, low IQs, bad genetics, or delinquent peers? Even if we could manage such manipulations, ethical constraints would prohibit them.

Thus, as criminologists, we may never be able to meet all the criteria of causality, so we are essentially stuck with building a case for the factors we think are causing crime by amassing as much support as we can regarding temporal ordering and covariance or correlation, and perhaps accounting for other factors in advanced statistical models. Ultimately, social science, especially criminology, is a difficult field in terms of establishing causality, and we shall see that the empirical validity of various criminological theories is hindered by such issues.

⬛ Measures of Crime

Crime can be measured in an infinite number of ways. To some extent, readers have measured crime by observing what has been happening in their own neighborhoods or reading or watching the news every day. However, some measures of crime go beyond these anecdotal or personal experiences, and these more exacting measures are what criminologists commonly use to gauge rates of crime across time and place.

Specifically, three major categories of crime measures are used by social scientists to examine crime. The first and most used measure is the **Uniform Crime Report** (UCR). Police send reports about certain crimes and arrests to the Federal Bureau of Investigation (FBI), which combines the many thousands they receive from across the nation and publishes the UCR annually.

The second measure is the **National Crime Victimization Survey** (NCVS; prior to the early 1990s, it was known as the National Crime Survey [NCS]). Like the UCR, the report is issued by the U.S. Department of Justice (DOJ), but the data are collected in an entirely different way. Specifically, interviews are conducted with a large, random sample of U.S. households asking how much crime they have experienced in half-year intervals. The NCVS is collected by the research branch of the DOJ called the Bureau of Justice Statistics (BJS) in conjunction with the U.S. Bureau of the Census, which is one of the earliest agencies to collect information about citizens and thus the most experienced at such endeavors.

The third measure, which is perhaps the most important for purposes of this book, is **self-report data (SRD),** which are primarily collected by independent academic scientists or think tank agencies, such as RAND Corporation. Participating in surveys or interviews, individuals report crimes against themselves or crimes they have committed. This measure is the most important for the purposes of this book because the UCR and NCVS do not provide in-depth information on the offenders or the victims, such as personality, biology or physiology, family life, and economic information. These factors are of the utmost importance for our purposes because there is a broad consensus that they cause people to commit crime, and yet, they are missing from the most commonly used measures of crime. Self-report data are the best, and in most cases the only, measure that can be used to figure out why some people offend and others do not. However, like the other measures, self-reports have numerous weaknesses (and strengths).

Each of these three measures is briefly examined here. Although the measures are not the primary emphasis of this book, it is important to understand their strengths and weaknesses.

⬛ The Uniform Crime Report

The UCR is the oldest and most used measure of crime rates in the United States for purposes of examining trends and distribution of crime. It began in the early 1930s, and although changes have been made along the way, it is relatively stable in terms of comparing various years and decades. As mentioned before, it is collected by many thousands of independent police agencies at federal, state, and local levels. These thousands of agencies send their reports of crimes and arrests to their respective state capitals, which then forward their synthesized reports to FBI headquarters, where all reports are combined to provide an overview of crime in the nation.

The FBI definitions of crimes often differ from state categorizations, and the way that they differentiate crimes are important in terms of future discussions in this chapter. The FBI concentrates on eight (four violent and four property) **index offenses**, or Part I offenses. The four violent crimes are murder and non-negligent manslaughter, forcible rape (not statutory), robbery, and aggravated assault (which involves intentions of serious injury to the victim). The four property offenses are burglary (which includes a breaking and entering or trespass), motor vehicle theft, larceny (which does not involve a trespass, for example, shoplifting), and arson (which was added to the crime index count in the late 1970s). All reports to police for these eight offenses are included in the crime index, whether or not they resulted in an arrest. This information is often referred to as crimes known to police (CKP).

The UCR also includes about two dozen other offenses known as **nonindex offenses**, or Part II offenses, which are reported only if an arrest is made in a case. These offenses range from violent crimes (such as simple assault), to embezzlement and fraud, to offenses that are considered violations of the law only if an individual is under 18 years of age (such as running away from home). The major problem with the estimates of these nonindex offenses is that the likelihood of arresting someone for such crimes is less than 10% of the actual occurrence, so the data regarding nonindex offenses are highly inaccurate. The official count from the FBI is missing at least 90% of the actual offenses that take place in the United States. Therefore, we will primarily concentrate on the index offenses for the purposes of our discussion.

▲ **Image 1.3**   The annual UCRs are produced by the FBI. Local, county, and state criminal justice agencies send their annual crime data to the J. Edgar Hoover Building in Washington, DC. UCR data are, by their nature, incomplete, as many crimes are never reported to the police at all. This **dark figure** of crime might be as high as 90% of all crime incidents.

Even the count of index offenses has numerous problems. The most important and chronic problem with using the UCR as a measure of crime is that, most of the time, victims fail to report crimes—yes, even aggravated assault, forcible rape, robbery, burglary, and larceny. Recent studies estimate that about 70% to 80% of these serious crimes are not reported to police. Criminologists call this missing amount of crime the dark figure because it never shows up in the police reports that get sent to the FBI.

There are many reasons why victims do not report these serious crimes to the police. One of the most important is that they consider it a personal matter. Many times, the offense is committed by a family member, a close friend, or an acquaintance. For instance, police are rarely informed about aggravated assaults among siblings. Rape victims are often assaulted on a date or by someone they know; they may feel that they contributed to their attack by choosing to go out with the offender, or they may believe that police won't take such a claim seriously. Regardless of the case, many crime victims prefer to handle it informally and not involve the police.

Another major reason why police are not called is that victims don't feel the crime is important enough to report. For example, thieves may steal a small item that the victim won't miss, so they don't see the need to report what the police or FBI would consider a serious crime. This is likely related to another major reason why people do not report crime to the police: They have no confidence that reporting the case to law enforcement will do any good. Many people, often residents of the neighborhoods that are most crime ridden, are likely to feel that the police are not competent or will not seriously investigate their charges.

There are many other reasons why people do not report their victimizations to police. Some may fear retaliation, for example, in cases involving gang activity; many cities, especially those with many gangs, have seen this occur even more in recent years. The victims may also fail to report a crime for fear that their own illegal activities will be exposed; an example is a prostitute who has been brutally beaten by her pimp. In U.S. society, much crime is committed against businesses, but they are very reluctant to report crimes because they do not want a reputation for being a hot spot for criminal activity. Sometimes, victims call the police or 911, but they leave the scene if the police fail to show up in a reasonable amount of time. This has become a chronic problem despite efforts by police departments to prioritize calls.

Perhaps the most chronic, most important, and often ignored failure to report crimes can be traced to U.S. school systems. Most studies of crime and victimization in schools show that much and maybe even most juvenile crimes occur in schools, but these offenses almost never get reported to police, even when school resource officers (SROs) are assigned to the school. Schools—and especially private schools—have a vested interest in not reporting crimes that occur on their premises to the police. After all, no school (or school system) wants to become known as crime-ridden.

Schools are predisposed to being crime ridden because the most likely offenders and the most likely victims—young people—interact there in close quarters for most of the day. The school is much happier, however, if teachers and administrators deal informally with the parties involved in an on-campus fight; the school does not want these activities reported to and by the media. In addition, the student parties involved in the fight do not want to be formally arrested and charged with the offense, so they are also happy with the case being handled informally. Finally, the parents of the students are also generally pleased with the informal process because they do not want their children involved with a formal legal case.

Even universities and colleges follow this model of not reporting crimes when they occur on campus. A good example can be seen on most of the websites of virtually all major colleges, which are required by federal law to report crimes on campus. Official reports of crimes, ranging from rapes to liquor law violations, are often in the single digits each year for campuses housing many thousands of students. Of course, some crimes may not be reported to the school, and others may be dealt with administratively rather than calling police. The absence of school data is a big weakness of the UCR.

Besides the dark figure, there are many other criticisms of the UCR as a measure of crime. For example, the way that crimes are counted can be misleading. Specifically, the UCR counts only the most serious crime that is committed in a given incident. For example, if a person or persons rob, rape, and murder a victim, only the murder would show up in the UCR; the robbery and rape would not be recorded. Furthermore, there are inconsistencies in how the UCR counts the incidents; for example, if a person walks into a bar and assaults eight people there, it would be counted as eight assaults, but if the same person walks into the bar and robs every person of wallets and purses, it is counted as one robbery. This makes little sense, but it is the official count policy traditionally used by the UCR.

Some other more important criticisms of the UCR involve political considerations, such as the fact that many police departments (such as in Philadelphia and New York City) have systematically altered the way that crimes are defined, for example, by manipulating the way that classifications and counts of crimes are recorded (e.g., the difference between aggravated assault [an index crime] and simple assault [a nonindex crime]), so that official estimates make it seem that major crimes have decreased in the city when, in fact, they may have actually increased.

It is important also to note the strengths of the UCR measure. First, it was started in 1930, making it the longest systematic measure of crime in the United States. This is a very important advantage if one wants to examine crime over most of the 20th century. We will see later that there have been extremely high crime rates at certain times (such as during the 1930s and the 1970s to 1980) and very low crime rates at other points (such as the early 1940s and recent years [late 1990s-present]). Other measures, such as the NCVS and national self-report data, did not come into use until much later, so the UCR is important for the fact that it started so early.

Another important strength of this measure is that two of the offenses that the UCR concentrates on are almost always reported and therefore overcome the weakness of the dark figure or lack of reporting to police. These two offenses are murder and non-negligent manslaughter and motor vehicle theft. Murder is almost always reported because a dead body is found; very few murders go unreported to authorities. Although a few may elude official recording, for example, if the body is transported elsewhere or carefully hidden, almost all murders are recorded. Similarly, motor vehicle thefts, a property crime, are almost always reported because any insurance claims must provide a police report. Most cars are worth thousands (or at least many hundreds) of dollars, so victims tend to report when their vehicle(s) have been stolen; this provides a rather valid estimate of property crime in specific areas. The rest of the offenses (yes, even the other index crimes) counted by the UCR are far less reliable. If someone is doing a study on homicide or motor vehicle theft, the UCR is likely the best source of data, but for any other crime, researchers should probably look elsewhere.

This is even further advised for studies examining nonindex offenses, which the UCR counts only when someone is arrested for a given offense. The vast majority of nonindex offenses do not result in an arrest. To shed some light on how much actual nonindex crime is not reported to police, it is useful to examine the **clearance rate** of the index offenses in the UCR, our best indicator of solving crimes. Even for the crimes that the FBI considers most serious, the clearance rate is about 21% of crimes reported to police, meaning that they took a report for a crime. Of course, the more violent offenses have higher clearance rates because (outside of murder) they inherently have a witness, namely the victim, and because police place a higher priority on solving violent crimes. However, for some of the index crimes, especially serious property offenses, the clearance rates are very low. Furthermore, it should be noted that the clearance rate of serious index crimes has not improved at all over the last few decades despite much more advanced resources and technology, such as DNA, fingerprints, and faster cars. These data on the clearance rates are only for the most serious, or index, crimes; thus, the reporting of the UCR regarding nonindex crimes is even more inaccurate because there is even less reporting (i.e., the dark figure) and less clearance of these less serious offenses. In other words, the data provided by the UCR regarding nonindex offenses are totally invalid and for the most part completely worthless.

Ultimately, the UCR is a good measure for (1) measuring the overall crime rate in the United States over time, (2) examining what crime was like prior to the 1970s, and (3) investigating murder and motor vehicle theft because they are almost always reported. Outside of these offenses, the UCR has serious problems, and fortunately, we have better measures to use for examining crime rates in the United States.

The National Crime Victimization Survey

Another commonly used measure of crime is the NCVS (the NCS until the early 1990s), which is distinguished from other key measures of crime because it concentrates on the victims of crime, whereas other measures tend to emphasize the offenders. In fact, that is the key reason why this measure was started in 1973 after several years of preparation and pretesting. To clarify, one of the key recommendations of Lyndon Johnson's President's Commission on Law Enforcement and Administration of Justice in the late 1960s was to learn more about the characteristics of victims of crime; at that time, virtually no studies had been done on the subject, whereas much research had been done on criminal offenders. So, the efforts of this commission set into motion the creation of the NCVS.

Since it began, the NCVS has been designed and collected by two agencies: the U.S. Bureau of the Census and the BJS, which is one of the key research branches of the U.S. DOJ. The NCVS is collected in a completely different way from the other commonly used measures of crimes; the researchers select tens of thousands of U.S. households, and each member over 12 years of age in these households is interviewed every 6 months about crime that occurred in that previous 6-month period (each selected household remains in the survey for 3 years, resulting in seven collection periods, which includes the initial interview). Although the selection of households is to some extent random, the way this sampling is designed guarantees that a certain proportion of houses selected have certain characteristics. For example, before the households are selected, they are first categorized according to factors such as region of the country, type of area (urban, suburban, rural), income level, and racial or ethnic composition. This type of sampling, called a multistage, stratified cluster sampling design, ensures that enough households are included in the survey to make conclusions regarding these important characteristics. As you will see later in this chapter, some of the most victimized groups (by rate) in the United States do not make up a large portion of the population or households. So, if the sampling design was not set up to select a certain number of people from certain groups, it is quite likely the researchers would not obtain enough cases to draw conclusions about them.

The data gathered from this sample are then adjusted, and statistical estimates are made about crime across the United States, with the NCVS estimates showing about three times more crime than the UCR rates. Some may doubt the ability of this selected sample to represent crime in the nation, but most studies find that its estimates are far more accurate than those provided by the UCR (with the exception of homicide and maybe motor vehicle theft). This is largely due to the expertise and professionalism of the agencies that collect and analyze the data, as well as the carefully thought out and well-administered survey design, which is indicated by the interview completion rates (typically more than 90%, higher than virtually all other crime and victimization surveys).

One of the biggest strengths of the NCVS is that it directly addresses the worst problem with the previously discussed measure, the UCR. Specifically, the greatest weakness of the UCR is the dark figure, or the crimes that victims fail to report, which happens most of the time (except in cases of homicide or motor vehicle theft). The NCVS interviews victims about crimes that happened to them, even those that were not reported to police. Thus, the NCVS captures far more crime events than the UCR, especially crimes that are highly personal (such as rape). So, the extent to which the NCVS captures much of this dark figure of crime is its greatest strength.

Despite this important strength, the NCVS, like the other measures of crime, has numerous weaknesses. Probably the biggest problem is that two of the most victimized groups in U.S. society are systematically not included in the NCVS. Specifically, homeless people are completely left out because they do not have a home and the participants are contacted through households; yet, they remain one of the most victimized groups per capita in our society.

Another highly victimized group in our society that is systematically not included in the NCVS is young children. Studies consistently show that the younger a person, the more likely he or she is to be victimized. Infants, in particular, especially in their first hours or days of life, face great risk of death or other sort of victimization, typically from parents or caregivers. This is not very surprising, especially in light of the fact that very young children cannot defend themselves or run away. They can't even tell anyone until they are old enough to speak, and then most are too afraid to do so or are not given an opportunity. Although, to some extent, it is understandable to exempt young children from such sensitive questions, the loss of this group is huge in terms of estimating victimization in the United States.

The NCVS also misses the crimes suffered by American businesses, which cumulatively is an enormous amount. In the early years of the NCVS, businesses were also sampled, but that was discontinued in the late 1970s. Had it continued, it would be invaluable information for social scientists and policy makers, not to mention the businesses that are losing billions of dollars each year as a result of crimes committed against them.

Many find it surprising that the NCVS does not collect data on homicide, which most people and agencies consider the most serious and important crime. Researchers studying murder cannot get information from the NCVS but must rely on the UCR, which is most accurate in its reporting for this crime type.

The NCVS also has issues with people accurately reporting the victimization that has occurred to them in the previous 6 months. However, studies show that their reports are surprisingly accurate most of the time. Often when participants report incidents inaccurately, they make unknowing mistakes rather than intentionally lying. Obviously, victims sometimes forget incidents that occur to them, probably because, most of the time, they know or are related to the person offending against them, so they never think of it as a crime per se but rather as a personal disagreement. When asked if they were victims of theft, they may not think to report the time that a brother or uncle borrowed a tool without asking and never returned it.

Although the NCVS researchers go to great lengths to prevent it, a common phenomenon known as **telescoping** tends to occur, which leads to overreporting of incidents. Telescoping is the human tendency to perceive events as having occurred more recently than they actually did. This is one of the key reasons why NCVS researchers interview the household subjects every 6 months, but telescoping still happens. For instance, a larceny may have occurred 8 months ago, but it seems like it happened just a few months ago to the participant, so it is reported to the researchers as occurring in the last 6 months when it really didn't. Thus, this inflates the estimates for that crime interval for the nation.

As mentioned before, an additional weakness is that the NCVS did not start until 1973, so it cannot provide any estimates of victimization prior to that time. A study of national crime rates prior to the 1970s has little choice but to use the UCR. Still, for most crimes, the NCVS has provided a more accurate estimate over the past three decades. Since the NCVS was created, the crime trends it reveals tend to be highly consistent with those shown by the UCR. For example, both measures show violent crime rates peaking at the same time (about 1980), and both agree on when the rates increased (the 1970s) and decreased (the last decade [late 1990s to present]) the most. This is very good, because if they did not agree, that would mean at least one (or both) were wrong. Before we discuss the national trends in crime rates, however, we will examine the strengths and weaknesses of a third measure of crime.

Self-Report Studies of Crime

The final measure of crime consists of various self-report studies of crime, in which individuals report (either in a written survey or interview) the extent of their own past criminal offending or victimization and other information. There is no one systematic study providing a yearly estimate of crime in the United States; rather, self-report studies tend to be conducted by independent researchers or institutes. Even when they do involve a national sample (such as the National Youth Survey), they almost never use such data to make estimates of the extent of crime or victimization across the nation.

This lack of a long-term, systematic study that can be used to estimate national crime rates may be the greatest weakness of self-report studies; however, this very weakness—not having a universal consistency in collection—is also its greatest strength. To clarify, researchers can develop their questionnaires to best fit the exact purposes of their study. For example, if researchers are doing a study on the relationship between a given personality trait (e.g., narcissism) and criminal offending, they can simply give participants a questionnaire that contains a narcissism scale and items that ask about past criminal behavior. Of course, these scales and items must be checked for their reliability and validity, but it is a relatively easy way to directly measure and test the hypotheses with which the researcher is most concerned.

Some question the accuracy of the self-report data because they believe participants typically lie, but most studies have concluded that participants generally tell the truth. Specifically, researchers have compared self-reported offenses to lie detector machine results, giving the same survey to the same individuals to see if they answer the same each time (called test-retest reliability) and cross-checking their self-reported arrests to police arrest data. All of these methods showed that most people tend to be quite truthful when answering surveys.[19]

The most important aspect of self-report surveys is that they are the only available source of data for determining the social and psychological reasons for why people commit crime. The UCR and NCVS have virtually no data on the personality, family life, biological development, or other characteristics of criminal offenders, which are generally considered key factors in the development of criminality. Therefore, although we examine the findings of all three measures in the next section, the vast majority of the content we cover in this book will be based on findings from self-report studies.

◪ What Do the Measures of Crime Show Regarding the Distribution of Crime?

It is important to examine the most aggregated trends of crime, namely the ups and downs of overall crime rates in the United States across different decades. We will start with crime in the early 1900s, largely because the best data started being collected during this era and also because the 20th century (especially the most recent decades) is most relevant to our understanding of the reasons for our current crime rates. However, most experts believe that the U.S. crime rate, whether in terms of violent or property offending, used to be far higher prior to the 20th century; this historians have concluded based on sporadic, poorly recorded documentation in the 18th and 19th centuries. By virtually all accounts, crime per capita (especially homicide) was far higher in the 1700s and 1800s than at any point after 1900, which is likely due to many factors, but perhaps most importantly because formal agencies of justice, such as police and corrections (i.e., prisons, parole, etc.), did not exist in most of the United States until the middle or end of the 1800s. Up to that time, it was up to individual communities or vigilantes to deal with offenders.

Therefore, there was little ability to investigate or apprehend criminals or a means to imprison them. But, as industrialization increased, the need to establish formal police agencies and correctional facilities evolved as a way to deal with people who offended in modern cities. By 1900, most existing states had formed police and prison systems, which is where our discussion will begin.

The level of crime in the United States, particularly homicide, was relatively low at the beginning of the 20th century, perhaps because of the formal justice agencies that had been created during the 19th century. For example, the first metropolitan U.S. police departments were formed in Boston and then New York

---

[19]Michael Hindelang, Travis Hirschi, and J. Weis, *Measuring Delinquency* (Beverly Hills: Sage, 1981).

during the 1830s; in the same decade but a bit earlier, the first state police department, the Texas Rangers, was established, and the federal marshals were founded still earlier. Although prisons started in the late 1790s, they did not begin to resemble their modern form or proliferate rapidly until the mid-1800s. The first juvenile court was formed in the Chicago area in 1899. The development of these formal law enforcement and justice agencies may have contributed to the low level of crime and homicide in the very early 1900s. (Note: our discussion of the crime rate in the early 1900s will primarily deal with homicide because it is the only valid record of crime at that time; UCR did not start until the 1930s. Most people consider this the most serious crime, and its frequency typically reflects the overall crime rate.)

The effect of the creation of these formal agencies did not persist long into the 20th century. Looking at the level of homicides that occurred in the United States, it is obvious that large increases took place between 1910 and 1920, which is likely due to extremely high increases in industrialization, as the U.S. economy moved from an agricultural to industrial emphasis. More important, population growth was rapid as a result of urbanization. Whenever high numbers of people move into an area (in this case, cities) and form a far more dense population (think of New York City at that time or Las Vegas in current times), it creates a crime problem. This is likely due to more opportunity for crime against others; after all, when people are crammed together, it creates a situation in which there are far more potential offenders in close proximity with far more potential victims. A good modern example of this is high schools, which studies show have higher crime rates than city subways or other crime-ridden areas, largely because they densely pack people together, and in such conditions, opportunities for crime are readily available. Thus, the rapid industrialization and urbanization of the early 1900s probably are the most important reasons for the increase in homicide and crime in general at that time in the United States.

The largest increases in U.S. homicide in the early 1900s occurred during the 1920s and early 1930s, with the peak level of homicide coming in the early 1930s. Criminologists and historians have concluded that this huge increase in the homicide rate was primarily due to two factors beyond the industrialization and urbanization that explained the increase prior to the 1920s. First, the U.S. Congress passed a constitutional amendment that banned the distribution and consumption of alcohol beginning in 1920. The period that followed is known as Prohibition. This legal action proved to be a disaster, at least in terms of crime, and Congress later agreed—but not until the 1930s—by passing another amendment to do away with the previous amendment that banned alcohol. For about 14 years, which notably recorded the highest U.S. rates of homicide and crime before 1950, the government attempted to stop people from drinking. Prior to Prohibition, gangsters were relatively passive and did not hold much power.

This political mandate gave the black market a lot of potential in terms of monetary profit and reasons for killing off competition. Some of the greatest massacres of rival gangs of organized crime syndicates (e.g., Italian Mafia) occurred during the Prohibition era. The impact on crime

▲ **Image 1.4**   Group portrait of a police department liquor squad posing with cases of confiscated alcohol and distilling equipment during Prohibition.

was likely only one of the many problems with Prohibition, but it was a very important and deadly one for our purposes. Once Prohibition ended in the early 1930s, the homicide and crime rates decreased significantly, which may have implications regarding modern drug policies. According to studies, many banned substances today are less violence producing than alcohol, which studies show is the most violence-causing substance. For example, most criminologists believe that the current War on Drugs may actually be causing far more crime than it seeks to prevent (even if it may be lowering the number of drug addicts), due to the black market it creates for drugs that are in demand, much like the case with alcohol during Prohibition.

Another major reason why the homicide rate and overall crime levels increased so much during the early 1930s was the Great Depression, which sent the United States into an unprecedented state of economic upheaval. Most historians and criminologists agree that the stock market crash of the late 1920s was a primary contributor to the large levels of homicide in the early 1930s. We will return to this subject later when we examine the strain theory of crime, which places an emphasis on economic structure and poverty as the primary causes of crime.

Although the homicide and crime rate experienced a significant drop after Prohibition was eliminated, another reason for this decrease likely was due to social policies of the New Deal, which was implemented by President Franklin D. Roosevelt. Such policies included creating new jobs for people hit hardest by the Depression through programs such as Job Corps and the Tennessee Valley Authority, both of which still exist today. Although such programs likely aided the economic (and thus crime) recovery for the United States, world events of the early 1940s provided the greatest reasons for the huge decreases that were seen at that time.

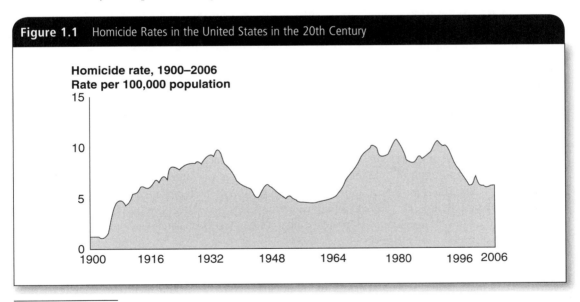

Figure 1.1   Homicide Rates in the United States in the 20th Century

Homicide rate, 1900–2006
Rate per 100,000 population

SOURCE: National Center for Health Statistics, Vital Statistics. Available from Bureau of Justice Statistics, http://www.google.com/imgres?imgurl=http:// bjs.ojp.usdoj.gov/content/glance/hmrt.gif&imgrefurl=http://bjs.ojp.usdoj.gov/content/glance/hmrt.cfm&usg=__AE1aBbgO72i4-jm8uUrCYdt7X6U= &h=314&w=372&sz=5&hl=en&start=0&zoom=1&tbnid=kLHaWLtLP8nPaM:&tbnh=143&tbnw=170&prev=/images%3Fq%3Dvital%2Bstatistics %2Bhomicide%26um%3D1%26hl%3Den%26client%3Dfirefox-a%26sa%3DN%26rls%3Dorg.mozilla:en-US:official%26channel%3Ds%26biw%3D12 80%26bih%3D607%26tbs%3Disch:1&um=1&itbs=1&iact=rc&dur=443&ei=UfnJTOmIG4-4sAOOirWPDg&oei=UfnJTOmIG4-4sAOOirWPDg&esq =1&page=1&ndsp=17&ved=1t:429,r:1,s:0&tx=96&ty=110. For related data about homicide trends, see *Homicide Trends in the U.S.*

NOTE: The 2001 rate includes deaths attributed to the 9/11 terrorist attacks.

The entry of the United States into World War II was probably the biggest contributor to decreasing U.S. crime in the early 20th century. As you will notice, the homicides decreased dramatically in the 4 years (1941–1945) that hundreds of thousands of young men (the most likely offenders and victims) were sent overseas to fight on two fronts, Europe and the South Pacific. Any time a society loses a huge portion of the most common offenders, namely young (teenage to twenties) males, they can expect a drop in crime like the one the United States experienced while these giant portions of men were elsewhere. However, at the end of 1945, most of these men returned and began making babies, which triggered the greatest increase in babies that U.S. society had ever seen. This generation of babies started what historians call the baby boom, which would have the greatest impact on crime levels that has ever been recorded in U.S. history.

Although crime rates increased after soldiers returned home from overseas in the late 1940s, they did not rise to anywhere near the levels of Prohibition and the Great Depression. Alcohol was legal, and the economy was doing relatively well after World War II. During the 1950s, the crime level remained relatively low and stable until the early 1960s, when the impact of the baby boom emerged in terms of the crime rate. If a very large share of the population is made up of young people, particularly teenage or early-20s males, the crime rate will almost inevitably go up. This is exactly what occurred in the United States, starting in the early 1960s, and it led to the largest 10-year increase in crime that the country has ever seen.

The Baby-Boom Effect

The UCR shows that the greatest single-decade increase in the crime rate occurred between 1965 and 1975. In that time, the overall crime rate more than doubled, an increase that was unprecedented. Notably, this increase occurred during the War on Poverty, which was set into motion by President Lyndon B. Johnson in a program he termed the Great Society; it turned many people and policy makers against having the government address economic issues for bettering society. However, the demographic factor that they did not take into account was that this was the era in which most people in society were in young age groups, which predisposed the nation for the high crime rates experienced at this time. In contrast, the following generation, called Generation X, which includes those individuals born between 1965 and 1978, had a very low birthrate, which may have contributed to the low crime rates observed in recent years.

The high level of young people in society was not the only societal trend going on during the late 1960s and early 1970s that would lead to higher crime rates. For example, a large number of people were being arrested as a result of their participation in the civil rights movement, the women's rights movement, and anti-Vietnam War activities. Perhaps most important, the 1970s had the highest levels of drug usage and favorable attitudes toward drugs since accurate national studies have been conducted. Virtually all measures of drug usage peaked during the 1970s or early 1980s.

So, ultimately, many things came together between the mid-1960s and the late 1970s that predisposed this generation of baby boomers to lead to the greatest increase in the crime rate that the United States has ever seen, culminating in the peak of crime and homicide in 1980. All of our measures agree that crime, especially homicide, reached their highest level about that year. Although other periods, such as the early 1990s, showed similar increases in crime, largely due to juvenile offending, no other period was higher than the 1980 period, and it was most likely due to the many baby boomers coming of age and high drug usage.

Crime levels declined somewhat in the early 1980s and then rose again in the late 1980s and early 1990s, but the crime and homicide rate never exceeded the 1980 peak. Furthermore, after 1994, the crime rate dropped drastically every year for about a decade to the point where it was as low as it had been in the early 1960s. Thus, the U.S. crime rate is currently about where it was about 40 years ago.

There are many reasons for this huge decrease over the last decade and a half. One of the biggest reasons is that the population has a relatively smaller proportion of young people than during the 1960s and 1970s, but obviously, there is more to the picture. Drug usage, as well as attitudes favorable toward drugs, has dropped a lot in recent years. Furthermore, the incarceration rate of prisoners is about 400% to 500% what it was in the early 1970s. Although many experts have claimed that locking up criminals does not reduce crime levels in a society, it is hard to deny that imprisoning five times more people will not result in catching many, if not most, of the habitual offenders. It makes sense: By rate, the United States currently locks up more citizens in prison than virtually any other developed country.

Almost all crime tends to be nonrandom. Consistent with this, the crime measures show a number of trends in which crime occurs among certain types of people, in certain places, at certain times, and so on. We turn to an examination of such concentrations of crime, starting with large, macro differences in crime rates across regions of the United States.

Regional and City Differences in Rates of Crime

Crime tends to be higher in certain regions of the country. According to the UCR, the United States is separated into four regions: Northeast, Midwest, South, and West. For the last few decades, crime rates (based on crime per capita) have been significantly higher in two of these regions: the South and the West. These two regions consistently have higher rates than the other regions, with one or the other having the highest rates for violence or property offenses or both each year. Some studies have found that, when poverty levels are accounted for, much of the regional differences are explained away. Although this is a very simple conclusion, the studies seem to be consistent in tying regional differences to variations in social factors, notably socioeconomic levels.

Regardless of the region, there seems to be extreme variation from high to low crime rates across states and cities within each of these large regions. For example, crime measures consistently show that certain U.S. states and jurisdictions have consistently high crime rates. The two standouts are Louisiana and the District of Columbia, with the latter having an extremely high rate, typically more than eight times the national average for homicide. It is quite ironic that arguably the most powerful city in the world has an extremely serious crime problem, certainly one of the worst in our nation.

Another question is why do crime rates in states or jurisdictions, or in cities and counties, vary drastically from one region to the next. For instance, Camden, New Jersey, one of the cities in the lower-rate Northeast region according to the UCR, had the highest rate of crime among all U.S. cities in the years 2004 and 2005. Detroit, Michigan, was second worst for both of these years; it used to be number one before Camden outdid it. At the same time, however, it is important to note that New Jersey also had some of the safest cities in the nation for these years—2 of the 10 safest cities are in New Jersey, which shows how much crime can vary from place to place, even those in relatively close proximity. Notably, in the most recent estimates from the FBI, the cities of St. Louis and New Orleans exhibited the highest rates of serious violent crimes in 2006 and 2007, respectively. An important factor regarding New Orleans was the devastation of the city's infrastructure after Hurricane Katrina, which essentially wiped out the city's criminal justice system and resources.

Crime has also been found to cluster within a given city, whether the overall city is relatively low crime or high crime. Virtually every area (whether urban, suburban, or rural) has what are known as **hot spots** or places that have high levels of crime activity. Such places are often bars or liquor stores or other types of businesses, such as bus stops and depots, convenience stores, shopping malls, motels and hotels, fast-food restaurants, check-cashing businesses, tattoo parlors, discount stores (such as dollar stores), and so on. However, hot spots can also be residential places, such as a home that police are constantly called to for domestic violence or certain apartment complexes that are crime ridden, something often seen in subsidized housing areas.

Even the nicest of cities or areas have hot spots, and even the worst cities or areas also tend to have most of their police calls coming from specific addresses or businesses. This is one of the best examples of how crime does not tend to be random. Many police agencies have begun using spatial software on computer systems to analyze where the hot spots in a given city are and to predict where certain crimes are likely to occur in the future. This allows more preventive patrols and proactive strategies in such zones. One criminological theory, **routine activities theory,** is largely based on explaining why hot spots exist.

Another way that crime is known to cluster is time of day. This varies greatly depending on the type of group one is examining. For example, juvenile delinquency and victimization, especially for violence, tends to peak sharply at 3:00 p.m. on school days (about half of the days of the year for children), which is the time that youths are most likely to lack supervision (i.e., children are let out of school and are often not supervised by adults until they get home). On the other hand, adult crime and victimization, especially for violence, tends to peak much later on, almost all days at about 11:00 p.m., which is a sharp contrast from the juvenile peak in midafternoon. These estimates are primarily based on FBI and UCR data.

To some extent, the peak hour for juveniles is misleading; other nonpolice-based estimates show that just as much crime is going on during school, but schools tend to not report it. This widespread lack of reporting by schools occurs because none of the parties involved want a formal police report taken. Typically, the youth doesn't want to be arrested, the child's parents do not want their son or daughter to be formally processed by police, and most importantly, no school wants to become known as a dangerous place. So, the police are typically called only in extreme cases, for example, if a student pulls a gun on another student or actually uses a weapon.

This also occurs in colleges and universities because such institutions largely depend on tuition for funding, and this goes down if enrollment levels decline. After all, no parents want to send their teenagers to a college that is high in crime. Federal law now requires virtually all colleges to report to the public their crime levels, so there is a lot at stake if police take formal reports on crime events. Thus, most colleges, like the K–12 school systems, have an informal process in place in which even violent crimes are often handled informally.

Crimes tend to peak significantly during the summer. Studies show that criminals tend to be highly opportunistic, meaning that they happen to be going about their normal activities when they see an opportunity to commit a crime, as compared to a more hydraulic model, in which an offender actually goes out looking to commit a crime. Because criminals are like everyone else in most ways, they tend to be out and about more in the summer, so they are more likely to see opportunities at that time. Furthermore, youths are typically out of school during the summer, so they are often bored and not supervised by adults as much as during the traditional school year. Burglary tends to rise 300% during the summer, an increase that may be linked to the fact that people go on vacation and leave their home vacant for weeks or months at a time. All of the factors come together to produce much higher rates in the summer than in any other season.

A couple of crimes, such as murder and robbery, tend to have a second peak in the winter, which most experts believe is due to high emotions during the holidays, additional social interaction (and often the drinking of alcohol) during the holidays, and an increase in wanting money or goods for gift giving, which would explain robbery increases at that time. These offenses are the exception, however, not the rule. Most offenses, including murder and robbery, tend to have their highest peaks during warmer summer months.

Age is perhaps the most important way that crime and victimization tend to cluster in certain groups. Almost no individuals get arrested before the age of 10; if they do, it is a huge predictor that the child is likely going to become a habitual, chronic offender. However, from the age of 10 to 17, the offending rate for all serious street crimes (i.e., FBI index crimes) goes up drastically, peaking at an average age of 17. Then, the offending rates begin decreasing significantly, such that by the time people reach the age of 20, the likelihood of being arrested has fallen in half as compared to the midteenage years. This offending level continues to decline throughout life for all of the serious index crimes, although other crimes are likely to be committed more often at later ages, such as white-collar crimes, tax evasion, and gambling.

The extraordinarily high levels of offending in the teenage years have implications for how we prevent and deal with juvenile delinquents and for why we are so bad at preventing habitual offenders from committing so many crimes. We are often good at predicting who the most chronic, serious offenders are by the time they are in their 20s, but that does little good because most offenders have committed most of their crimes before they hit 20 years old.

Another characteristic important in the way crime clusters is gender. In every society, at every time in recorded history, males commit far more serious street crimes (both violent and property) than do females. It appears there is almost no closing of this gap in offending, at least for FBI index crimes. Even in the most recent years, males are the offenders in 80% to 98% of all serious violent crimes (murder, robbery, aggravated assault, forcible rape) and make up the vast majority of offenders in property index crimes (burglary, motor vehicle theft, arson, and larceny). The last offense is often surprising to people because the most common type of larceny is shoplifting, which people often perceive as being done mostly by women. All studies show that men commit most of the larcenies in the United States. It is important to realize that males in all societies throughout the world commit the vast majority of offenses, and the more violent and serious the crimes, the more men are represented in such offenses.

However, there are a few nonindex crimes that females commit as much as, or more than, males. Specifically, in terms of property crimes, embezzlement and fraud are two offenses that females commit at comparable rates to men, which likely has to do with enhanced opportunities to commit such crimes. Most of the workforce is now female, which wasn't true in past decades, and many women work in banking and other businesses that tempt employees by having large amounts of money available for embezzling. In terms of public disorder, prostitution arrests tend to be mostly female, which is not too surprising.

The only other offense in which females are well represented is running away from home, which is a status offense (illegal for juveniles only). However, virtually all other sources and studies of offending rates (e.g., self-report studies) show that male juveniles run away far more than females, but because of societal norms and values, females get arrested far more than males. Feminist theories of the patriarchal model (in short, men are in charge and dominate or control females) and the chivalry model (simply, females are treated differently because they are seen as more innocent) argue that females are protected as a type of property. This may be important in light of the contrasting findings regarding female and male rates of running away, as compared to the opposite trends regarding female and male rates of being arrested for running away. The bottom line is that families are more likely to report the missing child and to press law enforcement agencies to pursue girls rather than boys who have run away. We will explore explanations for such differences in later chapters, particularly in the section in which we cover conflict and feminist theories of crime.

Victimization and offending rates are also clustered according to the density of a given area. All sources of crime data show that rates of offending and victimization are far higher per capita in urban areas than in suburban and rural regions. Furthermore, this trend is linear: The more rural an area, the lower the rates for crime and victimization. To clarify, urban areas have, by far, the highest rates for offending, followed by suburban areas; the least amount of crime and victimization is found in rural (e.g., farming) areas. This trend has been shown for many decades and is undisputed. Keep in mind that such rates are based on population in such areas, so even per capita of citizens in a given region, this trend holds true. This is likely due to enhanced opportunities to commit crime in urban and even suburban areas, as well as the fact that rural communities tend to have stronger **informal controls**, such as family, church, and community ties. Studies consistently show that informal controls are far more effective in preventing and solving crimes than are the **formal controls** of justice, which include all official aspects of law enforcement and justice, such as police, courts, and corrections (i.e., prisons, probation, parole).

A good example of the effectiveness of informal sanctions can be seen in the early formation of the U.S. colonies, such as the Massachusetts Bay Colony. In the early 1600s, crime per capita was at an all-time low in what would become the United States. It may surprise many that police and prisons did not exist then; rather, the low crime rate was due to high levels of informal controls: When people committed crimes, they were banished from the society or were shunned.

As in Nathaniel Hawthorne's novel *The Scarlet Letter*, even for what would now be considered a relatively minor offense (adultery), a person was forced to wear a large letter (such as an A for adultery or a T for theft), and they were shunned by all others in their social world. Such punishments were (and still are) highly effective punishments and deterrents for offenders, but they work only in communities with high levels of informal controls. Studies have shown that, in such societies, crime tends to be extremely low, in fact, so low that such communities may "invent" serious crimes so that they can have an identifiable group on which to blame societal problems. We see this occur in the Massachusetts Bay Colony, with the creation of a new offense, witchcraft, for which hundreds of people were put on trial and many were executed.

Such issues will be raised again later when we discuss the sociological theories of Emile Durkheim. Nevertheless, it is interesting to note that some judges and communities have gone back to such public shaming punishments, such as making people carry signs of their offenses and putting the names of offenders in the local newspapers. But, the ultimate conclusion at this point is that rural communities tend to have far higher informal controls, which keep their crime rates low. On the other hand, urban (especially inner-city) areas tend to have very low levels of informal controls, which lead to few attempts by the community to police itself.

Crime also tends to cluster according to social class, with the lower classes experiencing far more violent offending and victimization. This is now undisputed and consistently shown across all sources of data regarding criminal offending. Interestingly, the characteristics thus far examined for offending rates tend to be a mirror image for victimization rates. To clarify, young, poor urban males tend to have the highest rates of criminal offending—and this group also has the highest rates of victimization as a result of violent offending. This mirror image phenomenon is often referred to as the **equivalency hypothesis**. However, this equivalency hypothesis is not true regarding the relationship of social class and property crimes.

Specifically, members of middle- to upper-class households tend to experience more victimization for property crimes than do lower-class households, but the most likely offenders in most property crimes are from the lower class. This makes sense; offenders will tend to steal from the people and places that have the most property or money to steal. This tendency has been found since criminological data have been collected, even back to the early 1800s, and it is often found in present times, although not consistently shown each year (such as in the NCVS data). Nevertheless, the equivalency hypothesis still holds true for violent crimes: Lower-class individuals commit more violent crimes, and they are victimized more as a result of such violent crimes.

Race and Ethnicity Rates of Crime

Another important characteristic for how crimes are clustered in U.S. society is race or ethnicity. Regarding most violent crimes, the most victimized group by far in the United States is Native Americans or American Indians. According to NCVS data, Native Americans are victimized at almost twice the rate of any other racial or ethnic group. This is likely due to the extreme levels of poverty, unemployment, and so on, that exist on virtually all American Indian reservations. Although some Indian tribes have recently gained profits from operating gaming casinos on their lands, the vast majority of tribes in most states are not involved in such endeavors, so deprivation and poverty are still the norm.

Although there is little offending data regarding this group (the UCR does not adequately measure this group in their arrest data), it is generally assumed that Native Americans have the highest rates of offending as well. This is a fairly safe assumption because research has clearly shown that the vast majority of criminal offending or victimization is **intraracial**. This means that crime tends to occur within a race or ethnicity (e.g., Whites offending against Whites) as opposed to being **interracial**, or across race or ethnicity (e.g., Whites offending against Blacks).

Another major group that experiences an extremely high rate of victimization, particularly for homicide, is Blacks. (The term *Black* is used here, as opposed to African American because that is what most measures [UCR/NCVS] use and because many African Americans are not Black, for example, many citizens from Egypt or South Africa). According to UCR data for homicide, which the NCVS does not report, Blacks have by far the highest rates of victimization and offending. Again, this is likely due to the extreme levels of poverty experienced by this group, as well as the high levels of single-headed households among this population (which likely explains much of the poverty).

A good example of the extent of the high levels of homicide rates among Blacks, in terms of both offending and victimization, can be seen in previously examined rates of certain U.S. cities. Washington, DC, New Orleans, St. Louis, and Detroit are some U.S. cities with the highest murder rates. They also have some of the highest proportions of Black residents as compared to other U.S. cities. Notably, studies have shown that, when researchers control for poverty rates and single-headed households, the racial and ethnicity effects seem to be explained away.

For example, data show that Washington, DC, and the state of Louisiana are the top two jurisdictions for high rates of poverty, with most of the poor being children or teenagers, who are most prone to commit crimes, especially violent offenses (such as murder). Thus, the two highest racial and ethnic groups for violent crime—Native Americans and Blacks—both as victims and offenders, also tend to have the highest rates in the nation for poverty and broken families. It should also be noted that Hispanics, a common ethnic group in the United States, also have relatively high offending and victimization rates as compared to Whites and other minorities, such as Asians, who tend to experience low rates.

Although there are numerous other ways that crime tends to be clustered in our society, we have covered most of the major variables in this chapter. The rest of this book will now deal with examining the various theories for why crime tends to occur in such groups and individuals, and the way that we determine how accurate these theories are will directly relate to how well they explain the high rates of the groups that we have discussed.

Policy Implications

One of the key elements of a good theory is that it can help inform policy makers in making decisions about how to reduce crime. This is a theme that will be reviewed at the end of each chapter in this book. After all, a criminological theory is only truly useful in the real world if it helps to reduce criminal offending there. Many theories have been used as the basis of such changes in policy, and we will present examples of how theories of crime discussed in each chapter have guided policy making. We will also present empirical evidence for such policies, specifically whether or not such policies were successful (and many are not).

⊠ Chapter Summary

- Criminology is the scientific study of crime, which involves the use of the scientific method and testing hypotheses and distinguishes it from other perspectives of crime (e.g., journalism, religious, legal) because these fields are not based on science.
- Criminological theory seeks to do more than simply predict criminal behavior; rather, the goal is to more fully understand the reasons why certain individuals offend or do not.
- Definitions of crime vary drastically across time and place; most criminologists tend to focus on deviant behaviors whether or not they violate the criminal codes.
- There are various paradigms or unique perspectives of crime, which have contrasting assumptions and propositions. The Classical School of criminological theory assumes free will and choice among individuals, whereas the Positive School rejects the notion of free will and focuses on biological, social, and psychological factors that negate the notion of choice among criminal offenders. Other theoretical paradigms, such as conflict and critical offending and integrated models, also exist.
- Criminological theories can be classified by their fundamental assumptions as well as other factors, such as the unit of measure on which they focus—the individual or the group, micro or macro, respectively—or the assumptions they make regarding basic human nature: good, bad, or blank slate.
- There are more than a handful of characteristics that good criminological theories should have and that bad criminological theories do not have.
- Three criteria are required for determining causality, which is essential in determining whether or not an independent (predictive) variable (X) actually affects a dependent (consequent) variable (Y).
- This introduction also discussed the primary measures that are used for estimating the crime committed in the United States; these primary measures are police reports (e.g., the UCR), victimization surveys (e.g., NCVS), and offenders' self-reports of the crimes they have committed.
- We also discussed in this section what the measures of crime have shown regarding the distribution of criminal offending in terms of region, race and ethnicity, time of day, gender, and socioeconomic status. It is important to note that various theories presented in this book will be tested on the extent to which they can explain the clustering of crime among certain individuals and groups.

KEY TERMS

Classical School	Equivalency hypothesis	Logical consistency
Clearance rate	Formal controls	Macrolevel of analysis
Conflict perspective	Hot spots	*Mala in se*
Consensual perspective	Hypotheses	*Mala prohibita*
Correlation or covariation	Index offenses	Microlevel of analysis
Criminology	Informal controls	National Crime Victimization Survey (NCVS)
Dark figure	Interracial	Nonindex offenses
Deviance	Intraracial	Paradigm
Empirical validity	Legalistic approach	Parsimony

Policy implications

Positive School

Scientific method

Scope

Self-report data (SRD)

Social sciences

Spuriousness

Telescoping

Temporal ordering

Testability

Theory

Uniform Crime Report (UCR)

DISCUSSION QUESTIONS

1. How does criminology differ from other perspectives of crime?

2. How does a good theoretical explanation of crime go beyond simply predicting crime?

3. Should criminologists emphasize only crimes made illegal by law, or should they also study acts that are deviant but not illegal? Explain why you feel this way.

4. Even though you haven't been exposed to all the theories in this book, do you favor the classical or positive assumptions regarding criminal behavior? Explain why.

5. Which types of theories do you think are most important, those that explain individual behavior (micro) or those that explain criminality among groups (macro)? Explain your decision.

6. Do you tend to favor the idea that human beings are born bad, good, or blank slate? Discuss your decision and provide an example.

7. Which characteristics of a good theory do you find most important? Which are least important? Make sure to explain why you feel that way.

8. Of the three criteria for determining causality between a predictor variable and a consequent variable, which do you think is easiest to show, and which is hardest to show? Why?

9. Which of the measures of crime do you think is the best estimate of crime in the United States? Why? Also, which measure is the best to use for determining associations between crime and personality traits? Why?

10. Looking at what the measures of crime show, which finding surprises you the most? Explain why.

11. What types of policies do you feel reduce crime the most? Which do you think have little or no effect in reducing crime?

WEB RESOURCES

Measures of Crime

Overview of Criminological Theory (by Dr. Cecil Greek at Florida State University): http://www.criminology.fsu.edu/crimtheory/

FBI's UCR: http://www.fbi.gov/ucr/ucr.htm

NCVS: http://bjs.ojp.usdoj.gov/index.cfm?ty=dcdetail&iid=245

Preclassical and Classical Theories of Crime

This chapter will examine the earliest logical theories of rule breaking, namely explanations of criminal conduct that emphasize free will and the ability of individuals to make rational decisions regarding the consequences of their behavior. The natural capabilities of human beings to make decisions based on expected costs and benefits were acknowledged during the **Age of Enlightenment** in the 17th and 18th centuries. This understanding of human capabilities led to what is considered the first rational theory of criminal activity, **deterrence theory.** The importance and impact of this theory had a more profound impact on justice systems in the United States than any other perspective to date. Furthermore, virtually all criminal justice systems (e.g., policing, courts, corrections) are based on this theoretical model even today.

Such theories of human rationality were in stark contrast to the theories focused on religious or supernatural causes of crime, which had prevailed through most of human civilization up to the Age of Enlightenment. In addition, the Classical School theories of crime are distinguished from theories in subsequent chapters of this book by their emphasis on free will and rational decision making of individuals, which modern theories of crime tend to ignore. The theoretical perspectives discussed in this chapter all focus on the ability of human beings to choose their own behavior and destinies, whereas paradigms that existed before and after this period tend to emphasize the inability of individuals to control their behavior due to external factors. Therefore, the Classical School is perhaps the paradigm best suited for analysis of what types of calculations are going on in someone's head before they commit a crime.

The different Classical School theories presented in this chapter vary in many ways, most notably in what they propose as the primary constructs and processes individuals use to determine whether they are going to commit a crime. For example, some Classical School theories place an emphasis on the potential negative consequences of their actions, whereas others focus on the possible benefits of such activity. Still others concentrate on the opportunities and existing situations that predispose people to engage in criminal activity. Regardless of their differences, all of the theories examined in this chapter emphasize a common theme: Individuals commit crime because they identify certain situations and acts as beneficial due to the perceived lack of punishment and the perceived likelihood of profits, such as money or peer status. In other words, the potential offender weighs out the possible costs and pleasures of committing a given act and then behaves in a rational way based on this analysis.

The most important distinction of these Classical School theories, as opposed to those discussed in future chapters, is that they emphasize individuals making their own decisions regardless of extraneous influences, such as the economy or bonding with society. Although many extraneous factors may influence the ability of an individual to rationally consider offending situations, the Classical School assumes that the individual takes all these influences into account when making the decision about whether or not to engage in criminal behavior. Given the focus placed on individual responsibility, it is not surprising that Classical School theories are used as the basis for the U.S. policies on punishment for criminal activity. The Classical School theories are highly compatible and consistent with the conservative, get-tough movement that has existed since the mid-1970s. Thus, the Classical School still retains the highest importance in terms of policy and pragmatic punishment in the United States, as well as throughout all countries in the Western world.

As you will see, the Classical School theoretical paradigm was presented as early as the mid-1700s, and its prominence as a model of offending behavior in criminal justice systems is still dominant. Scientific and academic circles, however, have dismissed many of the claims of this perspective. For reasons we shall explore in this chapter, the assumptions and primary propositions of Classical School theories have been neglected by most recent theoretical models of criminologists. This dismissal is likely premature, given the impact that this perspective has had on understanding human nature, as well as the profound influence it has had on most criminal justice systems, especially in the United States.

◿ Preclassical Perspectives of Crime and Punishment

Over the long course of human civilization, people have mostly believed that criminal activity was caused by supernatural causes or religious factors. It has been documented that some primitive societies believed that crime increased during major thunderstorms or major droughts. Most primitive cultures believed that, when a person engaged in behavior that violated the tribe's or clan's rules, the devil or evil spirits were making them do it.[1] For example, in many societies at that time, if someone committed criminal activity, it was common to perform exorcisms or primitive surgery, such as breaking open the skull of the perpetrator to allow the demons to leave his or her head. Of course, this almost always resulted in the death of the accused person, but it was seen as a liberating experience for the offender.

This was just one form of dealing with criminal behavior, but it epitomizes how primitive cultures understood the causes of crime. As the movie *The Exorcist* revealed, exorcisms were still being performed on offenders by representatives of a number of religions, including Catholicism, in the 21st century to get the devil out of them, as commonly said. In June 2005, a Romanian monk and four nuns acknowledged engaging in an exorcism that led to the death of the victim, who was crucified, a towel stuffed in her mouth, and left without food for many days.[2] When this monk and the nuns were asked to explain why they did this, they defiantly said they were trying to take the devils out of the 23-year-old woman. Although they were prosecuted by Romanian authorities, many governments might not have done so because many societies around the world still believe in and condone such practices.

---

[1]Stephen E. Brown, Finn-Aage Esbensen, and Gilbert Geis, *Criminology: Explaining Crime and Its Context,* 5th ed. (Cincinnati: Anderson, 2004).
[2]Associated Press, "Five Charged in Deadly Exorcism," June 24, 2005.

Readers may be surprised to learn that the Roman Catholic Church still authorizes college-level courses on how to perform exorcisms. Specifically, recent news reports revealed that a Vatican-recognized university was offering a course in exorcism and demonic possession for a second year because of their concern about the "devil's lure."[3] In fact, Pope Benedict XVI welcomed a large group of Italian exorcists who visited the Vatican in September 2005 and encouraged them to carry on their work for the Catholic Church.[4] Furthermore, in 1999, the Roman Catholic Church issued their revised guidelines for conducting exorcisms, which recommend consulting physicians when exorcisms are performed; it also provides an 84-page description of the language (in Latin) to be used in such rituals. It should be noted that the use of such exorcisms is quite rare, especially in more developed nations. However, U.S. Catholic bishops recently (in November 2010) cited the need for more trained exorcists and even held a conference in Baltimore, Maryland, on how to conduct them. This two-day training session instructed clergy on evaluating evil possession, as well as reviewing the rituals that comprise an exorcism. More than 50 bishops and 60 priests attended this training session, despite the tendency for exorcists in U.S. dioceses to keep a very low profile.[5]

One of the most common supernatural beliefs in primitive cultures was that the full moon caused criminal activity. Then, as now, there was much truth to the full-moon theory. In primitive times, people believed that crime was related to the influence of higher powers, including the destructive influence of the moon itself. Modern studies have shown, however, that the increase in crime is primarily due to a Classical School theoretical model: There are simply more opportunities to commit crime when the moon is full because there is more light at night, which results in more people out on the streets. In any case, nighttime is well established as a high-risk period for adult crime, such as sexual assault.

Although some primitive theories had some validity in determining when crime was more common, virtually none of them accurately predicted who would commit the offenses. During the Middle Ages, just about everyone was from the lower classes, and only a minority of that group engaged in offending against the normative society. So, for most of human civilization, there was virtually no rational theoretical understanding for why individuals violate the laws of society; instead, people believed crime was caused by supernatural or religious factors of the devil-made-me-do-it variety.

Consistent with these views, the punishments related to offending during this period were quite harsh by modern standards. Given the assumption that evil spirits drove the motivations for criminal activity, the punishments of criminal acts, especially those deemed particularly offensive to the norms of a given society, were often quite inhumane. Common punishments at that time were beheading, torture, being burned alive at the stake, and being drowned, stoned, or quartered. Good discussions of such harsh examples of punishment, such as quartering, can be found in Brown et al.'s discussion of punishment.[6]

Although many would find the primitive forms of punishment and execution to be quite barbaric, some modern societies still practice them. For example, Islamic court systems, as well as other religious and ethnic cultures, are often allowed to carry out executions and other forms of corporal punishment. Fifteen individuals were whipped with a cane for gambling in Aceh, Indonesia, a highly conservative Muslim region. The caning was held in public and outside a mosque.[7] In the United States, gambling is a relatively minor crime—when it is not totally legal, as in many places in the United States. It is interesting to note, however,

---

[3] Associated Press, "University Offering Course on Exorcism," July 9, 2005.

[4] Reuters, "Pope Benedict XVI Warmly Greets Exorcist Convention," September 17, 2005.

[5] Associated Press, "Catholic Bishops: More Exorcists Needed," November 13, 2010.

[6] Brown et al., *Criminology*, 177–179 (see chap. 1, n.1).

[7] Irwan Firdaus, "Gamblers Whipped in Muslim Indonesia," Associated Press, June 25, 2005.

that a relatively recent Gallup poll regarding the use of caning (i.e., public whipping) of convicted individuals was supported by most of the American public.[8]

Compared to U.S. standards, the more extreme forms of corporal punishment, particularly public executions carried out by many religious courts and countries, are drawn-out and very painful. An example is stoning, in which people are buried up to the waist and local citizens throw small stones at them until they die (large stones are not allowed because they would lead to death too quickly). In most of the Western world, such brutal forms of punishment and execution were done away with in the 1700s due to the impact of the Age of Enlightenment.

The Age of Enlightenment

In the midst of the extremely draconian times of the mid-1600s, Thomas Hobbes, in his book *Leviathan* (1651), proposed a rational theory of why people are motivated to form democratic states of governance.[9] Hobbes started with a basic framework that all individuals are in a constant state of warfare with all other individuals. Hobbes used the extreme examples of primitive tribes and sects. He argued that the primitive state of man is selfish and greedy, so people live in a constant state of fear of everyone else. However, Hobbes also proclaimed that people are also rational, so they will rationally organize sound forms of governance, which can create rules to avoid this constant state of fear. Interestingly, once a government is created, the state of warfare mutates from one waged among individuals or families to one between nations. This can be seen in modern times; after all, it is quite rare that a person or family declares war against another (although gangs may be an exception), but we often hear of governments declaring war.

Hobbes stated that the primitive state of fear—of constant warfare of everyone against everyone else—was the motivation for entering into a contract with others in creating a common authority. At the same time, Hobbes also specified that it was this exact emotion—fear—that was needed to make citizens conform to the given rules or laws in society. Strangely, it appears that the very emotion that inspires people to enter into an agreement to form a government is the same emotion that inspires these same individuals to follow the rules of the government that is created. It is ironic, but it is quite true.

Given the social conditions during the 1600s, this model appears somewhat accurate; there was little sense of community regarding working toward progress as a group. It had not been that long since the Middle Ages, when a third of the world's population had died from sickness, and many cultures were severely deprived or in extreme states of poverty. Prior to the 1600s, the feudal system was the dominant model of governance in most of the Western world. During this feudal era, a very small group of aristocrats (less than 1% of the population) owned and operated the largely agricultural economy that existed. Virtually no rights were afforded to individuals in the Middle Ages or at the time Hobbes wrote his book. Most people had no individual rights regarding criminal justice, let alone a say in the existing governments of the time. Hobbes's book clearly took issue with this lack of say in the government, which has had profound implications regarding the justice systems of that time.

Hobbes clearly stated that, until the citizens were entitled to a certain degree of respect from their governing bodies, as well as their justice systems, they would never fully buy into the authority of government or the system of justice. Hobbes proposed a number of extraordinary ideas that came to define the Age

---

[8]Philip Shenon, "U.S. Youth in Singapore Loses Appeal on Flogging," *New York Times,* April 1, 1994.

[9]Thomas Hobbes, *Leviathan* (New York: Library of Liberal Arts, 1958; first published in 1651).

of Enlightenment. He presented a drastic paradigm shift to this new social structure, which had extreme implications for justice systems throughout the world.

Hobbes explicitly declared that people are rational beings who choose their destinies by creating societies. Hobbes further proposed that individuals in such societies democratically create rules of conduct that all members of that society must follow. These rules, which all citizens decide upon, become laws, and the result of not following the laws is punishment determined by the democratically instituted government. It is clear from Hobbes's statements that the government, as instructed by the citizens, not only has the authority to punish individuals who violate the rules of the society but, more importantly, has a duty to punish them. When such an authority fails to fulfill this duty, it can quickly result in a breakdown in the social order.

The arrangement of citizens promising to abide by the rules or laws set forth by a given society in return for protection is commonly referred to as the **social contract**. Hobbes introduced this idea, but it was also emphasized by all other Enlightenment theorists after him, such as Rousseau, Locke, Voltaire, Montesquieu, and others. The idea of the social contract is an extraordinarily important part of Enlightenment philosophy. Although Enlightenment philosophers had significant differences in what they believed, the one thing they had in common was the belief in the social contract: the idea that people invest in the laws of their society with the guarantee that they will be protected from others who violate such rules.

Another shared belief among Enlightenment philosophers was giving people a say in the government, especially the justice system. All of them emphasized fairness in determining who was guilty, as well as appropriate punishments or sentences. During the time in which Enlightenment philosophers wrote, individuals who stole a loaf of bread to feed their families were sentenced to death, whereas upper-class individuals who stole large sums of money or committed murder were pardoned. This goes against common sense; moreover, it violates the social contract. If citizens observe people being excused for violating the law, then their belief in the social contract breaks down. This same feeling can be applied to modern times. When the Los Angeles police officers who were filmed beating a suspect were acquitted of criminal charges, a massive riot erupted among the citizens of the community. This was a good example of the social contract breaking down when people realized that the government failed to punish members of the community (ironically and, significantly, police officers) who had violated its rules.

The concept of the social contract was likely the most important contribution of Enlightenment philosophers, but there were others. Another key concept of Enlightenment philosophers focused on democracy, emphasizing that every person in society should have a say via the government; specifically, they promoted the ideal of *one person, one vote*. Granted, at the time they wrote, this meant one vote for each White, land-owning male, and not for women, minorities, or the poor, but it was a step in the right direction. Until then, no individuals outside of the aristocracy had any say in government or the justice system.

The Enlightenment philosophers also talked about each individual's right to the pursuit of life, liberty, and happiness. This probably sounds familiar because it is contained in the U.S. Constitution. Until the Enlightenment, individuals were not considered to have these rights; rather, they were seen as instruments for serving totalitarian governments. Although most citizens of the Western world take these rights for granted, they did not exist prior to the Age of Enlightenment.

Perhaps the most relevant concept that Enlightenment philosophers emphasized, as mentioned above, was the idea that human beings are rational and therefore have free will. The philosophers of this age focused on the ability of individuals to consider the consequences of their actions, and they assumed that people freely choose their behavior (or lack thereof), especially in regard to criminal activity. The father of criminal justice made this assumption in his formulation of what is considered to be the first bona fide theory of why people commit crime.

⚅ The Classical School of Criminology

The foundation of the Classical School of criminological theorizing is typically traced to the Enlightenment philosophers, but the specific origin of the Classical School is considered to be the 1764 publication of *On Crimes and Punishments* by Italian scholar Cesare Bonesana, Marchese de Beccaria (1738–1794), commonly known as Cesare Beccaria. Amazingly, he wrote this book at the age of 26 and published it anonymously, but its almost instant popularity persuaded him to come forward as the author. Due to this significant work, most experts consider Beccaria the father of criminal justice and the father of the Classical School of criminology, but perhaps most importantly, the father of deterrence theory. This section provides a comprehensive survey of the ideas and impact of Cesare Beccaria and the Classical School.

Influences on Beccaria and His Writings

The Enlightenment philosophers had a profound impact on the social and political climate of the late 1600s and 1700s. Growing up in this time period, Beccaria was a child of the Enlightenment, and as such, he was highly influenced by the concepts and propositions that these great thinkers proposed. The Enlightenment philosophy is readily evident in Beccaria's essay, and he incorporates much of its assumptions into his work. As a student of law, Beccaria had a good background for determining what was and was not rational in legal policy. But his loyalty to the Enlightenment ideal was ever-present throughout his work.

Beccaria places an emphasis on the concept of the social contract and incorporates the idea that citizens give up certain rights in exchange for the state's or government's protection. He also asserts that acts or punishments by the government that violate the overall sense of unity will not be accepted by the populace, largely due to the need for the social contract to be a fair deal. Beccaria explicitly stated that laws are compacts of free individuals in a society. In addition, he specifically notes his appeal to the ideal of the greatest happiness shared by the greatest number, which is otherwise known as **utilitarianism**. This, too, was a focus of Enlightenment philosophers. Finally, the emphasis on free will and individual choice is key to his propositions and theorizing. Indeed, as we shall see, Enlightenment philosophy is present in virtually all of his propositions; he directly cites Hobbes, Montesquieu, and other Enlightenment thinkers in his work.[10]

Beccaria's Proposed Reforms and Ideas of Justice

When Beccaria wrote, authoritarian governments ruled the justice systems, which were actually quite unjust during that time. For example, it was not uncommon that a person who stole a loaf of bread to feed his or her family would be imprisoned for life or executed. A good example of this is seen in the story of *Les Misérables:* The protagonist, Jean Valjean, gets a lengthy prison sentence for stealing food for his starving loved ones. On the other hand, a judge might excuse a person who had committed several murders because the confessed killer was from a prominent family.

▲ **Image 2.1**   Cesare Beccaria (1738–1794)

---

[10]Hobbes, *Leviathan;* Charles Louis de Secondat, Baron de la Brede et de Montesquieu, *The Spirit of the Laws* (New York: Library of Liberal Arts, 1949; first published in 1748).

Beccaria sought to rid the justice system of such arbitrary acts on the part of individual judges. Specifically, Beccaria claimed in his essay that "only laws can decree punishments for crimes . . . judges in criminal cases cannot have the authority to interpret laws."[11] Rather, he believed that legislatures, elected by the citizens, must define crimes and the specific punishment for each criminal offense. One of his main goals was to prevent a single person from assigning an overly harsh sentence on a defendant and allowing another defendant in a similar case to walk free for the same criminal act, which was quite common at that time. Thus, Beccaria's writing calls for a set punishment for a given offense without consideration of the presiding judge's personal attitudes or the defendant's background.

Beccaria believed that "the true measure of crimes is namely the harm done to society."[12] Thus, anyone at all who committed a given act against the society should face the same consequence. He was very clear that the law should impose a specific punishment for a given act, regardless of the contextual circumstances. One aspect of this principle was that it ignored the intent the offender had in committing the crime. Obviously, this is not true of most modern justice systems; intent often plays a key role in the charges and sentencing of defendants in many types of crimes. Most notably, the different degrees of homicide in most U.S. jurisdictions include first-degree murder, which requires proof of planning or *malice aforethought;* second-degree murder, which typically involves no evidence of planning but rather a spontaneous act of killing; and various degrees of manslaughter, which generally include some level of provocation on the part of the victim. This is just one example of how important intent, legally known as **mens rea** (literally, *guilty mind*), is in most modern justice systems. Many types of offending are graded by degree of intent, as opposed to the focus on only the act itself, known legally as **actus reus** (literally, *guilty act*). Beccaria's propositions focus on only the *actus reus* because he claimed that an act against society was just as harmful, regardless of the intent, or *mens rea.* Despite his recommendations, most societies factor in the intent of the offender in criminal activity. Still, this proposal of *a given act should be given equal punishment* certainly seemed to represent a significant improvement over the arbitrary punishments in the regimes and justice systems of the 1700s.

Another important reform that Beccaria proposed was to do away with practices that were common in "justice" systems of the time—the word *justice* is in quotes because they were largely systems of injustice. Specifically, Beccaria claimed that secret accusations should not be permitted; rather, defendants should be able to confront and cross-examine witnesses. Writing about secret accusations, he says "their customary use makes men false and deceptive"; he asks, "Who can defend himself against calumny when it comes armed with tyranny's strongest shield, *secrecy?*"[13] Although some modern countries still accept and use secret accusations and disallow the cross-examination of witnesses, Beccaria set the standard in guaranteeing such rights to defendants in the United States and most Western societies.

In addition, Beccaria argued that torture should not be used against defendants:

> A cruelty consecrated by the practice of most nations is torture of the accused . . . either to make him confess the crime or to clear up contradictory statements, or to discover accomplices . . . to discover other crimes of which he might be guilty but of which he is not accused.[14]

---

[11]Cesare Beccaria, *On Crimes and Punishments,* trans. Henry Paolucci (New York: MacMillan, 1963), 14.

[12]Ibid., 64.

[13]Ibid., 25–26.

[14]Ibid., 30.

Although some countries, such as Israel and Mexico, currently allow the use of torture for eliciting information or confessions, most countries abstain from the practice. There has been wide discussion about a memo, written by former U.S. Attorney General Alberto Gonzales when he was President George W. Bush's lead counsel at the White House, claiming that the U.S. military could use torture against terrorist suspects. However, the United States has traditionally agreed with Beccaria, who believed that any information or oaths obtained under torture were relatively worthless; our country apparently agrees, at least in terms of domestic criminal defendants. Beccaria's belief in the worthlessness of torture is further seen in his statement that "it is useless to reveal the author of a crime that lies deeply buried in darkness."[15]

It is likely that Beccaria believed the use of torture was one of the worst aspects of the criminal justice systems of his time and a horrible manifestation of the truly barbaric acts common in the Middle Ages. This is seen in his further elaboration of torture:

> This infamous crucible of truth is a still-standing memorial of the ancient and barbarous legislation of a time when trials by fire and by boiling water, as well as the uncertain outcomes of duels, were called "judgments of God."[16]

Beccaria also expressed his doubt of the relevance of any information received via method of torture:

> Thus the impression of pain may become so great that, filling the entire sensory capacity of the tortured person, it leaves him free only to choose what for the moment is the shortest way of escape from pain.[17]

As Beccaria sees it, the policy implications from such use of torture are that "of two men, equally innocent or equally guilty, the strong and courageous will be acquitted, the weak and timid condemned."[18]

Beccaria also claimed that defendants should be tried by fellow citizens or peers, not by judges: "I consider an excellent law that which assigns popular jurors, taken by lot, to assist the chief judge . . . that each man ought to be judged by his peers."[19] It is clear that Beccaria felt that the responsibility of determining the facts of a case should be placed in the hands of more than one person, a belief driven by his Enlightenment beliefs about democratic philosophy, in which citizens of the society should have a voice and serve in judging the facts and deciding the verdicts of criminal cases. This proposition is representative of Beccaria's overall philosophy toward fairness and democratic processes, which Enlightenment philosophers shared.

Today, U.S. citizens often take for granted the right to have a trial by a jury of their peers. It may surprise some readers to know that some modern, developed countries have not provided this right. For example, Russia just recently held its first jury trials in 85 years. When Vladimir Lenin was in charge of Russia, he banished jury trials. Over the course of several decades, the bench trials in Russia produced a 99.6% rate of convictions. This means that virtually every person in Russia who was accused of a crime was found guilty. Given the relatively high percentage of defendants found to be innocent of crimes in the United States, not to mention the numerous people who have been released from death row after DNA analysis showed they were not guilty, it is rather frightening to think of how many falsely accused individuals were convicted and unjustly sentenced in Russia over the last century.

---

[15]Ibid., 30–31.

[16]Ibid., 31.

[17]Ibid., 32.

[18]Ibid., 32.

[19]Ibid., 21.

Another important area of Beccaria's reforms involves the emphasis on making the justice system, particularly its laws and decisions, more public and better understood. This fits the Enlightenment assumption that individuals are rational: If people know the consequences of their actions, they will act accordingly. Beccaria stated that "when the number of those who can understand the sacred code of laws and hold it in their hands increases, the frequency of crimes will be found to decrease."[20] At the time, the laws were often unknown to the populace, in part because of widespread illiteracy but perhaps more as a result of the failure to publicly declare what the laws were. Even when laws were posted, they were often in languages (e.g., Latin) that the citizens did not read or speak. So, Beccaria stressed the need for society to ensure that the citizens were educated about what the laws were; he believed that this alone would lead to a significant decrease in law violations.

Furthermore, Beccaria believed that the important stages and decision-making processes of any justice system should be public knowledge rather than being held in secret or decided behind closed doors. As Beccaria stated, "Punishment ... must be essentially public."[21] This has a highly democratic and Enlightenment feel to it in the sense that citizens of a society have the right to know what vital judgments are being made. After all, in a democratic society, citizens give the government the deeply profound responsibility of distributing punishment for crimes against the society. The citizens are entitled to know what decisions their government officials are making, particularly regarding justice. Besides providing knowledge and an understanding of what is going on, it also ensures a form of checks and balances on what is happening. Furthermore, the public nature of trials and punishments inherently produces a form of deterrence for those individuals who may be considering such criminal activity.

One of Beccaria's most profound and important proposed reforms is one of the least noted. Beccaria said, "The surest but most difficult way to prevent crimes is by perfecting education."[22] We know of no other review of his work that notes this hypothesis, which is quite amazing, because most of the reviews are done for an educational audience. Furthermore, this emphasis on education makes sense given Beccaria's emphasis on knowledge of laws and consequences of criminal activity, as well as his focus on deterrence.

Beccaria's Ideas Regarding the Death Penalty

Another primary area of Beccaria's reforms dealt with the use—and, in his day, the abuse—of the death penalty. First, let it be said that Beccaria was against the use of capital punishment. (Interestingly, he was not against corporal punishment, which he explicitly stated was appropriate for violent offenders.) Perhaps this was due to the times in which he wrote, in which a large number of people were put to death, often by harsh methods. Still, Beccaria had several rational reasons for why he felt the death penalty was not an efficient and effective punishment.

First, Beccaria argued that the use of capital punishment inherently violated the social contract:

Is it conceivable that the least sacrifice of each person's liberty should include sacrifice of the greatest of all goods, life? . . . The punishment of death, therefore, is not a right, for I have demonstrated that it cannot be such; but it is the war of a nation against a citizen whose destruction it judges to be necessary or useful.[23]

[20]Ibid., 17.

[21]Ibid., 99.

[22]Ibid., 98.

[23]Ibid., 45.

In a related theme, the second reason that Beccaria felt that the death penalty was an inappropriate form of punishment was that, if the government endorsed the death of a citizen, it would provide a negative example to the rest of society. He said, "The death penalty cannot be useful, because of the example of barbarity it gives men."[24] Although some studies show some evidence that use of the death penalty in the United States deters crime,[25] most studies show no effect or even a positive effect on homicides.[26] Researchers have called this increase of homicides after executions the **brutalization effect,** and a similar phenomenon can be seen at numerous sporting events when violence breaks out among spectators at boxing matches, hockey games, soccer or football games, and so on. There have even been incidents in recent years at youth sporting events.

To further complicate the possible contradictory effects of capital punishment, some analyses show that both deterrence and brutalization occur at the same time for different types of murder or crime, depending on the level of planning or spontaneity of a given act. For example, a sophisticated analysis of homicide data from California examined the effects of a high-profile execution in 1992, largely because it was the first one in the state in 25 years.[27] As predicted, the authors found that nonstranger felony murders, which typically involve some planning, significantly decreased after the high-profile execution, whereas the level of argument-based, stranger murders, which are typically more spontaneous, significantly increased during the same time period. Thus, both the effects of deterrence and brutalization were observed at the same time and location following a given execution.

Another primary reason that Beccaria was against the use of capital punishment was that he believed it was an ineffective deterrent. Specifically, he thought that a punishment that was quick, such as the death penalty, could not be an effective deterrent as compared to a drawn-out penalty. As Beccaria stated, "It is not the intensity of punishment that has the greatest effect on the human spirit, but its duration."[28] It is likely that many readers can relate to this type of argument, not that they necessarily agree with it; the idea of spending the rest of one's life in a cell is a very scary concept to most people. To many people, such a concept is more frightening than death, which supports Beccaria's idea that the duration of the punishment may be more of a deterrent than the short, albeit extremely intense, punishment of execution.

---

[24]Ibid., 50.

[25]For a review and analysis showing a deterrent effect of capital punishment, see Steven Stack, "The Effect of Publicized Executions on Homicides in California," *Journal of Crime and Justice 21* (1998): 1–16. Also see Isaac Ehrlich, "The Deterrent Effect of Capital Punishment: A Question of Life and Death," *American Economic Review 65* (1975): 397–417; Isaac Ehrlich, "Capital Punishment and Deterrence," *Journal of Political Economy 85* (1977): 741–788; Stephen K. Layson, "Homicide and Deterrence: A Reexamination of United States Time-Series Evidence," *Southern Economic Journal 52* (1985): 68–89; David P. Phillips, "The Deterrent Effect of Capital Punishment: Evidence on an Old Controversy," *American Journal of Sociology 86* (1980): 139–148; Steven Stack, "Publicized Executions and Homicide, 1950–1980," *American Sociological Review 52* (1987): 532–540; Steven Stack, "Execution Publicity and Homicide in South Carolina," *Sociological Quarterly 31* (1990): 599–611; Steven Stack, "The Impact of Publicized Executions on Homicide," *Criminal Justice and Behavior 22* (1995): 172–186.

[26]William C. Bailey, "The Deterrent Effect of the Death Penalty for Murder in California," *Southern California Law Review 52* (1979): 743–764; David Lester, "The Deterrent Effect of Execution on Homicide," *Psychological Reports 64* (1989): 306–314; William J. Bowers, "The Effect of Execution is Brutalization, Not Deterrence," *Capital Punishment: Legal and Social Science Approaches,* ed. Kenneth C. Haas and James A. Inciardi (Newbury Park: Sage, 1988), 49–89; William C. Bailey and Ruth D. Peterson, "Murder and Capital Punishment: A Monthly Time Series Analysis of Execution Publicity," *American Sociological Review 54* (1989): 722–743; James A. Fox and Michael L. Radelet, "Persistent Flaws in Econometric Studies of the Deterrent Effect of the Death Penalty," *Loyola of Los Angeles Law Review 23* (1990): 29–44; John K. Cochran, Mitchell Chamlin, and Mark Seth, "Deterrence or Brutalization? An Impact Assessment of Oklahoma's Return to Capital Punishment," *Criminology 32* (1994): 107–134; for a review, see William C. Bailey and Ruth D. Peterson, "Capital Punishment, Homicide, and Deterrence: An Assessment of the Evidence," in *Studying and Preventing Homicide,* ed. M. Dwayne Smith and Margaret A. Zahn (Thousand Oaks: Sage, 1999), 223–245.

[27]John K. Cochran and Mitchell B. Chamlin, "Deterrence and Brutalization: The Dual Effects of Executions," *Justice Quarterly* (2000): 685–706.

[28]Beccaria, *On Crimes,* 46–47.

Beccaria's Concept of Deterrence and the Three Key Elements of Punishment

Beccaria is generally considered the father of deterrence theory for good reason. Beccaria was the first known scholar to write a work that summarized such extravagant ideas regarding the direction of human behavior toward choice as opposed to fate or destiny. Prior to his work, the common wisdom on the issue of human destiny was that it was chosen by the gods or God. At that time, governments and societies generally believed that people were born either good or bad. Beccaria, as a child of the Enlightenment, defied this belief in proclaiming that people freely choose their destinies and thus their decisions to commit or not commit criminal behavior.

Beccaria suggested three characteristics of punishment that would make a significant difference in whether an individual decides to commit a criminal act. These vital deterrent characteristics of punishment included celerity (swiftness), certainty, and severity.

Swiftness

The first of these characteristics was celerity, which we will refer to as **swiftness of punishment.** Beccaria saw two reasons why swiftness of punishment was important. At the time he wrote, some defendants were spending many years awaiting trial. Often, this was a longer time than they would have been locked up as punishment for their alleged offenses, even if the maximum penalty was imposed. As Beccaria stated, "The more promptly and the more closely punishment follows upon the commission of a crime, the more just and useful will it be."[29] Thus, the first reason that Beccaria recommended swiftness of punishment was to reform a system that was slow to respond to offenders.

The second reason that Beccaria emphasized swift sentencing was related to the deterrence aspect of punishment. A swift trial and punishment was important, Beccaria said, "because of privation of liberty, being itself a punishment, should not precede the sentence."[30] This not only was unjust, in the sense that some of these defendants would not have been incarcerated for such a long period even if they were convicted and sentenced to the maximum for the charges they were accused of committing, but it was detrimental because the individual would not link the sanction with the violation(s) they committed. Specifically, Beccaria believed that people build an association between the pains of punishment and their criminal acts. He asserted:

> Promptness of punishments is more useful because when the length of time that passes between the punishment and the misdeed is less, so much the stronger and more lasting in the human mind is the association of these two ideas, crime and punishment; they then come insensibly to be considered, one as the cause, the other as the necessary inevitable effect. It has been demonstrated that the association of ideas is the cement that forms the entire fabric of the human intellect.[31]

An analogy can be made with training animals or children; you have to catch them in the act, or soon after, or the punishment does not matter because the offender does not know why he or she is being punished. Ultimately, Beccaria argued that, for both reform and deterrent reasons, punishment should occur quickly after the act. Despite the commonsense aspects of making punishments swift, this has not been

---

[29]Ibid., 55.
[30]Ibid., 55.
[31]Ibid., 56.

examined by modern empirical research and therefore is the most neglected of the three elements of punishment Beccaria emphasized.

Certainty

The second characteristic that Beccaria felt was vital to the effectiveness of deterrence was **certainty of punishment.** Beccaria considered this the most important quality of punishment: "Even the least of evils, when they are certain, always terrify men's minds."[32] He also said, "The certainty of punishment, even if it be moderate, will always make a stronger impression than the fear of another which is more terrible but combined with the hope of impunity."[33] As scientific studies later showed, Beccaria was accurate in his assumption that perceived certainty or risk of punishment was the most important aspect of deterrence.

It is interesting to note that certainty is the least likely characteristic of punishment to be enhanced in modern criminal justice policy. Over the last few decades, the risk of criminals being caught and arrested has not increased. Law enforcement officials have been able to clear only about 21% of known felonies. Such clearance rates are based on the rate at which known suspects are apprehended for crimes that are reported to police. Law enforcement officials are no better at solving serious crimes known to police than they were in past decades, despite increased knowledge and resources put toward solving such crimes.

Severity

The third characteristic that Beccaria emphasized was **severity of punishment.** Specifically, Beccaria claimed that, for a punishment to be effective, the possible penalty must outweigh the potential benefits (e.g., financial payoff) of a given crime. However, this came with a caveat. This aspect of punishment was perhaps the most complicated part of Beccaria's philosophy, primarily because he thought that too much severity would lead to more crime, but the punishment must exceed any benefits expected from the crime. Beccaria said:

> For a punishment to attain its end, the evil which it inflicts has only to exceed the advantage derivable from the crime; in this excess of evil one should include the . . . loss of the good which the crime might have produced. All beyond this is superfluous and for that reason tyrannical.[34]

Beccaria makes clear in this statement that punishments should equal or outweigh any benefits of a crime to deter individuals from engaging in such acts. However, he also explicitly states that any punishments that largely exceed the reasonable punishment for a given crime are inhumane and may lead to further criminality.

A modern example of how punishments can be taken to an extreme and thereby cause more crime rather than deter it is the current *three-strikes-you're-out* approach to sentencing. Such laws have become common in many states, such as California. In such jurisdictions, individuals who have committed two prior felonies can be sentenced to life imprisonment for committing a crime, even a nonviolent crime, that the

---

[32]Ibid., 58.
[33]Ibid., 58.
[34]Ibid., 43.

state statutes consider a *serious felony*. Such laws have been known to drive some relatively minor property offenders to become violent when they know that they will be incarcerated for life when they are caught. A number of offenders have even wounded or killed people to avoid apprehension, knowing that they would face life imprisonment even for a relatively minor property offense. In a recent study, the authors analyzed the impact of three-strikes laws in 188 large cities in the 25 states that have such laws and concluded that there was no significant reduction in crime rates as a result. Furthermore, the areas with three-strikes laws typically had higher rates of homicide.[35]

Ultimately, Beccaria's philosophy on the three characteristics of good punishment in terms of deterrence—swiftness, certainty, and severity—is still highly respected and followed in most Western criminal justice systems. Despite its contemporary flaws and caveats, perhaps no other traditional framework is so widely adopted. With only one exception, Beccaria's concepts and propositions are still considered the ideal in virtually all Western criminal justice systems.

Beccaria's Conceptualization of Specific and General Deterrence

Beccaria also defined two identifiable forms of deterrence: specific and general. Although these two forms of deterrence tend to overlap in most sentences by judges, they can be distinguished in terms of the intended target of the punishment. Sometimes, the emphasis is clearly on one or the other, as Beccaria noted in his work.

Although Beccaria did not coin the terms **specific** and **general deterrence,** he clearly makes the case that both are important. Regarding punishment, he said, "The purpose can only be to prevent the criminal from inflicting new injuries on its citizens and to deter others from similar acts."[36] The first portion of this statement—preventing the criminal from reoffending—focuses on the defendant and the defendant alone, regardless of any possible offending by others. Punishments that focus primarily on the individual are considered specific deterrence, also referred to as special or individual deterrence. This concept is appropriately labeled because the emphasis is on the specific individual who offended. On the other hand, the latter portion of Beccaria's quotation emphasizes the deterrence of others, regardless of whether the individual criminal is deterred. Punishments that focus primarily on other potential criminals and not on the actual criminal are referred to as general deterrence.

Readers may wonder how a punishment would not be inherently both a specific and general deterrent. After all, in today's society, virtually all criminal punishments given to individuals (i.e., specific deterrence) in our society are done in court, a public venue, so people are somewhat aware of the sanctions (i.e., general deterrence). However, when Beccaria wrote in the 18th century, many if not most of sentencing was done behind closed doors and was not known to the public and had no way to deter other potential offenders. Therefore, Beccaria saw much utility in letting the public know what punishments were handed out for given crimes. This fulfilled the goal of general deterrence, which was essentially scaring others into not committing such criminal acts, while it also furthered his reforms of letting the public know if fair and balanced justice was being administered.

Despite the obvious overlap, there are some identifiable distinctions between specific and general deterrence seen in modern sentencing strategy. For example, some judges have chosen to hand out punishments

---

[35]Tomislav V. Kovandzic, John J. Sloan III, and Lynne M. Vieraitis, "'Striking Out' as Crime Reduction Policy: The Impact of 'Three-Strikes' Laws on Crime Rates in the U.S. Cities," *Justice Quarterly 21* (2004): 207–240.
[36]Ibid., 42.

to defendants in which they are obligated, as a condition of their probation or parole, to walk along their towns' main streets while wearing signs that say "Convicted Child Molester" or "Convicted Shoplifter." Other cities have implemented policies in which pictures and identifying information of those individuals who are arrested, such as prostitutes or men who solicit them, are put in newspapers or placed on billboards.

These punishment strategies are not likely to be much of a specific deterrent. Having now been labeled, these individuals may actually be psychologically encouraged to engage in doing what the public expects them to do. The specific deterrent effect may not be particularly strong. However, authorities are hoping for a strong general deterrent effect in most of these cases. They expect that many of the people who see these sign-laden individuals on the streets or in public pictures are going to be frightened from engaging in similar activity.

There are also numerous diversion programs, particularly for juvenile, first-time, and minor offenders, which seek to punish offenders without engaging them in public hearings or trials. The goal of such programs is to hold the individuals accountable and have them fulfill certain obligations without having them dragged through the system, which is often public. Thus, the goal is obviously to instill specific deterrence without using the person as a poster child for the public, which obviously negates any aspects of general deterrence.

Although most judges invoke both specific and general deterrence in many of the criminal sentences that they hand out, there are notable cases in which either specific or general deterrence is emphasized, sometime exclusively. Ultimately, Beccaria seemed to emphasize general deterrence and overall crime prevention, which is suggested by his statement, "It is better to prevent crimes than to punish them. This is the ultimate end of every good legislation."[37] This claim implies that it is better to deter potential offenders before they offend rather than imposing sanctions on already convicted criminals. Beccaria's emphasis on prevention (over reaction) and general deterrence is also evident in his claim that education is likely the best way to reduce crime. After all, the more educated an individual is regarding the law and potential punishments, as well as public cases in which offenders have been punished, the less likely he or she will be to engage in such activity. Beccaria's identification of the differential emphases in terms of punishment was a key element in his work that continues to be important in modern times.

Summary of Beccaria's Ideas and His Influence on Policy

Ultimately, Beccaria summarized his ideas on reforms and deterrence with this statement:

> In order for punishment not to be, in every instance, an act of violence of one or of many against a private citizen, it must be essentially public, prompt, necessary, the least possible in the given circumstances, proportionate to the crimes, dictated by the laws.[38]

In this statement, Beccaria is saying that the processing and punishment administered by justice systems must be known to the public, which delegates to the state the authority to make such decisions. Furthermore, he asserts that the punishment must be appropriately swift, certain (i.e., necessary), and appropriately severe, which fits his concept of deterrence. Finally, he reiterates the need to have equal punishment for a given criminal act, as opposed to having arbitrary punishments imposed by one judge.

---

[37]Ibid., 93.
[38]Beccaria, *On Crimes*, 99.

These are just some of many ideas that Beccaria proposed, but he apparently saw these points as being the most important.

Although we, as U.S. citizens, take for granted the rights proposed by Beccaria, they were quite unique concepts during the 18th century. In fact, the ideas proposed by Beccaria were so unusual and revolutionary then that he published his book anonymously. It is obvious that Beccaria was considerably worried about being accused of blasphemy by the church and of being persecuted by governments for his views.

Regarding the first claim, Beccaria was right; the Roman Catholic Church excommunicated Beccaria when it became known that he wrote the book. In fact, his book remained on the list of condemned works until relatively recently (the 1960s). On the other hand, government officials of the time surprisingly embraced his work. The Italian government and most European and other world officials, particularly dictators, embraced his work as well. Beccaria was invited to visit many other country capitals, even those of the most authoritarian states at that time, to help reform their criminal justice systems. For example, Beccaria was invited to meet with Catherine the Great, the czarina of Russia during the late 1700s, to help revise and improve their justice system. Most historical records suggest that Beccaria was not a great diplomat or representative of his ideas, largely because he was not physically or socially adequate for such endeavors. However, his ideas were strong and stood on their own merit.

Dictators and authoritarian governments may have liked his reform framework so much because it explicitly stated that treason was the most serious crime. As Beccaria said, "The first class of crime, which are the gravest because most injurious, are those known as crimes of *lese majesty* [high treason].... Every crime ... injures society, but it is not every crime that aims at its immediate destruction."[39] According to Enlightenment philosophy, violations of law are criminal acts not only against the direct victims but also against the entire society because they break the social contract. As Beccaria stated, the most heinous criminal acts are those that directly violate the social contract, which would be treason and espionage.

In his reform proposals, dictators may have seen a chance to pacify revolutionary citizens, who might be aiming to overthrow their governments. In many cases, reforms were only a temporary solution. After all, the American Revolution occurred in the 1770s, the French Revolution occurred in the 1780s, and other revolutions occurred soon after this period.

Governments that tried to apply his ideas to the letter experienced problems, but generally, most European (and American) societies that incorporated Beccaria's ideas had fairer and more democratic justice systems than they had before Beccaria. This is why he is to this day considered the father of criminal justice.

Impact of Beccaria's Work on Other Theorists

Beccaria's work had an immediate impact on the political and philosophical state of affairs in the late 18th century. He was invited to many other countries to reform their justice systems, and his propositions and theoretical model of deterrence were incorporated into many of the newly formed constitutions of countries, most formed after major revolutions. The most notable of these were the Constitution and Bill of Rights of the United States.

It is quite obvious that the many founding documents constructed before and during the American Revolution in the late 1700s were heavily influenced by Beccaria and other Enlightenment philosophers. Specifically, the concept that the U.S. government is "of the people, by the people, and for the people" makes it clear that the Enlightenment idea of democracy and voice in government is of utmost importance. Another clear example is the emphasis on due process and individual rights in the U.S. Bill of Rights. Among

---

[39]Ibid., 68.

the important concepts derived from Beccaria's work are the right to trial by jury, the right to confront and cross-examine witnesses, the right to a speedy trial, and the right to be informed about decisions of the justice system (such as charges, pleas, trials, verdicts, sentences, etc.).

The impact of Beccaria's ideas on the working ideology of our system of justice cannot be overstated. The public nature of our justice system comes from Beccaria, as does the emphasis on deterrence. The United States, as well as virtually all Western countries, incorporates in its justice system the certainty and severity of punishment to reduce crime. This system of deterrence remains the dominant model in criminal justice: The goal is to deter potential and previous offenders from committing crime by enforcing punishments that will make them reconsider the next time they think about engaging in such activity. This model assumes a rational thinking human being, as described by Enlightenment philosophy, who can learn from past experiences or from seeing others punished for offenses that he or she is rationally thinking about committing. Thus, Beccaria's work has had a profound impact on the existing philosophy and working of most justice systems throughout the world.

Beyond this, Beccaria also had a large impact on further theorizing about human decision making related to committing criminal behavior. One of the more notable theorists inspired by Beccaria's ideas was Jeremy Bentham (1748–1832) of England, who has become a well-known classical theorist in his own right, perhaps because he helped spread the Enlightenment/Beccarian philosophy to Britain. His influence in the development of classical theorizing is debated, with a number of major texts not covering his writings at all.[40] Although he did not add a significant amount of theorizing beyond Beccaria's propositions regarding reform and deterrence, Bentham did further refine the ideas presented by previous theorists, and his legacy is well known.

One of the more important contributions of Bentham was the concept of *hedonistic calculus,* which is essentially the weighing of pleasure versus pain. This, of course, is strongly based on the Enlightenment/Beccarian concept of rational choice and utility. After all, if the expected pain outweighs the expected benefit of doing a given act, the rational individual is far less likely to do it. On the other hand, if the expected pleasure outweighs the expected pain, a rational person will engage in the act. Bentham listed a set of criteria that he thought would go into the decision making of a rational individual. An analogy would be an imagined two-sided balance scale in which the pros and cons of crimes are considered, and then the individual makes a rational decision about whether to commit the crime.

Beyond the idea of hedonistic calculus, Bentham's contributions to the overall assumptions of classical theorizing did not revise the theoretical model in a significant way. Perhaps the most important contribution he made to the Classical School was helping to popularize the framework in Britain. In fact, Bentham became

▲ **Image 2.2** Jeremy Bentham, often credited as the founder of University College London, insisted that his body be put on display there after his death. You can visit a replica of it today.

[40]George B. Vold, Thomas J. Bernard, and Jeffrey B. Snipes, *Theoretical Criminology,* 5th ed. (New York: Oxford University Press, 2002).

more known for his design of a prison structure, known as the *panopticon,* which was used in several countries and in early Pennsylvania penitentiaries. This model of prisons involved using a type of wagon wheel design, in which a post at the center allowed 360-degree visual observation of the various spokes, hallways that contained all of the inmate cells.

The Neoclassical School of Criminology

A number of governments, including the newly formed United States, incorporated Beccaria's concepts and propositions in the development of their justice systems. The government that most strictly applied Beccaria's ideas—France after the revolution of the late 1780s—found that it worked pretty well except for one concept. Beccaria believed that every individual who committed a certain act against the law should be punished the same way. Although equality in punishment sounds like a good philosophy, the French realized very quickly that not everyone should be punished equally for a certain act.

The French system found that sentencing a first-time offender the same as a repeat offender did not make much sense, especially when the first-time offender was a juvenile. Furthermore, there were many circumstances in which a defendant appeared to be unmalicious in doing an act, such as someone with limited mental capacity or a person who acted out of necessity. Perhaps most important, Beccaria's framework specifically dismisses the intent (i.e., *mens rea*) of criminal offenders, while focusing only on the harm done to society by a given act (i.e., *actus reus*). French society, as well as most modern societies such as the United States, deviated from Beccaria's framework in taking the intent of offenders into account, often in a very important way, such as in determining what type of charges should be filed against those accused of homicide. Therefore, a new school of thought regarding the classical or deterrence model developed, which became known as the **Neoclassical School** of criminology.

The only significant difference between the Neoclassical School and the Classical School of criminology is that the Neoclassical (*neo* means new) School takes into account contextual circumstances of the individual or situation that allow for increases or decreases in the punishment. For example, would a society want to punish a 12-year-old, first-time offender the same way as they would punish a 35-year-old, previous offender for shoplifting the same item? In addition, does a society want to punish a mentally challenged person for stealing a car one time as much as a normal person who has been convicted of stealing more than a dozen cars? The answer is probably not—at least that is what most modern criminal justice authorities have decided, including those in the United States.

This was also the conclusion of French society, which quickly realized that, in this respect, Beccaria's system was neither fair nor effective in terms of deterrence. They came to acknowledge that circumstantial factors play an important part in how malicious or guilty a certain defendant is in a given crime. The French revised their laws to take into account both mitigating and aggravating circumstances. This neoclassical concept became the standard in all Western justice systems.

The United States also followed this model and considers such contextual factors in virtually all of its charges and sentencing decisions. For example, juvenile defendants are actually processed in completely different courts. Furthermore, defendants who are first-time offenders are generally given options for diversion programs or probation as long as their offenses are not serious.

The Neoclassical School adds an important caveat to the previously important Classical School, but it assumes virtually all other concepts and propositions of the Classical School: the social contract, due process rights, and the idea that rational beings will be deterred by the certainty, swiftness, and severity of punishment. This neoclassical framework had, and continues to have, an extremely important impact on the world.

Loss of Dominance of Classical and Neoclassical Theory

For about 100 years after Beccaria wrote his book, the Classical and Neoclassical Schools were dominant in criminological theorizing. During this time, most governments, especially those in the Western world, shifted their justice frameworks toward the neoclassical model. This has not changed, even in modern times. For example, when officials attempt to reduce certain illegal behaviors, they increase the punishments or put more effort into catching such offenders.

▲ **Image 2.3**   Charles Darwin (1809–1882), Author of Evolutionary Theory

However, the classical and neoclassical framework lost dominance among academics and scientists in the 19th century, especially after Darwin's publication in the 1860s of *The Origin of Species*, which introduced the concept of evolution and natural selection. This perspective shed new light on other influences on human behavior beyond free will and rational choice (e.g., genetics, psychological deficits). Despite this shift in emphasis among academic and scientific circles, it remains true that the actual workings of justice systems of most Western societies still retain the framework of classical and neoclassical models as their model of justice.

Three-strikes laws are an example; others include police department gang units and injunctions that condemn any observed loitering by or gathering among gang members in a specified region. Furthermore, some jurisdictions, such as California, have created gang enhancements for sentencing; after the jury decides whether the defendant is guilty of a given crime, it then considers whether the person is a gang member. If a jury in California decides that the defendant is a gang member, which is usually determined by evidence provided by local police gang units, it automatically adds more time to any sentence the judge gives. These are just some examples of how Western justice systems still rely primarily on deterrence of criminal activity through increased enforcement and enhanced sentencing. The bottom line is that modern justice systems still base most of their policies on classical or neoclassical theoretical frameworks that fell out of favor among scientists and philosophers in the late 1800s.

Policy Implications

Many policies are based on deterrence theory: the premise that increasing punishment sanctions will deter crime.[41] This is seen throughout the system of law enforcement, courts, and corrections. This is rather interesting given the fact that classical deterrence theory has not been the dominant explanatory model among criminologists for decades. In fact, a recent poll of close to 400 criminologists in the nation ranked classical theory as 22 out of 24 theories in terms of being the most valid explanation of serious and

---

[41]Stephen G. Tibbetts and Craig Hemmens, *Criminological Theory: A Text Reader.* (Thousand Oaks: Sage, 2010), 67–69.

persistent offending.[42] Still, given the dominance of classical deterrence theory in most criminal justice policies, it is very important to discuss the most common strategies.

First, the death penalty is used as a general deterrent for committing crime. As the father of deterrence theory predicted, most studies show that capital punishment has a negligible effect on criminality. A recent review of the extant literature concluded that "the death penalty does not deter crime."[43] In fact, some studies show evidence for a brutalization effect, an increase in homicides after a high-profile execution.[44] Although the evidence is somewhat mixed, it is safe to say that the death penalty is not a consistent deterrent.

Another policy flowing from classical and neoclassical models is adding more police officers to deter crime in a given area. A recent review of the existing literature concluded that simply "adding more police officers will not reduce crime."[45] Rather, it is generally up to communities to police themselves via informal factors of control (such as family, church, community ties, etc.). However, this same review did find that, sometimes, when police engage in problem-solving activities at a specific location, it can reduce crime, but at that point, the strategy is not based on deterrence.[46] Furthermore, a recent report concluded that proactive arrests for drunk driving have consistently been found to reduce such behavior, as does arresting offenders for domestic violence, but only if these measures are employed consistently.[47]

Regarding court and correctional strategies, one example is the scared straight approach that became popular several decades ago.[48] These programs essentially sought to scare or deter juvenile offenders into going straight by showing them the harshness and realities of prison life. However, nearly all evaluations of these programs showed that they were ineffective, and some evaluations indicated that these programs led to higher rates of recidivism.[49] There seem to be few successful deterrent policies in the court and corrections components of the criminal justice system. A recent review found that one of the only court-mandated policies that seem promising is providing protection orders for battered women.[50]

The policies, programs, and strategies based on classical deterrence theory will be examined more thoroughly in the final chapter of this book. To sum up, however, most of these strategies don't seem to work consistently to deter. This is because such a model assumes people are rational and think carefully before choosing their behavior, whereas most research findings suggest that people often engage in behaviors that they know are irrational or that offenders tend to engage in behaviors without rational decision making,[51] which criminologists often refer to as *bounded rationality*.[52] Therefore, it is not surprising that many attempts by police and other criminal justice authorities to deter potential offenders do not seem to

---

[42]Lee Ellis, Jonathan Cooper, and Anthony Walsh. "Criminologists' Opinions About Causes and Theories of Crime and Delinquency: A Follow-up," *The Criminologist 33* (2008): 23–26.

[43]Samuel Walker, *Sense and Nonsense about Crime and Drugs,* 6th ed. (Belmont: West/Wadsworth, 2005), 102.

[44]Bowers, "The Effect of Execution."

[45]Walker, *Sense and Nonsense,* 79.

[46]Ibid., 81.

[47]Lawrence Sherman, Denise Gottfredson, Doris MacKenzie, John Eck, Peter Reuter, and Shawn Bushway, *Preventing Crime: What Works, What Doesn't, What's Promising: A Report to the United States Congress* (Washington, DC: US Department of Justice, 1997).

[48]For a review of these programs and evaluations of them, see Richard Lundman, *Prevention and Control of Juvenile Delinquency,* 2nd ed. (Oxford: Oxford University Press, 1993).

[49]Ibid.

[50]Sherman et al., *Preventing Crime.*

[51]For a review of the extant research on this topic, see Alex Piquero and Stephen Tibbetts, *Rational Choice and Criminal Behavior* (New York: Routledge, 2002).

[52]For a more recent discussion on the complexity of developing policies based on deterrence and rational choice models, see Travis Pratt, "Rational Choice Theory, Crime Control Policy, and Criminological Relevance," *Criminology and Public Policy 7* (2008): 43–52.

have much effect in preventing crime. This explanation will be more fully discussed in the final chapter of the book.

⊠ Conclusion

This chapter examined the earliest period of theorizing about criminological theory, which evolved from the Enlightenment. The Classical School of criminology evolved out of ideas from the Enlightenment era in the mid- to late 18th century. This school of thought emphasized free will and rational choices that individuals make, with an emphasis on making choices regarding criminal behavior based on the potential costs and benefits that could result from such behavior. This chapter also explored the concepts and propositions of the father of the Classical School, which built the framework on which deterrence theory is based. We also discussed the various reforms that Beccaria proposed, many of which were adopted in the formation of the U.S. Constitution and Bill of Rights. The significance of the Classical School in both theorizing about crime and in actual administration of justice in the United States cannot be overestimated. The Classical and Neoclassical Schools of criminology remain to this day the primary framework in which justice is administered, despite the fact that scientific researchers and academics have, for the most part, moved past this perspective to consider social and economic factors.

⊠ Chapter Summary

- The dominant theory of criminal behavior for most of the history of human civilization used demonic, supernatural, or other metaphysical explanations of behavior.
- The Age of Enlightenment was important because it brought a new logic and rationality to understanding human behavior, especially regarding the ability of individual human beings to think for themselves. Hobbes and Rousseau were two of the more important Enlightenment philosophers, and both stressed the importance of the social contract.
- Cesare Beccaria, who is generally considered the father of criminal justice, laid out a series of recommendations for reforming the brutal justice systems that existed throughout the world in the 1700s.
- Beccaria is also widely considered the father of Classical School or deterrence theory; he based virtually all of his theoretical framework on the work of Enlightenment philosophers, especially their emphasis on humans as rational beings who consider perceived risks and benefits before committing criminal behaviors. This is the fundamental assumption of deterrence models of crime reduction.
- Beccaria discussed three key elements that punishments should have to be effective deterrents; a punishment should be certain, swift, and severe.
- Specific deterrence involves sanctioning an individual to deter that particular individual from offending in the future. General deterrence involves sanctioning an individual to deter other potential offenders by making an example out of the individual being punished.
- The Neoclassical School was formed because societies found it nearly impossible to punish offenders equally for a given offense. The significant difference between the Classical and Neoclassical Schools is that the neoclassical model takes aggravating and mitigating circumstances into account when an individual is sentenced.

- Jeremy Bentham helped reinforce and popularize Beccaria's ideas in the English-speaking world, and he further developed the theory by proposing hedonistic calculus, a formula for understanding criminal behavior.
- Despite falling out of favor among most criminologists in the late 1800s, the classical and neoclassical framework remains the dominant model and philosophy of all modern Western justice systems.

KEY TERMS

Actus reus

Age of Enlightenment

Brutalization effect

Certainty of punishment

Classical School

Deterrence theory

General deterrence

Mens rea

Neoclassical School

Severity of punishment

Social contract

Specific deterrence

Swiftness of punishment

Utilitarianism

DISCUSSION QUESTIONS

1. Do you see any validity to the supernatural or religious explanations of criminal behavior? Provide examples of why you feel the way you do. Is your position supported by scientific research?

2. Which portions of Enlightenment thought do you believe are most valid in modern times? Which portions do you find least valid?

3. Of all of Beccaria's reforms, which do you think made the most significant improvement to modern criminal justice systems and why? Which do you think had the least impact and why?

4. Of the three elements of deterrence that Beccaria described, which do you think has the most important impact on deterring individuals from committing crime? Which of the three do you think has the least impact on deterring potential criminals? Back up your selections with personal experience.

5. Between general and specific deterrence, which do you think is more important for a judge to consider when sentencing a convicted individual? Why do you feel that way?

6. Provide examples of general and specific deterrence in your local community or state. Use the Internet if you can't find examples from your local community. Do you think such deterrence is effective?

7. Given the modern interpretation by the U.S. government of the definition of torture in context with what Beccaria thought about this issue, do you think that the father of criminal justice and deterrence would agree with the interrogation policies of the Bush administration during the Iraq War, which indisputably violated the guidelines set by the Geneva Conventions? Explain your position.

8. Regarding the use of the death penalty, list and explain at least three reasons why the father of criminal justice and deterrence felt the way he did. Which of these arguments do you agree with the most? Which argument do you disagree with the most? Ultimately, are you more strongly for or against the death penalty after reading the arguments of Beccaria?

9. Regarding the Neoclassical School, which mitigating factors do you think should reduce the punishment of a criminal defendant the most? Which aggravating circumstances do you think should increase the sentence of a criminal defendant the most? Do you believe that all persons who commit the same act should be punished exactly the same, regardless of age, experience, or gender?

10. What types of policy strategies based on classical and deterrence theory do you support? Which don't you support? Why?

WEB RESOURCES

Age of Enlightenment

http://history-world.org/age_of_enlightenment.htm

Cesare Beccaria

http://www.constitution.org/cb/beccaria_bio.htm

http://cepa.newschool.edu/het/profiles/beccaria.htm

http://www.criminology.fsu.edu/crimtheory/beccaria.htm

Jeremy Bentham

http://www.utilitarianism.com/bentham.htm

http://cepa.newschool.edu/het/profiles/bentham.htm

Demonic Theories of Crime

http://www.criminology.fsu.edu/crimtheory/week2.htm

http://www.salemweb.com/memorial/

Deterrence

http://www.associatedcontent.com/article/32600/evolution_of_deterrence_crime_theory.html

http://www.deathpenaltyinfo.org/article.php?scid=12&did=167

Thomas Hobbes

http://www.philosophypages.com/hy/3x.htm

http://oregonstate.edu/instruct/phl302/philosophers/hobbes.html

Neoclassical School

http://www.answers.com/topic/neo-classical-school

Rousseau

http://www.philosophypages.com/ph/rous.htm

http://www.rousseauassociation.org/

3

Modern Applications of the Classical Perspective

Deterrence, Rational Choice, and Routine Activities or Lifestyle Theories of Crime

This chapter will discuss the early aggregate studies of deterrence in the late 1960s, then the perceptual studies of the 1970s, and finally, the longitudinal and scenario studies of the 1980s and 1990s to present. Other policy applications, such as increased penalties toward drunk driving, white-collar crime, and so on, will also be examined. The chapter will also discuss the development of rational choice theory in economics and its later application to crime. Finally, it will examine the use of **routine activities** or **lifestyle theory** as a framework for modern research and applications for reducing criminal activity.

In Chapter 2, we discussed the early development of the Classical and Neoclassical Schools of criminological thought. This theoretical perspective has been the dominant framework used by judges and practitioners in the practice of administering justice and punishment even in current times, but beginning in the late 19th century, criminological researchers dismissed the classical and neoclassical frameworks. Rather, criminological research and theorizing began emphasizing factors other than free will and deterrence. Instead, an emphasis was placed on social, biological, or other factors that go beyond free will and deterrence theory. These theories will be discussed in later sections, but first, we will examine the recent rebirth of classical and neoclassical theory and deterrence.

✉ Rebirth of Deterrence Theory and Contemporary Research

As discussed above, the Classical and Neoclassical School frameworks fell out of favor among scientists and philosophers in the late 19th century, largely due to the introduction of Darwin's ideas about evolution and natural selection. However, virtually all Western criminal systems retained the classical and neoclassical frameworks for their model of justice, particularly the United States. Nevertheless, the ideology of Beccaria's work was largely dismissed by academics and theorists after the presentation of Darwin's theory of evolution in the 1860s. Therefore, the Classical and Neoclassical Schools fell out of favor in terms of criminological theorizing for about 100 years. However, in the 1960s, the Beccarian model of offending experienced a rebirth.

In the late 1960s, several studies using aggregate measures of crime and punishment were published that used a deterrence model for explaining why individuals engage in criminal behavior. These studies revealed a new interest in the deterrent aspects of criminal behavior and further supported the importance of certainty and severity of punishment in deterring individuals from committing crime, particularly homicide. In particular, evidence was presented that showed that increased risk or certainty of punishment was associated with less crime for most serious offenses. Plus, it is a fact that most offenders who are arrested once never get arrested again, which lends some basic support for deterrence.

Many of these studies used statistical formulas to measure the degree of certainty and severity of punishment in given jurisdictions. One measure used the ratio of crimes reported to police as compared to the number of arrests in a given jurisdiction. Another measure of certainty of punishment was the ratio of arrests to convictions, or findings of guilt, in criminal cases. Other measures were also employed. Most of the studies showed the same result: The higher the likelihood of arrest compared with reports of crime, or the higher the conviction rate compared to the arrest rate, the lower the crime rate was in a jurisdiction. On the other hand, the scientific evidence regarding measures of severity, which such studies generally indicated by the length of sentence for comparable crimes or a similar type of measure, did not show much impact on crime.

Additional aggregate studies examined the prevalence and influence of capital punishment on the crime rate in given states.[1] The evidence showed that the states with death penalty statutes also had higher murder rates than nondeath penalty states. Furthermore, the studies showed that murderers in death penalty states who were not executed actually served less time than murderers in nondeath penalty states. Thus, the evidence regarding increased sanctions, including capital punishment, was mixed. Still, a review of the early deterrence studies by the National Academy of Sciences concluded that, overall, there was more evidence for a deterrent effect than against it, although the finding was reported in a tone that lacked confidence, perhaps cautious of what future studies would show.[2]

It was not long before critics noted that studies incorporating aggregate (i.e., macrolevel) statistics are not adequate indicators or valid measures of the deterrence theoretical framework largely because the model emphasizes the perceptions of individuals. Using aggregate or group statistics is flawed because

---

[1]Daniel Glaser and Max S. Zeigler, "Use of the Death Penalty v. Outrage at Murder," *Crime and Delinquency 20* (1974): 333–338; Charles Tittle, Franklin E. Zimring, and Gordon J. Hawkins, *Deterrence—The Legal Threat in Crime Control* (Chicago: University of Chicago Press, 1973); Johannes Andenaes, *Punishment and Deterrence* (Ann Arbor: University of Michigan Press, 1974); Jack P. Gibbs, *Crime, Punishment and Deterrence* (New York: Elsevier, 1975).

[2]Alfred Blumstein, Jacqueline Cohen, and Daniel Nagin, eds., *Deterrence and Incapacitation: Estimating the Effects of Criminal Sanctions on Crime Rates* (Washington, DC: National Academy of Sciences, 1978).

different regions may have higher or lower crime rates than others, thereby creating bias in the level of ratios for certainty or severity of punishment. Furthermore, the group measures produced by these studies provide virtually no information on the degree to which individuals in those regions perceive sanctions as being certain, severe, or swift. Therefore, the emphasis on the unit of analysis in deterrence research shifted from the aggregate level to that of a more micro, individual level.

The following phase of deterrence research focused on individual perceptions of certainty and severity of sanctions, primarily drawn at one point in time, known as **cross-sectional studies.** A number of cross-sectional studies of individual perceptions of deterrence showed that perceptions of the risk or certainty of punishment were strongly associated with intentions to commit future crimes, but individual perceptions of severity of crimes were mixed. Furthermore, it readily became evident that it was not clear whether perceptions were causing changes in behavior or whether behavior was causing changes in perception. This led to the next wave of research, longitudinal studies of individual perceptions and deterrence, which measured perceptions of risk and severity, as well as behavior, over time.[3]

One of the primary concepts revealed by longitudinal research was that behavior was influencing perceptions of the risk and severity of punishment more than perceptions were influencing behavior. This was referred to as the **experiential effect,** which is appropriately named because people's previous experience highly influences their expectations regarding their chances of being caught and suffering the resulting penalties. A common example is that of people who drive under the influence of alcohol (or other substances).

Studies show that if you ask people who have never driven drunk how likely they would be caught if they drove home drunk, most predict an unrealistically high chance of getting caught. However, if you ask people who have been arrested for driving drunk, even those who have been arrested several times for this offense, they typically predict that the chance is very low. The reason for this is that these chronic drunk drivers have typically been driving under the influence for many years, mostly without being caught. It is estimated that more than 1 million miles are driven collectively by drunk drivers before one person is arrested.[4] If anything, this is likely a conservative estimate. Thus, people who drive drunk, with some doing so every day, are not likely to be deterred even when they are arrested more than once because they have done so for years. In fact, perhaps the most notable experts on the deterrence of drunk drivers, H. L. Ross and his colleagues, concluded that drunk drivers who "perceive a severe punishment if caught, but a near-zero chance of being caught, are being rational in ignoring the threat."[5] It is obvious that even the most respected scholars in the area admit that sanctions against drunk driving are nowhere near certain enough, even if they are growing in severity.

Another common example is seen with white-collar criminals. Some researchers have theorized that being caught by authorities for violating government rules enforced by the Securities and Exchange Commission (SEC) will make these organizations less likely to commit future offenses.[6] However, business

---

[3]Raymond Paternoster, Linda E. Saltzman, Gordon P. Waldo, and Theodore G. Chiricos, "Perceived Risk and Social Control: Do Sanctions Really Deter?" *Law and Society Review 17* (1983): 457–80; Raymond Paternoster, "The Deterrent Effect of the Perceived Certainty and Severity of Punishment: A Review of the Evidence and Issues," *Justice Quarterly 4* (1987): 173–217.

[4]H. Laurence Ross, *Deterring the Drunk Driver: Legal Policy and Social Control* (Lexington: Lexington Books, 1982); H. Laurence Ross, *Confronting Drunk Driving: Social Policy for Saving Lives* (New Haven: Yale University Press, 1992); H. Laurence Ross, "Sobriety Checkpoints, American Style," *Journal of Criminal Justice 22* (1994): 437–44; H. Laurence Ross, Richard McCleary, and Gary LaFree, "Can Mandatory Jail Laws Deter Drunk Driving? The Arizona Case," *Journal of Criminal Law and Criminology 81* (1990): 156–170.

[5]H. Laurence Ross, "Sobriety Checkpoints," 164.

[6]Sally Simpson and Christopher S. Koper, "Deterring Corporate Crime," *Criminology 30* (1992): 347–376.

organizations have been in violation of established practices for years before getting caught, so it is likely that they will continue to ignore the rules in the future more than organizations that have never violated the rules. As it was with drunk drivers, the certainty of punishment for white-collar violations is so low—and many would argue the severity is also quite low—that it is quite rational for businesses and business professionals to take the risk of engaging in white-collar crime.

It is interesting to note that white-collar criminals and drunk drivers are two types of offenders who are most likely to be deterred because they are mostly of the middle- to upper-level socioeconomic class. The extant research on deterrence has shown that individuals who have something to lose are the most likely to be deterred by sanctions. This makes sense: Those who are unemployed or poor or both do not have much to lose, and for them, or for some minorities, incarceration may not present a significant departure from the deprived lives that they lead.

The fact that official sanctions have limitations in deterring individuals from drunk driving and white-collar crime is not a good indication of the effectiveness of deterrence-based policies. Their usefulness becomes even more questionable when other populations are considered, particularly the offenders in most predatory street crimes (e.g., robbery, burglary, etc.), in which offenders typically have nothing to lose because they come from poverty-stricken areas and are often unemployed. One recent study showed that being arrested had little effect on perceptions of the certainty of punishment; offending actually corresponded with decreases in such perceptions.[7]

Some people don't see incarceration as that much of a step down in life, given the three meals a day, shelter, and relative stability provided by such punishment. This fact epitomizes one of the most notable paradoxes we have in criminology: The individuals we most want to deter are the least likely to be deterred, primarily because they have nothing to fear. In early Enlightenment thought, Hobbes asserted that, although fear was the tool used to enforce the social contract, people who weren't afraid of punishment could not effectively be deterred. That remains true in modern days.

Along these same lines, studies have consistently shown that for young male offenders—at higher risk, with low emotional or moral inhibitions, low self-control, and high impulsivity—official deterrence is highly ineffective in preventing crimes with immediate payoffs.[8] Thus, many factors go into the extent to which official sanctions can deter. As we have seen, even among those offenders who are in theory the most

---

[7]Greg Pogarsky, KiDeuk Kim, and Raymond Paternoster, "Perceptual Change in the National Youth Survey: Lessons for Deterrence Theory and Offender Decision-Making," *Justice Quarterly 22* (2005): 1–29.

[8]For a review, see Stephen Brown, Finn Esbensen, and Gilbert Geis, *Criminology*, 6th ed. (Cincinnati: LexisNexis, 2007), 201–204; Nancy Finley and Harold Grasmick, "Gender Roles and Social Control," *Sociological Spectrum 5* (1985): 317–330; Harold Grasmick, Robert Bursik, and Karla Kinsey, "Shame and Embarrassment as Deterrents to Noncompliance With the Law: The Case of an Antilittering Campaign," *Environment and Behavior 23* (1991): 233–251; Harold Grasmick, Brenda Sims Blackwell, and Robert Bursik, "Changes in the Sex Patterning of Perceived Threats of Sanctions," *Law and Society Review 27* (1993): 679–705; Pamela Richards and Charles Tittle, "Gender and Perceived Chances of Arrest," *Social Forces 59* (1981): 1182–1199; George Loewenstein, Daniel Nagin, and Raymond Paternoster, "The Effect of Sexual Arousal on Expectations of Sexual Forcefulness," *Journal of Research in Crime and Delinquency 34* (1997): 209–228; Toni Makkai and John Braithwaite, "The Dialects of Corporate Deterrence," *Journal of Research in Crime and Delinquency 31* (1994): 347–373; Daniel Nagin and Raymond Paternoster, "Enduring Individual Differences and Rational Choice Theories of Crime," *Law and Society Review 27* (1993): 467–496; Alex Piquero and Stephen Tibbetts, "Specifying the Direct and Indirect Effects of Low Self-Control and Situational Factors in Offenders' Decision Making: Toward a More Complete Model of Rational Offending," *Justice Quarterly 13* (1996): 481–510; Raymond Paternoster and Sally Simpson, "Sanction Threats and Appeals to Morality: Testing a Rational Choice Model of Corporate Crime," *Law and Society Review 30* (1996): 549–583; Daniel Nagin and Greg Pogarsky, "Integrating Celerity, Impulsivity, and Extralegal Sanction Threats Into a Model of General Deterrence: Theory and Evidence," *Criminology 39* (2001): 404–430; Alex Piquero and Greg Pogarsky, "Beyond Stanford and Warr's Reconceptualization of Deterrence: Personal and Vicarious Experiences, Impulsivity, and Offending Behavior," *Journal of Research in Crime and Delinquency 39* (2002): 153–186; for a recent review and an altered explanation of these conclusions, see Greg Pogarsky, "Identifying 'Deterrable' Offenders: Implications for Research on Deterrence," *Justice Quarterly 19* (2002): 431–452.

deterrable, official sanctions have little impact because their experience of not being caught weakens the value of deterrence.

The identification and understanding of the experiential effect had a profound effect on the evidence regarding the impact of deterrence. Researchers saw that, to account for such an experiential effect, any estimation of the influence of perceived certainty or severity of punishment must control for previous behaviors and experiences engaging in such behavior. The identification of the experiential effect was the primary contribution of the longitudinal studies of deterrence, but such studies faced even further criticism.

Longitudinal studies of deterrence provided a significant improvement over the cross-sectional studies that preceded this advanced methodology. However, such longitudinal studies typically involved designs in which measures of perceptions of certainty and severity of punishment were collected at points in time that were separated by up to a year apart, including long stretches between when the crime was committed and when the offenders were asked about their perceptions of punishment. Psychological studies have clearly established that perceptions of the likelihood and severity of sanctions vary significantly from day to day, let alone month to month or year to year.[9] Therefore, in the late 1980s and early 1990s, a new wave of deterrence research evolved, which asked study participants to estimate their immediate intent to commit a criminal act in a given situation, as well as their immediate perceptions of certainty and severity of punishment in this same situation. This wave of research was known as **scenario (vignette) research**.[10]

Scenario research (i.e., vignette design) was created to deal with the limitations of previous methodological strategies for studying the effects of deterrence on criminal offending, specifically, the criticism that individuals' perceptions of the certainty and severity of punishment changed drastically from time to time and across different situations. The scenario method dealt with this criticism directly by a providing a specific, realistic (albeit hypothetical) situation, in which a person engages in a criminal act. The participant in the study is then asked to estimate the chance that he or she would engage in such activity in the given circumstances and to respond to questions regarding perceptions of the risk of getting caught (i.e., certainty of punishment) and the degree of severity of punishment they expected.

Another important and valuable aspect of scenario research was that it promoted a contemporaneous (i.e., instantaneous) response about perceptions of risk and the severity of perceived sanctions. In comparison, previous studies (e.g., aggregate, cross-sectional, longitudinal) had always relied on either group or individual measures of perceptions over long periods of time. However, some argue that intentions to commit a crime given a hypothetical situation are not accurate measures of what one would do in reality. Studies have shown an extremely high correlation between what people report doing in a given scenario and what they

[9]Icek Ajzen and Martin Fishbein, *Understanding Attitudes and Predicting Social Behavior* (Englewood Cliffs: Prentice Hall, 1980); Martin Fishbein and Icek Ajzen, *Belief, Attitude, Intention, and Behavior* (Reading: Addison-Wesley, 1975); Icek Ajzen and Martin Fishbein, "Attitude-Behavior Relations: A Theoretical Analysis and Review of Empirical Research," *Psychological Bulletin 84* (1977): 888–918; for a recent review, see Pogarsky et al., "Perceptual Change."

[10]Loewenstein et al., "The Effect of Sexual Arousal"; Nagin and Paternoster, "Enduring Individual Differences"; Piquero and Tibbetts, "Specifying the Direct"; Paternoster and Simpson, "Sanction Threats"; Ronet Bachman, Raymond Paternoster, and Sally Ward, "The Rationality of Sexual Offending: Testing a Deterrence/Rational Choice Conception of Sexual Assault," *Law and Society Review 26* (1992): 343–372; Harold Grasmick and Robert Bursik, "Conscience, Significant Others, and Rational Choice: Extending the Deterrence Model," *Law and Society Review 24* (1990): 837–861; Harold Grasmick and Donald E. Green, "Legal Punishment, Social Disapproval, and Internalization as Inhibitors of Illegal Behavior," *Journal of Criminal Law and Criminology 71* (1980): 325–335; Stephen Klepper and Daniel Nagin, "The Deterrent Effects of Perceived Certainty and Severity of Punishment Revisited, *Criminology 27* (1989): 721–746; Stephen Tibbetts and Denise Herz, "Gender Differences in Students' Rational Decisions to Cheat," *Deviant Behavior 18* (1996): 393–414; Stephen Tibbetts and David Myers, "Low Self-Control, Rational Choice, and Student Test Cheating," *American Journal of Criminal Justice 23* (1999): 179–200; Stephen Tibbetts, "Shame and Rational Choice in Offending Decisions," *Criminal Justice and Behavior 24* (1997): 234–255.

would do in real life.[11] A recent review of criticisms of this research method showed that one weakness was that it did not allow respondents to develop their own perceptions and costs associated with each offense.[12] Despite such criticisms, the scenario method appears to be the most accurate that we have to date to estimate the effects of individual perceptions on the likelihood of such individuals engaging in given criminal activity at a given point in time. This is something that the previous waves of deterrence research—aggregate, cross-sectional, and longitudinal studies—could not estimate.

Ultimately, the studies using the scenario method showed that participants were more affected by perceptions of certainty and less so, albeit sometimes significant, perceptions of severity. This finding supported previous methods of estimating the effects of *formal* or *official deterrence*, meaning the deterrent effects of three general groups: law enforcement, courts, and corrections (i.e., prisons and probation or parole). So, the overall conclusion regarding the effects of official sanctions on individual decision making remained unaltered. However, one of the more interesting aspects of the scenario method research is that it helped solidify the importance of extralegal variables in deterring criminal behavior, variables that had been neglected by previous methods.

These extralegal or informal deterrence variables, which include any factors beyond the formal sanctions of police, courts, and corrections—such as employment, family, friends, or community—are typically known as informal or unofficial sanctions. These studies helped show that these informal sanctions provided most of the deterrent effect—if there was any. These findings coincided with the advent of a new model of deterrence, which became commonly known as *rational choice theory*.

Rational Choice Theory

Rational choice theory is a perspective that criminologists adapted from economists, who used it to explain a variety of individual decisions regarding a variety of behaviors. This framework emphasizes all the important factors that go into a person's decision to engage or not engage in a particular act. In terms of criminological research, the rational choice model emphasized both official or formal forms of deterrence, as well as the informal factors that influence individual decisions for criminal behavior. This represented a profound advance in the understanding of human behavior. After all, as studies showed, most individuals are more affected by informal factors than they are by official or formal factors.

Although there were several previous attempts to apply the rational choice model to the understanding of criminal activity, the most significant work, which brought rational choice theory into the mainstream of criminological research, was Cornish and Clarke's *The Reasoning Criminal: Rational Choice Perspectives on Offending* in 1986.[13] Furthermore, in 1988, Katz published his work *Seductions of Crime*, which, for the first time, placed an emphasis on the benefits (mostly the inherent physiological pleasure) in committing crime;[14] before Katz's publication, virtually no attention had been paid to the benefits of offending, let alone the fun that people feel when they engage in criminal behavior. A recent study showed that the publication

---

[11]Donald Green, "Measures of Illegal Behavior in Individual Behavior in Individual-Level Deterrence Research," *Journal of Research in Crime and Delinquency* 26 (1989): 253–275; Ajzen and Fishbein, *Understanding Attitudes*; I. Ajzen, "From Intentions to Actions: A Theory of Planned Behavior," in *Action-Control: From Cognition to Behavior*, eds. Julius Kuhl and Jurgen Beckmann (New York: Springer, 1985), 11–39; Icek Ajzen and Martin Fishbein, "The Prediction of Behavioral Intentions in a Choice Situation," *Journal of Experimental Psychology* 5 (1969): 400–416.

[12]Jeffrey A. Bouffard, "Methodological and Theoretical Implications of Using Subject-Generated Consequences in Tests of Rational Choice Theory," *Justice Quarterly* 19 (2002): 747–771.

[13]Derek Cornish and Ron Clarke, *The Reasoning Criminal: Rational Choice Perspectives on Offending* (New York: Springer-Verlag, 1986).

[14]Jack Katz, *Seductions of Crime* (New York: Basic Books, 1988).

of Cornish and Clarke's book, as well as the timing of other publications, such as Katz's, led to an influx of criminological studies in the late 1980s to mid-1990s based on the rational choice model.[15]

These studies on rational choice showed that while official or formal sanctions tend to have some effect on individuals' decisions to commit crime, they almost always are relatively unimportant compared to extralegal or informal factors. The effects of people's perceptions of how much shame or loss of self-esteem they would experience, even if no one else found out that they committed the crime, was one of the most important variables in determining whether or not they would do so.[16] Additional evidence indicated that females were more influenced by the effects of shame and moral beliefs in this regard than were males.[17] Recent studies have shown that levels of personality traits, especially low self-control and empathy, are likely the reasons why males and females differ so much in engaging in criminal activity.[18] Finally, the influence of peers has a profound impact on individual perceptions of the pros and cons of offending by significantly decreasing the perceived risk of punishment if people see their friends get away with crimes.[19]

Another area of rational choice research dealt with the influence that an individual's behavior would have on those around them. A recent review and test of perceived social disapproval showed that this was one of the most important variables in decisions to commit crime.[20] In addition to self-sanctions, such as feelings of shame and embarrassment, the perceived likelihood of how loved ones and friends, as well as employers, would respond is perhaps the most important factor that goes into a person's decision to engage in criminal activity. These are the people we deal with every day and may be the source of our livelihoods, so it should not be too surprising that our perceptions of how they will react affect strongly how we behave.

Perhaps the most important finding of rational choice research was that the expected benefits, particularly the pleasure offenders would get from offending, had one of the most significant effects on their decisions to offend. Many other conclusions have been made regarding the influence of extralegal or informal factors on criminal offending, but the ultimate conclusion that can be made is that these informal deterrent variables typically hold more influence on individual decision making regarding deviant activity than the official or formal factors that were emphasized by traditional Classical School models of behavior.

The rational choice model of criminal offending became the modern framework of deterrence. Official authorities acknowledged the influence of extralegal or informal factors, which is seen in modern efforts to incorporate the family, employment, and community in rehabilitation efforts. Such efforts are highly consistent with the current state of understanding regarding the Classical School or rational choice frameworks, namely that individuals are more deterred by the impact of their actions on informal aspects of their lives as opposed to the formal punishments they face by doing illegal acts.

---

[15]Stephen Tibbetts and Chris Gibson, "Individual Propensities and Rational Decision-Making: Recent Findings and Promising Approaches," in *Rational Choice and Criminal Behavior*, eds. Alex Piquero and Stephen Tibbetts (New York: Routledge, 2002), 3–24.

[16]Grasmick and Bursik, "Conscience"; Pogarsky, "Identifying 'Deterrable' Offenders"; Tibbetts, "Shame and Rational Choice"; Nagin and Paternoster, "Enduring Individual Differences"; Tibbetts and Myers, "Low Self-Control"; Tibbetts and Herz, "Gender Differences"; Harold Grasmick, Brenda Sims Blackwell, and Robert Bursik, "Changes Over Time in Gender Differences in Perceived Risk of Sanctions," *Law and Society Review 27* (1993): 679–705; Harold Grasmick, Robert Bursik, and Bruce Arneklev, "Reduction in Drunk Driving as a Response to Increased Threats of Shame, Embarrassment, and Legal Sanctions," *Criminology 31* (1993): 41–67; Stephen Tibbetts, "Self-Conscious Emotions and Criminal Offending," *Psychological Reports 93* (2004): 101–131.

[17]Tibbetts and Herz, "Gender Differences"; Grasmick et al., "Changes in the Sex Patterning"; Finley and Grasmick, "Gender Roles"; Pogarsky et al., "Perceptual Change"; Stephen Tibbetts, "Gender Differences in Students' Rational Decisions to Cheat," *Deviant Behavior 18* (1997): 393–414.

[18]Nagin and Paternoster, "Enduring Individual Differences"; Grasmick et al., "Changes Over Time"; Tibbetts, "Self-Conscious Emotions."

[19]Pogarsky et al., "Perceptual Change."

[20]Pogarsky, "Identifying 'Deterrable' Offenders."

✉ Routine Activities Theory

Routine activities theory is another contemporary form of the Classical School framework in the sense that it assumes a rational decision-making offender. The general model of routine activities theory was originally presented by Lawrence Cohen and Marcus Felson in 1979.[21] This theoretical framework emphasized the presence of three factors that come together in time and place to create a high likelihood for crime and victimization. These three factors are: motivated offender(s), suitable target(s), and lack of guardianship. Overall, the theory is appropriately named, in the sense that it assumes that most crime occurs in the daily routine of people who happen to see—and then seize—tempting opportunities to commit crime. Studies tend to support this idea, as opposed to the idea that most offenders leave their home knowing they are going to commit a crime; the latter offenders are called *hydraulic* and are relatively rare compared to the opportunistic type.

▲ **Image 3.1**   Marcus Felson, 1947– , Rutgers University, Author of Routine Activities Theory

Regarding the first factor noted as being important for increasing the likelihood of criminal activity—a motivated offender—the routine activities theory does not provide much insight. Rather, the model simply assumes that some individuals tend to be motivated and leaves it at that. Fortunately, we have many other theories that can fill this notable absence. Instead, the strength of routine activities theory is in the elaboration of the other two aspects of a crime-prone environment: suitable targets and lack of guardianship.

Suitable targets can include a variety of situations. For example, a very suitable target can be a vacant house in the suburbs, which the family has left for summer vacation. Data clearly show that burglaries more than double in the summer when many families are on vacation. Other forms of suitable targets range from an unlocked car to a female alone at a shopping mall carrying a lot of cash and credit cards or purchased goods. Other likely targets are bars or other places that serve alcohol. Offenders have traditionally targeted drunk persons because they are less likely to be able to defend themselves, as illustrated in a history of rolling drunks for their wallets that extends back to the early part of the 20th century. This is only a short list of the many types of suitable targets that are available to motivated offenders in everyday life.

The third and final aspect of the routine activities model for increased likelihood of criminal activity is the lack of guardianship. Guardianship is often thought of as a police officer or security guard, which often is the case. There are many other forms of guardianship, however, such as owning a dog to protect a house, which studies demonstrate can be quite effective. Just having a car or house alarm constitutes a form of guardianship. Furthermore, the presence of an adult, neighbor, or teacher can be an effective type of guarding the area against crime. In fact, recent studies show that the presence of increased lighting in the area can prevent a significant amount of crime, with one study showing a 20% reduction in overall crime in areas randomly chosen to receive improved lighting as compared to control areas that did not.[22] Regardless of the type of guardianship, it is the absence of adequate guardianship that sets the stage for crime; on the other

---

[21]Lawrence Cohen and Marcus Felson, "Social Change and Crime Rates: A Routine Activities Approach," *American Sociological Review 44* (1979): 214–241.

[22]David P. Farrington and Brandon C. Welsh, "Improved Street Lighting and Crime Prevention," *Justice Quarterly 19* (2002): 313–343.

hand, each step taken toward protecting a place or person is likely to deter offenders from choosing the target in relation to others. Locations that have a high convergence of motivated offenders, suitable targets, and lack of guardianship are typically referred to as *hot spots.*

Perhaps the most supportive evidence for routine activities theory and hot spots was the study of 911 calls for service during 1 year in Minneapolis, Minnesota.[23] This study examined all serious calls (as well as total calls) to police for a 1-year period. Half of the top 10 places from which police were called were bars or locations where alcohol was served. As mentioned above, establishments that serve alcohol are often targeted by motivated offenders for their high proportion of suitable targets. Furthermore, a number of bars tend to have a low level of guardianship in relation to the number of people they serve. Readers of this book may well relate to this situation. Most college towns and cities have certain drinking establishments that are known as being hot spots for crime.

Still, the Minneapolis hot spot study showed other types of establishments that made the top 10 rankings. These included bus depots, convenience stores, run-down motels and hotels, downtown malls and strip malls, fast-food restaurants, towing companies, and so on. The common theme linking these locations and the bars was the convergence of the three aspects described by routine activities theory as being predictive of criminal activity. Specifically, these are places that attract motivated offenders, largely because they have a lot of vulnerable targets and lack sufficient levels of security or guardianship. The routine activities framework has been applied in many contexts and places, many of them international.[24]

Modern applications of routine activities theory include geographic profiling, which uses satellite positioning systems in perhaps the most attractive and marketable aspect of criminological research in contemporary times. Essentially, such research incorporates computer software for global positioning systems (GPS) for identifying the exact location of every crime that takes place in a given jurisdiction. Such information has been used to solve or predict various crimes to the point where serial killers have been caught because the sites where the victims were found were triangulated to show the most likely place where the killer lived.

Some theorists have proposed a theoretical model based on individuals' lifestyles, which has a large overlap with routine activities theory, as shown in previous studies reviewed.[25] It only makes sense that a person who lives a more risky lifestyle, for example, by frequenting bars or living in a high-crime area, will be at more risk by being close to various hot spots identified by routine activities theory. Although some criminologists label this phenomenon a lifestyle perspective, it is virtually synonymous with the routine activities model because such lifestyles incorporate the same conceptual and causal factors in routine activities.

---

[23]Lawrence Sherman, Patrick R. Gartin, and Michael Buerger, "Hot Spots of Predatory Crime: Routine Activities and the Criminology of Place," *Criminology 27* (1989): 27–56.

[24]Jon Gunnar Bernburg and Thorolfur Thorlindsson, "Routine Activities in Social Context: A Closer Look at the Role of Opportunity in Deviant Behavior," *Justice Quarterly 18* (2001): 543–567; see also Richard Bennett, "Routine Activity: A Cross-National Assessment of a Criminological Perspective," *Social Forces 70* (1991): 147–163; James Hawdon, "Deviant Lifestyles: The Social Control of Routine Activities," *Youth and Society 28* (1996): 162–188; James L. Massey, Marvin Krohn, and Lisa Bonati, "Property Crime and the Routine Activities of Individuals," *Journal of Research in Crime and Delinquency 26* (1989): 378–400; Terrance Miethe, Mark Stafford, and J. Scott Long, "Social Differences in Criminological Victimization: A Test of Routine Activities/Lifestyles Theories," *American Sociological Review 52* (1987): 184–194; Elizabeth Mustaine and Richard Tewksbury, "Predicting Risks of Largency Theft Victimization: A Routine Activity Analysis Using Refined Lifestyle Measures," *Criminology 36* (1998): 829–857; D. Wayne Osgood, Janet Wilson, Patrick M. O'Malley, Jerald Bachman, and Lloyd Johnston, "Routine Activities and Individual Deviant Behavior," *American Sociological Review 61* (1996): 635–655; Dennis Roncek and Pamela Maier, "Bars, Blocks, and Crimes Revisited: Linking the Theory of Routine Activities to the Empiricism of Hot Spots," *Criminology 29* (1991): 725–753; Robert Sampson and John Wooldredge, "Linking the Micro- and Macro-Level Dimensions of Lifestyle-Routine Activity and Opportunity Models of Predatory Victimization," *Journal of Quantitative Criminology 3* (1987): 371–393.

[25]Hawdon, "Deviant Lifestyles"; Sampson and Wooldredge, "Linking the Micro"; Brown et al., *Criminology.*

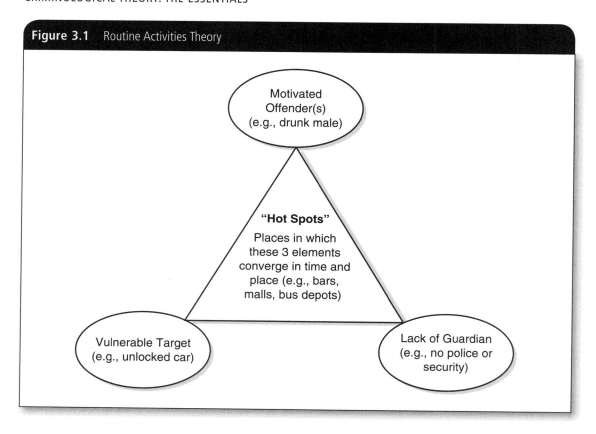

Figure 3.1   Routine Activities Theory

Policy Implications

There are numerous policy implications that can be derived from the theories and scientific findings in this chapter. Here, we will concentrate on some of the most important policies. First, we look at the policy of broken windows, which has many assumptions similar to those of routine activities and rational choice theories. The broken windows perspective emphasizes the need for police to crack down on more minor offenses to reduce more major crimes.[26] Although many cities have claimed reductions in serious crimes by using this theory (such as New York and Los Angeles), the fact is that crime was reduced by the same amount across most cities during the same time (the late 1990s to mid 2000s).

Still, other policies that can be derived from theories in this chapter include the *three-strikes-you're-out* policy, which assumes that offenders will make a rational choice not to commit future offenses because they could go to prison for life if they commit three felonies; the negatives certainly outweigh the expected benefits for the third crime. For deterrence to be extremely effective, punishment must be swift, certain, and severe. Where does the three-strikes policy fall into this equation? The bottom line is that it is much more

---

[26]James Q. Wilson and George Kelling, "Broken Windows: The Police and Neighborhood Safety," *Atlantic Monthly* (March 1982): 29–38.

severe than it is swift or certain. Given the Beccaria's philosophy (see Chapter 1), this policy will probably not work because it is not certain or swift. However, it is severe in the sense that a person can be sentenced to life if they commit three felony offenses over time.

A controversial three-strikes law was passed by voter initiative in California, and other states have adopted similar types of laws.[27] It sends third-time felons to prison for the rest of their lives regardless of the nature of that third felony. California first requires convictions for two *strikeable* felonies, crimes like murder, rape, aggravated assault, burglary, drug offenses, and so on. Then, any third felony can trigger a life sentence. The stories about some nonviolent offenders going to prison for the rest of their lives for stealing pieces of pizza or shoplifting DVDs, while rare, are quite true.

The question we are concerned with here is, does the three-strikes policy work? As a specific deterrent, the answer is clearly yes; offenders who are in prison for the rest of their lives cannot commit more crimes on the streets. In that regard, three-strikes works very well. Some people feel, however, that laws like three-strikes need to have a general deterrent effect to be considered successful, meaning that this law should deter everyone from engaging in multiple crimes. So, is three-strikes a general deterrent? Unfortunately, there are no easy answers to this question because laws vary from state to state, the laws are used at different rates across the counties in a given state, and so forth. There is at least some consensus in the literature, however.

One study from California suggests that three-strikes reduced crime,[28] but the remaining studies show that three-strikes either has no effect on crime or that it actually increases crime.[29] How could three-strikes increase crime? The authors attributed the increase in homicide, following three-strikes, to the possibility that third strikers have an incentive to kill victims and any witnesses in an effort to avoid apprehension. Although this argument is tentative, it may be true.[30] This is just one of the many policy implications that can be derived from this chapter. We expect that readers of this book will come up with many more policy implications, but it is vital that they examine the empirical literature in determining these policies' usefulness in reducing criminal activity. Other policy implications regarding the theories and findings discussed in this chapter will be discussed in the final section of this book.

In a strategy that is also strongly based on the rational choice model, a number of judges have started using shaming strategies to deter offenders from recidivating.[31] They have ordered everything from publicly posting pictures of men arrested for soliciting prostitutes to forcing offenders to walk down main streets of towns wearing signs that announce they committed crimes. These are just two examples of an increasing trend that emphasizes the informal or community factors required to deter crime. Unfortunately, to date, there have been virtually no empirical evaluations of the effectiveness of such shaming penalties, although studies of expected shame for doing an act consistently show a deterrent effect.[32]

---

[27]David Shichor and Dale K. Sechrest, eds., *Three Strikes and You're Out: Vengeance as Social Policy* (Thousand Oaks: Sage, 1996).

[28]Joanna M. Shepherd, "Fear of the First Strike: The Full Deterrent Effect of California's Two- and Three-Strikes Legislation," *Journal of Legal Studies 31* (2002): 159–201.

[29]See Lisa Stolzenberg and Stewart J. D'Alessio, "Three Strikes and You're Out: The Impact of California's New Mandatory Sentencing Law on Serious Crime Rates," *Crime and Delinquency 43* (1997): 457–469; Mike Males and Dan Macallair, "Striking Out: The Failure of California's 'Three-Strikes and You're Out Law," *Stanford Law and Policy Review 11* (1999): 65–72.

[30]Thomas B. Marvell and Carlisle E. Moody, "The Lethal Effects of Three-Strikes Laws," *Journal of Legal Studies 30* (2001): 89–106; see also Tomislav Kovandzic, John J. Sloan III, and Lynne M. Vieraitis, "Unintended Consequences of Politically Popular Sentencing Policy: The Homicide-Promoting Effects of 'Three Strikes' in U.S. Cities (1980-1999)," *Criminology and Public Policy 1* (2002): 399–424. For a review of empirical evaluations of three-strikes laws, see John Worrall, "The Effect of Three-Strikes Legislation on Serious Crime in California," *Journal of Criminal Justice, 32* (2004): 283–296.

[31]Piquero and Tibbetts, *Rational Choice.*

[32]Tibbetts, "Gender Differences in Students' Rational."

✉ Conclusion

This chapter reviewed the more recent forms of classical and deterrence theory, such as rational choice theory, which emphasizes the effects of informal sanctions (e.g., family, friends, employment) and benefits and costs of offending, and a framework called routine activities theory, which explains why victimization tends to occur far more often in certain locations (i.e., hot spots) due to the convergence of three key elements in time and place—motivated offender(s), vulnerable target(s), and lack of guardianship—which create attractive opportunities for crime as individuals go about their typical, everyday activities. The common element across all of these perspectives is the underlying assumption that individuals are rational beings who have free will and thus choose their behavior based on assessment of a given situation, such as possible risks versus potential payoff. Although the studies examined in this chapter lend support for many of the assumptions and propositions of the classical framework, it is also clear that there is a lot more involved in explaining criminal human behavior than the individual decision making that goes on before a person engages in rule violation. After all, human beings are often not rational and often do things spontaneously without considering the potential risks beforehand, especially chronic offenders. So, despite the use of the classical and neoclassical models in most systems of justice in the modern world, such theoretical models of criminal activity largely fell out of favor among experts in the mid-19th century, when an entirely new paradigm of human behavior became dominant. This new perspective became known as the Positive School, and we will discuss the origin and development of this paradigm in the following chapter.

✉ Chapter Summary

- After 100 years of neglect by criminologists, the classical and deterrence models experienced a rebirth in the late 1960s.
- The seminal studies in the late 1960s and early 1970s were largely based on aggregate and group rates of crime, as well as group rates of certainty and severity of punishment, which showed that levels of actual punishment, especially certainty of punishment, were associated with lower levels of crime.
- A subsequent wave of deterrence research, cross-sectional surveys, which were collected at one time, supported previous findings that perceptions of certainty of punishment had a strong, inverse association with offending, whereas findings regarding severity were mixed.
- Longitudinal studies showed that much of the observed association between perceived levels of punishment and offending could be explained by the experiential effect, which is the phenomena of behavior affecting perceptions as opposed to deterrence (i.e., perceptions affecting behavior).
- Scenario studies addressed the experiential effect by supplying a specific context through presenting a detailed vignette and then asking what subjects would do in that specific circumstance and what their perceptions of the event were.
- Rational choice theory emphasizes not only the formal and official aspects of criminal sanctions but also the informal or unofficial aspects, such as family and community.
- Whereas traditional classical deterrence theory ignored them, rational choice theory emphasizes the benefits of offending, such as the thrill it produces, as well as the social benefits of committing crime.
- Routine activities theory provides a theoretical model that explains why certain places have far more crime than others and why some locations have hundreds of calls to police each year, whereas others have none.

- Lifestyle theories of crime reveal that the way people live may predispose them to both crime and victimization.
- Routine activities theory and the lifestyle perspective are becoming key in one of the most modern approaches toward reducing or predicting crime and victimization. Specifically, GPS and other forms of geographical mapping of crime events have contributed to an elevated level of research and attention given to these theoretical models, due to their importance in specifically documenting where crime occurs and, in some cases, predicting where future crimes will occur.
- All of the theoretical models and studies in this section were based on classical and deterrence models, which assume that individuals consider the potential benefits and costs of punishment and then make their decisions to engage (or not) in the criminal act.

KEY TERMS

Cross-sectional studies	Rational choice theory	Scenario (vignette) research
Experiential effect	Routine activities theory (lifestyle theory)	

DISCUSSION QUESTIONS

1. Do you think the deterrence model should have been reborn, or do you think it should have just been left for dead? Explain why you feel this way.

2. Regarding the aggregate level of research in deterrence studies, do you find such studies valid? Explain why or why not.

3. In the comparison of longitudinal studies versus scenario (vignette) studies, which do you think offers the most valid method for examining individual perceptions regarding the costs and benefits in offending situations? Explain why you feel this way.

4. Can you relate to the experiential effect? If you can't, do you know someone who seems to resemble the behavior that results from this phenomenon? Make sure to articulate what the experiential effect is.

5. Regarding rational choice theory, would you rather be subject to formal sanctions if none of your family, friends, or employers found out that you engaged in shoplifting, or would you rather face the informal sanctions with no formal punishment (other than being arrested) for such a crime? Explain your decision.

6. As a teenager, did you or family or friends get a rush out of doing things that were deviant or wrong? If so, did that feeling seem to outweigh any legal or informal consequences that may have deterred you or people you know?

7. Regarding routine activities theory, which places, residences, or areas of your hometown do you feel fit this idea that certain places have more crime than others (i.e., hot spots)? Explain how you, friends, or others (including police) in your community deal with such areas. Does it work?

8. Regarding routine activities theory, which of its three elements do you feel is the most important to address in efforts to reduce crime in the hot spots?

9. What type of lifestyle characteristics lead to the highest criminal or victimizing rates? List at least five factors that lead to such propensities.

10. Find at least one study that uses mapping and geographical (GPS) data, and report the conclusions of that study. Do the findings and conclusions fit the routine activities theoretical framework or not? Why?

11. What types of policy strategies derived from rational choice and routine activities theories do you think would be most effective? Least effective?

WEB RESOURCES

Modern Testing of Deterrence

http://www.deathpenaltyinfo.org/

Rational Choice Theory

http://www.answers.com/topic/rational-choice-theory-criminology

Routine Activities and Lifestyle Theory

http://www.popcenter.org/learning/pam/help/theory.cfm

CHAPTER

4

Early Positive School Perspectives of Criminality

In this chapter, we will discuss the dramatic differences in assumptions between the Classical and Positive Schools of criminological thought. We will also touch on the pre-Darwinian perspectives of human behavior (e.g., **phrenology**), as well as the influence that Darwin had on perspectives of all social sciences, particularly criminology. Finally, we will discuss the theories and methods used by early positivists, particularly Cesare Lombroso, IQ theorists, and body-type researchers, with an emphasis on the criticisms of these perspectives, methodologies, and resulting policies.

After many decades of dominance by the Classical School (see Chapters 2 and 3), academics and scientists were becoming aware that the deterrence framework did not explain the distribution of crime. Their restlessness led to new explanatory models of crime and behavior. Most of these perspectives focused on the fact that certain individuals or groups tend to offend more than others and that such "inferior" individuals should be controlled or even eliminated. This ideological framework fit a more general stance toward **eugenics,** which is the study of and policies related to the improvement of the human race via control over reproduction, which we will see was explicitly mandated for certain groups. Thus, the conclusion was that there must be notable variations across individuals and groups that can help determine who are most at risk of offending.

So, in the early to mid-1800s, several perspectives were offered regarding how to determine which individuals or groups were most likely to commit crime. Many of these theoretical frameworks were made to distinguish the more "superior" individuals or groups from the "inferior" individuals or groups. Such intentions were likely related to the increased use of slavery in the world during the 1800s, as well as the fight of imperialism over rebellions at that time. For example, slavery was at its peak in the United States during this period, and many European countries controlled many dozens of colonies, which they were trying to retain for profit and domain.

Perhaps the first example of this belief was represented by **craniometry.** Craniometry was the belief that the size of the brain or skull represented the superiority or inferiority of certain individuals or ethnic or racial groups.[1] The size of the brain and the skull were considered because, at that time, it was believed that a person's skull perfectly conformed to brain structure; thus, the size of the skull was believed to reflect the size of the brain. Modern science has challenged this assumption, but there actually is a significant correlation between the size of the skull and the size of the brain. Still, even according to the assumptions of the craniometrists, it is unlikely that much can be gathered from the overall size of the brain, and certainly the skull, from simple measurements of mass.

The scientists who studied this model, if they were dealing with living subjects, would measure the various sizes or circumferences of the skulls. If they were dealing with recently dead subjects, then they would actually measure the brain weight or volume of the participants. When dealing with subjects who had died long before, craniometrists would measure the volume of skulls by pouring seeds inside and then pouring those that fit into graduated cylinders. Later, when these scientists realized that seeds were not a valid measure of volume, they moved toward using buckshot or ball bearings.

Most studies by the craniometrists tended to show that the subjects of White or Western European descent were superior to other ethnic groups in terms of larger circumference or volume in brain or skull size. Furthermore, the front portion of the brain (i.e., genu) was thought to be larger in the superior individuals or groups, and the hind portion of the brain or skull (i.e., splenium) was predicted to be larger in the lesser individuals or groups. Notably, these researchers typically knew which brains or skulls belonged to which ethnic or racial group before measurements were taken, making for an unethical and improper methodology. Such biased measurements continued throughout the 19th century and into the early 1900s.[2] These examinations were largely done with the intention of furthering the assumptions of eugenics, which aimed to prove under the banner of science that certain individuals and ethnic or racial groups are inferior to others. The fact that this was their intent is underscored by subsequent tests using the same subjects but without knowing which skulls or brains were from certain ethnic or racial groups; these later studies showed only a small correlation between size of the skull or brain and certain behaviors or personalities.[3]

Furthermore, once some of the early practitioners of craniometry died, their brains had volumes that were less than average or average. The brain of K. F. Gauss, for example, was relatively small but more convoluted, with more gyri and fissures. Craniometrists then switched their postulates to say that more convoluted or complex brain structure, with more fissures and gyri, indicated superior brains.[4] However, this argument was even more tentative and vague than the former hypotheses of craniometrists and thus did not last long. The same was true of craniometry, thanks to its noticeable lack of validity. However, it is important to note that modern studies show that people who have significantly larger brains tend to score higher on intelligence tests.[5]

---

[1]For a review, see Nicole Rafter, "The Murderous Dutch Fiddler," *Theoretical Criminology* 9, No. 1 (2005): 65–96.

[2]For example, see the comparison made between Robert Bean, "Some Racial Peculiarities of the Negro Brain," *American Journal of Anatomy* 5 (1906): 353–432, which showed a distinct difference in the brains across race when brains were identified by race before comparison, and Franklin P. Mall, "On Several Anatomical Characters of the Human Brain, Said to Vary According to Race and Sex, with Especial Reference to the Weight of the Frontal Lobe," *American Journal of Anatomy* 9 (1909): 1–32, which showed virtually no differences among the same brains when comparisons were made without knowing the races of the brains prior to comparison; see discussion in Stephen Jay Gould, *The Mismeasure of Man*, 2nd ed. (New York: Norton, 1996).

[3]See Mall, "On Several Anatomical Characters"; much of the discussion in this section is taken from Gould, *The Mismeasure*.

[4]Edward A. Spitska, "A Study of the Brains of Six Eminent Scientists and Scholars Belonging to the Anthropological Society, Together with a Description of the Skull of Professor E. D. Cope," *Transactions of the American Philosophical Society* 21 (1907): 175–308.

[5]Stanley Coren, *The Left-Hander Syndrome* (New York: Vintage, 1993); James Kalat, *Biological Psychology*, 8th ed. (New York: Wadsworth, 2004).

Despite the failure of craniometry to explain the difference between criminals and noncriminals, scientists were not ready to give up the assumption that criminal behavior could be explained by visual differences in the skull (or brain), and they certainly weren't ready to give up the assumption that certain ethnic or racial groups were superior or inferior relative to others. Therefore, the experts of the time created phrenology. Phrenology is the science of determining human dispositions based on distinctions (e.g., bumps) in the skull, which are believed to conform to the shape of the brain.[6] Readers should keep in mind that much of the theorizing by the phrenologists still aimed to support the assumptions of eugenics and show that certain individuals and groups of people were inferior or superior to others.

It is important to keep in mind that, like the craniometrists, phrenologists assumed that the shape of the skull conformed to the shape of the brain. Thus, a bump or other abnormality on the skull directly related to an abnormality in the brain at that spot. Such assumptions have been refuted by modern scientific evidence, so it is not surprising that phrenology fell out of favor in criminological thought rather quickly. Like its predecessor, however, phrenology got some things right. Certain parts of the brain are indeed responsible for specific tasks.

For example, in the original phrenological map, destructiveness was indicated by abnormalities above the left ear. Modern scientific studies show that the most vital part of the brain in terms of criminality associated with trauma is the left temporal lobe, the area above the left ear.[7] Furthermore, phrenologists were right in other ways. Most readers know that specific portions of the brain govern the operation of different physical activities; one area governs the action of our hands, whereas other areas govern our arms, legs, and so on. So, the phrenologists had a few things right but were completely wrong about the extent to which bumps on the skull could indicate who would be most disposed to criminal behavior.

Once phrenology fell out of favor among scientists, researchers and society in general did not want to depart from the assumption that certain individuals or ethnic groups were inferior to others. Therefore, another discipline, known as **physiognomy,** became popular in the mid-1800s. Physiognomy is the study of facial and other bodily aspects to indicate developmental problems, such as criminality. Not surprisingly, the early physiognomy studies focused on contrasting various racial or ethnic groups to prove that some were superior or inferior to others.[8]

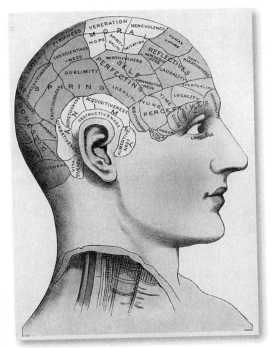

▲ **Image 4.1** This diagram shows the sections of the brain, as detailed by 19th-century phrenologists, who believed that each section was responsible for a particular human personality trait. If a section were enlarged or shrunken, the personality would be likewise abnormal. Doctors, particularly those doing entry examinations at American prisons, would examine each new inmate's head for bumps or cavities to develop a criminal profile. For example, if the section of brain responsible for acquisitiveness was enlarged, the offender probably was a thief. Lombroso and his school combined phrenology with other models that included external physical appearance traits to single out criminals from the general population.

---

[6]Orson S. Fowler, *Fowler's Practical Phrenology: Giving a Concise Elementary View of Phrenology* (New York: O. S. Fowler, 1842).

[7]For a review, see Adrian Raine, *The Psychopathology of Crime* (San Diego: Academic Press, 1993).

[8]Josiah C. Nott and George R. Gliddon, *Types of Mankind* (Philadelphia: Lippincott, Grambo, and Company, 1854); Josiah C. Nott and George R. Gliddon, *Indigenous Races on Earth* (Philadelphia: Lippincott, 1868).

Using modern understandings of science, it is not surprising that physiognomy did not last long as a respected scientific perspective of criminality. At any time other than the late 1800s, their ideas would not have been accepted for long, if at all. However, the theory emerged at an auspicious time. Specifically, Darwin published his work *The Origin of Species* in the late 1800s and made a huge impact on societal views regarding the rank order of groups in societies.

Darwin's model outlined a vague framework suggesting that humans had evolved from more primitive beings and that the human species (like all others) evolved from a number of adaptations preferred by natural selection. In other words, some species are selected by their ability to adapt to the environment, whereas others do not adapt as well and die off or at least become inferior in terms of dominance. This assumption of Darwin's work, which was quickly and widely accepted by both society and scientists throughout the world, falsely led to an inclination to believe that certain ethnic or racial groups are inferior or superior to other groups. Despite a backlash by many religious authorities, who were likely threatened by the popularity of a theory that promoted natural design as opposed to a higher being or creator, Darwin had created a scientific snowball that spread like a wildfire across virtually all scientific disciplines, particularly criminology.

Darwin was not a criminologist, so he is not considered a father or theorist in the field. However, he did set the stage for what followed in criminological thought. Specifically, Darwin's theory laid the groundwork for what would become the first major scientific theory of crime, namely Cesare Lombroso's theory of born criminals, which also tied together the assumptions and propositions of craniometry, phrenology, and physiognomy.

▲ **Image 4.2**   Cesare Lombroso (1836–1909)

⊠ Lombroso's Theory of Atavism and Born Criminals

Basing his work on Darwin's theory of natural selection, Cesare Lombroso (1835–1909) created what is widely considered the first attempt toward scientific theory in criminological thought. Most previous theorists were not scientists; Beccaria, for example, was trained in law and never tested his propositions. Unlike the craniometrists and phrenologists, Beccaria's goal was not to explain levels of criminality. However, Lombroso was trained in medical science, and he aimed to document his observations and use scientific methodology. Furthermore, timing was on his side in the sense that Darwin's theory was published 15 years prior to Lombroso's major work, and in that time, the idea of evolution had become immensely popular with both scientists and the public.

Lombroso's Theory of Crime

The first edition of Lombroso's work *The Criminal Man* was published in 1876 and created an immediate response in most

Western societies, influencing both their ideas and policies related to crime and justice.[9] In this work, Lombroso outlined a theory of crime that largely brought together the pre-Darwinian theories of craniometry, phrenology, and physiognomy. Furthermore, Lombroso thought that certain groups and individuals were atavistic and likely to be born to commit crime. **Atavistic** or **atavism** means that a person or feature of an individual is a throwback to an earlier stage of evolutionary development. In other words, he thought serious criminals were lower forms of humanity in terms of evolutionary progression. For example, Lombroso would likely suggest that chronic offenders are more like earlier stages of humankind, like a *missing link,* than they are like modern humans.

Lombroso noted other types of offenders, such as the mentally ill and *criminaloids,* who committed minor offenses due to external or environmental circumstances, but he argued that the *born criminals* should be the target in addressing crime, insisting that they were the most serious and violent criminals in any society. These are what most criminologists now refer to as *chronic offenders.* Furthermore, Lombroso claimed that born criminals cannot be stopped from their natural tendencies to be antisocial.

On the other hand, Lombroso claimed that, although it was their nature to commit crime, born criminals could be stopped, or at least partially deterred by society. According to Lombroso, societies could identify born criminals, even early in life, through their **stigmata.** Stigmata are physical manifestations of the atavism of an individual, features that indicate a prior evolutionary stage of development.

Lombroso's List of Stigmata

According to Lombroso, more than five stigmata indicate that an individual is atavistic and inevitably will be a born criminal. Understandably, readers may be wondering what these stigmata are, given their importance. This is a great question, but the answer varies. In the beginning, this list was largely based on Lombroso's work as a physician; it included features such as large eyes, large ears, and so on. Lombroso changed this list as he went along, however, which might be considered poor science, even in the last edition of his book published well into the 1900s.

For the most part, stigmata consisted of facial and bodily features that deviated from the norm. In other words, abnormally small or large noses, abnormally small or large ears, abnormally small or large eyes, abnormally small or large jaws—almost anything that went outside the bell curve on normal human physical development. Lombroso also threw in some extraphysiological features, such as tattoos and a family history of epilepsy and other disorders.[10] Although tattoos may be somewhat correlated to crime and delinquency, is it likely that they caused such antisocial behavior? Given Lombroso's model that people are born criminal, it is quite unlikely that these factors are causally linked to criminality. How many babies are born with tattoos? Ignoring the illogical nature of many of the stigmata, Lombroso professed that people who had more than five of these physical features were born criminals and that something should be done to prevent the inevitable in their future offending career.

As a physician working for the Italian army, Lombroso examined the bodies of war criminals who were captured and brought in for analysis. According to Lombroso, he first came to the realization of the nature of the criminal when a particular individual was brought in for him to examine:

---

[9]Cesare Lombroso, *The Criminal Man (L'uomo Delinquente),* 1st ed. (Milan: Hoepli, 1876), 2nd ed. (Turin: Bocca, 1878).

[10]Gould, *The Mismeasure,* 153.

This was not merely an idea, but a flash of inspiration. At the sight of that skull, I seemed to see all of a sudden, lighted up as a vast plain under a flaming sky, the problem of the nature of the criminal—an atavistic being who reproduces in his person the ferocious instincts of primitive humanity and the inferior animals.[11]

This was Lombroso's first exposure to such a criminal and his acknowledgment of the theory that he created. He further expanded on this theory by specifying some of the physical features he observed from this individual:

Thus were explained anatomically the enormous jaws, high cheek bones . . . solitary lines in the palms, extreme size of the orbits, handle-shaped ears found in criminals, savages and apes, insensibility to pain, extremely acute sight, tattooing, excessive idleness, love of orgies, and the irresponsible craving of evil for its own sake, the desire not only to extinguish life in the victim, but to mutilate the corpse, tear its flesh and drink its blood.[12]

Although most people may now laugh at his words, at the time he wrote this description, it would have rung true to most readers, which is likely why his book was the dominant text for many decades in the criminological field. In this description, Lombroso incorporates many of the core principles of his theory: the idea of criminals being atavistic or biological throwbacks in evolution, as well as the stigmata that became so important in his theoretical model.

A good example of the popular acceptance of Lombroso's "scientific" stigmata was that Bram Stoker used it in the 1896 novel *Dracula*, creating a character based on Lombrosian traits of a villain, such as a high bridge of a thin nose, arched nostrils, massive eyebrows, pointed ears, and so on. This novel was published in the late 1800s, when Lombroso's theory was highly dominant in society and science. Lombroso's ideas became quite popular among academics, scientists, philosophers, fiction writers, and those responsible for criminal justice policy.

Beyond identifying born criminals by their stigmata, Lombroso said he could associate the stigmata with certain types of criminals, for example, anarchists, burglars, murderers, shoplifters, and so on. Of course, his work is quite invalid by modern research standards.

Lombroso as the Father of Criminology and the Father of the Positive School

Lombroso's theory came a decade and a half after Darwin's work had been published and had spread rapidly throughout the Western world. Also, Lombroso's model supported what were then the Western world's views on slavery, deportation, and so on. Due to this timing and the fact that Lombroso became known as the first individual who actually tested his hypotheses through observation, Lombroso is widely considered the father of criminology. This title does not indicate respect for his theory, which has been largely rejected, or for his methods, which are considered highly invalid by modern standards. It is deserved, however, in the sense that he was the first person to gain recognition in testing his theoretical propositions. Furthermore, his theory coincided with a political movement that became popular at that time: the Fascist and Nazi movements of the early 1900s.

Beyond being considered the father of criminology, Lombroso is also considered the father of the **Positive School** of criminology because he was the first to gain prominence in identifying factors beyond

---

[11]Lombroso, *Criminal Man,* as cited and discussed by Ian Taylor, Paul Walton, and Jock Young, *The New Criminology: For a Social Theory of Deviance* (London: Routledge, 1973), 41.

[12]Taylor et al., *New Criminology,* 41–42.

free will or free choice, which the Classical School said were the sole cause of crime. Although previous theorists had presented perspectives that went beyond free will, such as craniometrists and phrenologists, Lombroso was the first to gain widespread attention. Lombroso's perspective gained almost immediate support in all developed countries of that time, which is the most likely reason for why Lombroso is considered the father of the Positive School of criminology.

It is important to understand the assumptions of *positivism,* which most experts consider somewhat synonymous with the term **determinism.** Determinism is the assumption that most human behavior is determined by factors beyond free will and free choice. In other words, determinism (i.e., the Positive School) assumes that human beings do not decide how they will act by logically thinking through the costs and benefits of a given situation. Rather, the Positive School attributes all kinds of behavior, especially crime, on biological, psychological, and sociological variables.

Many readers probably feel they chose their career paths and made most other key decisions in their lives. However, scientific evidence shows otherwise. For example, studies clearly show that far more than 90% of the world's population has adopted the religious affiliation (e.g., Baptist, Buddhist, Catholic, Judaism, etc.) of their parents or caretakers. Therefore, what most people consider an extremely important decision, the choice of what they believe regarding a higher being or force, is almost completely determined by the environment in which they were brought up. Almost no one sits down and studies various religions before deciding which one suits him or her the best. Rather, in almost all cases, religion is determined by culture, and this finding goes against the Classical School's assumption that free will rules. The same type of argument can be made about the clothes we wear, the food we prefer, and the activities that give us pleasure.

Another way of distinguishing positivism and determinism from the Classical School lies in the way scientists view human behavior, which can be seen best in an analogy with chemistry. Specifically, a chemist assumes that, if a certain element is subjected to certain temperatures, pressures, or mixtures with other elements, it will result in a predicted response to such influences. In a highly comparable way, a positivist assumes that, when human beings are subjected to poverty, delinquent peers, low intelligence, or other factors, they will react and behave in predictable ways. Therefore, there is virtually no difference in how a chemist feels about particles and elements and how a positivistic scientist feels about how humans react when exposed to biological and social factors.

In Lombroso's case, the deterministic factor was the biological makeup of the individual. However, we shall see in the next several sections that positivistic theories focus on a large range of variables, from biology to psychology to social aspects. For example, many readers may believe that bad parenting, poverty, and associating with delinquent peers are some of the most important factors in predicting criminality. If you believe that such variables have a significant influence on decisions to commit crime, then you are likely a positive theorist; you believe that crime is caused by factors above and beyond free choice or free will.

Lombroso's Policy Implications

Beyond the theoretical aspects of Lombroso's theory of criminality, it is important to realize that his perspective had profound consequences for policy. Lombroso was called to testify in numerous criminal processes and trials to determine the guilt or innocence of suspects. Under the banner of science (comparable to what we consider DNA or fingerprint analysis in modern times), Lombroso was asked to specify whether or not a suspect had committed a crime.[13] Lombroso based such judgments on the visual stigmata that he

---

[13]See Gould, *The Mismeasure.*

could see among suspects.[14] Lombroso documented many of his experiences as an expert witness at criminal trials. Here is one example: "[one suspect] was, in fact, the most perfect type of the born criminal: enormous jaws, frontal sinuses, and zygomata, thin upper lip, huge incisors, unusually large head. . . . [H]e was convicted."[15]

When Lombroso was not available for such "scientific" determinations of the guilty persons, his colleagues or students (often referred to as lieutenants) were often sent. Some of these students, such as Enrico Ferri and Raphael Garrofalo, became quite active in the fascist regime of Italy in the early 1900s. This model of government, like the Nazi Party of Germany, sought to remove the "inferior" groups from society.

Another policy implication in some parts of the world was identifying young children on the basis of observed stigmata, which often become noticeable in the first 5 to 10 years of life. This led to tracking or isolating certain children, largely based on physiological features. Although many readers may consider such policies ridiculous, modern medicine has supported the identification, documentation, and importance of what are termed **minor physical anomalies (MPAs),** which it holds may indicate high risk of developmental problems. Some of these MPAs include

- Head circumference out of the normal range
- Fine, "electric" hair
- More than one hair whorl
- Epicanthus, which is observed as a fold of skin from the lower eyelids to the nose, which appears as droopy eyelids
- Hypertelorism (orbital), which represents an increased interorbital distance
- Malformed ears
- Low-set ears
- Excessively large gap between the first and second toes
- Webbing between toes or fingers
- No earlobes
- Curved fifth finger
- Third toe is longer than the second toe
- Asymmetrical ears
- Furrowed tongue
- Simian crease[16]

Given that such visible physical aspects are still correlated with developmental problems, including criminality, it is obvious that Lombroso's model of stigmata in predicting antisocial problems has implications to the present day. Such implications are more accepted by modern medical science than they are in criminological literature. Furthermore, some modern scientific studies have shown that being unattractive predicts criminal offending, which somewhat supports Lombroso's theory of crime.[17]

---

[14]For a review of such identifications, see Gould, *The Mismeasure.*

[15]Cesare Lombroso, *Crime: Its Causes and Remedies* (Boston: Little & Brown, 1911), 436.

[16]Taken from Mary Waldrop and Charles Halverson, "Minor Physical Anomalies and Hyperactive Behavior in Young Children," in *Exceptional Infant: Studies in Abnormalities,* ed. Jerome Hellmuth (New York: Brunner/Mazel, 1971), 343–381, as cited and reviewed by Diana Fishbein, *Biobehavioral Perspectives on Criminology* (Belmont: Wadsworth, 2001).

[17]Robert Agnew, "Appearance and Delinquency," *Criminology 22* (1984): 421–440.

About three decades after Lombroso's original work was published, and after a long period of dominance, criminologists began to question his theory of atavism and stigmata. Furthermore, it became clear that more was involved in criminality than just the way people looked, such as psychological aspects of individuals. However, scientists and societies were not ready to depart from theories like Lombroso's, which assumed that certain people or groups of people were inferior to others, so they simply chose another factor to emphasize: intelligence or IQ.

The IQ Testing Era

Despite the evidence against Lombroso that was presented, his theorizing remained dominant until the early 1900s, when criminologists realized that stigmata and the idea of born criminals were not valid. However, even at that time, theorists and researchers were not ready to give up on the eugenics assumption that certain ethnic or racial groups were superior or inferior to others. Thus, a new theory emerged based on a more quantified measure that was originated, with benevolent intentions, by Alfred Binet in France. This new measure was IQ, short for *intelligence quotient*. At that time, IQ was calculated as chronological age divided by mental age, which was then multiplied by 100, with average score being 100. This scale was changed enormously over time, but the bottom line was using the test to determine whether someone was above or below the base score (100).

As mentioned above, Binet had good intentions: He created IQ scores to identify youth who were not performing up to par on educational skills. Binet was explicit in stating that IQ could be changed, which is why he proposed a score to identify slow learners so that they could be trained to increase their IQs.[18] However, when Binet's work was brought over to the United States, his basic assumptions and propositions were twisted. One of the most prominent individuals who used Binet's IQ test in the United States for purposes of deporting, incapacitating, sterilizing, and otherwise ridding society of low-IQ individuals was H. H. Goddard.

Goddard is generally considered the leading authority on the use and interpretation of IQ testing in the United States.[19] He adapted Binet's model to examine immigrants who were coming into the United States from foreign lands. It is important to note that Goddard proposed quite different assumptions regarding intelligence or IQ than did Binet. Goddard asserted that IQ was static or innate, meaning that such levels could not be changed, even with training. His assumption was that intelligence was passed from generation to generation, inherited from parents.

Goddard labeled low IQ as **feeblemindedness,** which actually became a technical, scientific term in the early 1900s meaning those who had significantly below-average levels of intelligence. Of course, being a scientist, Goddard specified certain levels of feeblemindedness, which were ranked based on the degree to which the score of an individual was below average. Ranking from the best to the lowest intelligence, the first group were the *morons,* the second-lowest group were the *imbeciles,* and the lowest-intelligence group were the *idiots.*

According to Goddard, from a eugenics point of view, the biggest threat to the progress of humanity was not the idiots but the morons. In Goddard's words, "The idiot is not our greatest problem. . . . He does not continue the race with a line of children like himself. . . . It is the moron type that makes for us our great problem."[20] Thus, the moron is the one group out of the three categories of feebleminded that is smart enough to slip through the cracks and reproduce.

---

[18]Gould, *The Mismeasure.*

[19]Again, most of this discussion is taken from Gould, *The Mismeasure,* because his review is perhaps the best known in the current literature.

[20]Henry H. Goddard, *The Kallikak Family, a Study of the Heredity of Feeble-Mindedness* (New York: MacMillan, 1912).

Goddard received many grants to fund his research to identify the feebleminded. Goddard took his research team to the major immigration center at Ellis Island in the early 1900s to identify the feebleminded as they attempted to enter the United States. Many members of his team were women, who he felt were better at distinguishing the feebleminded by sight:

> The people who are best at this work, and who I believe should do this work, are women. Women seem to have closer observation than men. It was quite impossible for others to see how . . . women could pick out the feeble-minded without the aid of the Binet test at all.[21]

Goddard was proud of the increase in the deportation of potential immigrants to the United States, enthusiastically reporting that deportations for the reason of mental deficiency increased by 350% in 1913 and 570% in 1914 over the average of the preceding 5 years.[22] However, over time, Goddard realized that his policy recommendations of deportation, incarceration, and sterilization were not based on accurate science.

▲ **Image 4.3**   Ellis Island

After consistently validating his IQ test on immigrants and mental patients, Goddard finally tested his intelligence scale on a relatively representative cross-section of American citizens, namely draftees for military service during World War I. The results showed that many of these recruits would score as feebleminded (i.e., lower than a mental age of 12) on the IQ test. Therefore, Goddard lowered the criterion of what determined a person of feeblemindedness from the mental age of 12 to 8. Although this appears to be a clear admission that his scientific method was inaccurate, Goddard continued to promote his model of the feebleminded for many years, and societies used his ideas.

However, toward the end of his career, Goddard admitted that intelligence could be improved, despite his earlier assumptions that it was innate and static.[23] In fact, Goddard actually claimed that he had "gone over to the enemy."[24] However, despite Goddard's admission that his assumptions and testing were not determinant of individuals' intelligence levels, the snowball had been rolling for too long and had gathered too much strength to fight even the most notable theorists' admonishments of the perspective.

Sterilization of individuals, mostly females, continued in the United States based on scores of intelligence tests. Often the justification was not a person's intelligence scores but those of his or her mother or father. Goddard had proclaimed that the *germ-plasm* determining feeblemindedness was passed on from

---

[21]Henry H. Goddard, "The Binet Tests in Relation to Immigration," *Journal of Psycho-Asthenics 18* (1913): 105–107; quote taken from page 106, as cited in Gould, *The Mismeasure.*

[22]Gould, *The Mismeasure*, 198.

[23]Henry H. Goddard, "Feeblemindedness: A Question of Definition," *Journal of Psycho-Asthenics 33* (1928): 219–227, as discussed in Gould, *The Mismeasure.*

[24]Goddard, "Feeblemindedness," 224.

one generation to the next, so it inevitably resulted in offspring being feebleminded as well. Thus, the U.S. government typically sterilized individuals, typically women, based on the IQ scores of their parents.

The case of Buck v. Bell, brought to the U.S. Supreme Court in 1927, dealt with the issue of sterilizing individuals who had scored, or whose parents had scored, as mentally deficient on intelligence scales. The majority opinion, written by one of the court's most respected jurists, Oliver Wendell Holmes, Jr., stated that:

> We have seen more than once that the public welfare may call upon the best citizens for their lives. It would be strange if it could not call upon those who already sap the strength of the state for these lesser sacrifices.... Three generations of imbeciles are enough.[25]

Thus, the highest court in the United States upheld the use of sterilization for the purposes of limiting reproduction among individuals who were deemed as being feebleminded according to an IQ score. Such sterilizations continued until the 1970s, when the practice was finally halted. Governors of many states, such as North Carolina, Virginia, and California, have given public apologies for what was done. For example, in 2002, the governor of California, Gray Davis, apologized for the state law passed almost a century earlier that had resulted in the sterilization of about 19,000 women in California.

Although this aspect of U.S. history is often hidden from the public, it did occur, and it is important to acknowledge this blot on our nation's history, especially at a time when we were fighting abuses of civil rights by the Nazis and other regimes. Ultimately, the sterilizations, deportations, and incarcerations based on IQ testing are an embarrassing episode in the history of the United States.

For decades, the issue of IQ was not researched or discussed much in the literature. However, in the 1970s, a very important study was published in which Travis Hirschi and Michael Hindelang examined the effect of intelligence on youths' behaviors.[26] Hirschi and Hindelang found that, among youth of the same race and social class, intelligence had a significant effect on delinquency and criminality among individuals. This study, as well as others, showed that the IQ of delinquents or criminals is about 10 points lower than the scores of noncriminals.[27]

This study led to a rebirth in research regarding intelligence testing within the criminological perspective. A number of recent studies have shown that certain types of intelligence are more important than others. For example, several studies have shown that having a low verbal intelligence has the most significant impact on predicting delinquent and criminal behavior.[28]

This tendency makes sense because verbal skills are important for virtually all aspects of life, from everyday interactions with significant others to filling out forms at work to dealing with people via employment. In contrast, most people do not have to perform advanced math or quantitative skills at their jobs or in day-to-day experiences, let alone spatial and other forms of intelligence that are more abstract. Thus, low verbal IQ is the type of intelligence that represents the most direct prediction for criminality, and this is most likely due to the

---

[25]As quoted in Gould, *The Mismeasure,* 365.

[26]Travis Hirschi and Michael Hindelang, "Intelligence and Delinquency: A Revisionist Review," *American Sociological Review* 42 (1977): 571–587.

[27]For a review, see Raymond Paternoster and Ronet Bachman, *Explaining Criminals and Crime* (Los Angeles: Roxbury, 2001).

[28]For a review, see Chris L. Gibson, Alex R. Piquero, and Stephen G. Tibbetts, "The Contribution of Family Adversity and Verbal IQ to Criminal Behavior," *International Journal of Offender Therapy and Comparative Criminology* 45 (2001): 574–592; see also, Hirschi and Hindelang, "Intelligence and Delinquency"; Terrie Moffitt, "The Neuropsychology of Delinquency: A Critical Review of Theory and Research," in *Crime and Justice: An Annual Review of Research,* Vol. 12, eds. Michael Tonry and Norval Morris (Chicago: University of Chicago Press, 1990): 99–169; Terrie Moffitt and Bill Henry, "Neuropsychological Studies of Juvenile Delinquency and Juvenile Violence," in *The Neuropsychology of Aggression,* ed. Joel S. Milner (Boston: Kluwer, 1991): 67–91; also see conclusion by Paternoster and Bachman, *Explaining Criminals,* 51.

general need for such skills in routine daily activities. After all, people who lack such communication skills will likely find it hard to obtain or retain employment, as well as deal with family and social problems due to their lack of being able to communicate in daily life.

This rebirth in studies regarding the link between intelligence and criminality seemed to reach a peak with the publication of Richard Herrnstein and Charles Murray's *The Bell Curve* in 1994.[29] Although this publication changed the terminology of *moron, imbecile,* and *idiot* to relatively benign terms (e.g., *cognitively disadvantaged*), their argument was consistent with that of the feeblemindedness researchers of the early 20th century. Herrnstein and Murray argued that people with low IQ scores are somewhat destined to be unsuccessful in school, become unemployed, produce illegitimate children, and commit crime. They also suggest that IQ or intelligence is primarily innate, or genetically determined, with little chance of improving it. These authors also noted that African Americans tend to score the lowest, whereas Asians and Jewish people tend to score highest, and they offer results from social indicators that support their argument that these intelligence levels result in relative success in life in terms of group-level statistics.

This book produced a public outcry, resulting in symposiums at major universities and other venues in which the authors' postulates were largely condemned. As noted by other reviews of the impact of this work, some professors at public institutions were sued in court because they used this book in their classes.[30] The book received blistering reviews from fellow scientists.[31] However, none of these scientific critics has fully addressed the undisputed fact that African Americans consistently score low on intelligence tests and that Asians and Jews score higher on these examinations. Furthermore, none have adequately addressed the issue that, even within these populations, low IQ scores (especially on verbal tests) predict criminality. For example, in samples of African Americans, the group that scores lowest on verbal intelligence consistently commits more crime and is more likely to become delinquent or criminal. So, despite the harsh criticism of *The Bell Curve*, it is apparent that there is some validity to the authors' arguments.

With the popularity of intelligence testing and IQ scores in the early 20th century, it is not surprising that this was also the period when other psychological models of deviance and criminality became popular. However, one of the most popular involved body-type theories.

Body-Type Theory: Sheldon's Model of Somatotyping

Although there were numerous theories based on body types in the late 1800s and early 1900s, such as Lombroso's and those of others who called themselves criminal anthropologists, none of these perspectives had a more enduring impact than that of William Sheldon. In the mid-1940s, a new theoretical perspective was proposed that merged the concepts of biology and psychology. Sheldon claimed that, in the embryonic and fetal stages of development, individuals tend to have an emphasis on the development of certain tissue layers.[32] According to Sheldon, these varying degrees of emphasis were largely due to heredity, which led to the development of certain body types and temperaments or personalities. This became the best-known body-type theory, also known as **somatotyping.**

[29]Richard J. Herrnstein and Charles Murray, *The Bell Curve: Intelligence and Class Structure in the United States* (New York: Free Press, 1994).

[30]See Stephen Brown et al., *Criminology* (see chap. 2, n. 1), 260.

[31]J. Blaine Hudson, "Scientific Racism: The Politics of Tests, Race, and Genes," *Black Scholar* 25 (1995): 1–10; David Perkins, *Outsmarting IQ: The Emerging Science of Learnable Intelligence* (New York: The Free Press, 1995); Francis Cullen, Paul Gendreau, G. Roger Jarjoura, and John P. Wright, "Crime and the Bell Curve: Lessons From Intelligent Criminology," *Crime and Delinquency* 43 (1997): 387–411; Robert Hauser, "Review of the Bell Curve," *Contemporary Sociology* 24 (1995): 149–153; Howard Taylor, "Book Review, *The Bell Curve*," *Contemporary Sociology* 24 (1995): 153–58; Gould, *The Mismeasure*, 367–390.

[32]William Sheldon, E. M. Hartl, and E. McDermott, *Varieties of Delinquent Youth* (New York: Harper, 1949).

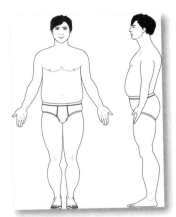

▲ **Image 4.4a**   Endomorph. Physical traits: soft body, underdeveloped muscles, round shape, overdeveloped digestive system. Associated personality traits: love of food, tolerance, evenness of emotions, love of comfort, sociability, good humor, relaxed mood, need for affection.

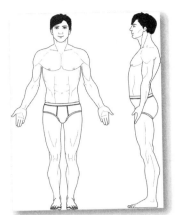

▲ **Image 4.4b**   Mesomorph. Physical traits: hard, muscular body; overly mature appearance; rectangular shape; thick skin; upright posture. Associated personality traits: love of adventure, desire for power and dominance, courage, indifference to what others think or want, assertive mien, boldness, zest for physical activity, competitive nature, love of risk and chance.

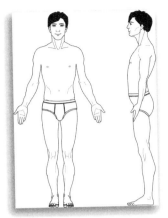

▲ **Image 4.4c**   Ectomorph. Physical traits: thin, flat chest, delicate build, young appearance, lightly muscled, stoop-shouldered, large brain. Associated personality traits: self-conscious, preference for privacy, introverted, inhibited, socially anxious, artistic, mentally intense, emotionally restrained.

According to Sheldon, all embryos must develop three distinct tissue layers, and this much is still acknowledged by perinatal medical researchers. The first layer of tissue is the **endoderm,** which is the inner layer of tissues and includes the internal organs, such as the stomach, large intestine, small intestine, and so on. The middle layer of tissue, called the **mesoderm,** includes the muscles, bones, ligaments, and tendons. The **ectoderm** is the outer layer of tissue, which includes the skin, capillaries, and much of the nervous system's sensors.

Sheldon used these medical facts regarding various tissue layers to propose that certain individuals tend to emphasize certain tissue layers relative to others, typically due to inherited dispositions. In turn, Sheldon believed that such emphases led to certain body types in an individual, such that people who had a focus on their endoderms in development would inevitably become **endomorphic,** or obese (see Image 4.4a). Individuals who had an emphasis on the middle layer of tissue typically became **mesomorphic,** or of an athletic or muscular build (see Image 4.4b). Finally, someone who had an emphasis on the outer layer of tissue in embryonic development would end up being an **ectomorphic** build, or thin (see Image 4.4c).

Sheldon and his research team graded each subject on three dimensions corresponding to these body types. Each body type was measured on a scale of 1 to 7, with 7 being the highest score. Obviously, no one could score a 0 for any body type because all tissue layers are needed for survival; we all need our internal organs, bone and muscular structure, and outer systems (e.g., skin, capillaries, etc.). Each somatotype always had the following order: endomorphy, mesomorphy, ectomorphy.

So, the scores on a typical somatotype might be 3-6-2, which would indicate that this person scored a 3 (a little lower than average) on endomorphy, a 6 (high) on mesomorphy, and a 2 (relatively low) on ectomorphy. According to Sheldon's theory, this hypothetical subject would be a likely candidate for criminality because of the relatively high score on mesomorphy. In fact, the results from his data, as well as all studies that have examined the association of body types and delinquency or criminality, would support this prediction.

Perhaps most important, Sheldon proposed that these body types matched personality traits or temperaments. Individuals who were endomorphic (obese), Sheldon claimed, tended to be more jolly or azy. The technical term for this temperament is referred to as **viscerotonic.** In contrast, people who were mesomorphic (muscular) typically had risk-taking and aggressive temperaments, called **somotonic.** Last, individuals who were ectomorphic (thin) tended to have introverted or shy personalities, which is referred to as **cerebrotonic.** Obviously, members of the middle group, the mesomorphs, had the highest propensity toward criminality because they were disposed toward a risk-taking and aggressive personality.

Interestingly, many politicians were subjects in Sheldon's research. Most entering freshmen at Ivy League schools, especially Harvard, were asked to pose for photos for Sheldon's studies. The Smithsonian Institution still retains a collection of nude photos, which includes George W. Bush, Hillary Rodham Clinton, and many other notable figures.[33]

Sheldon used poor methodology to test his theory. He based his measures of subjects' body types on what he subjectively judged from viewing three perspectives of each subject and often from only three pictures taken of each subject in the three poses. He also had his trained staff view many of the photos and make their determinations of how these individuals scored on each category of body type. The evidence of the reliability among these scorings was shown to be weak, meaning that the trained staff did not tend to agree with Sheldon or among themselves on the somatotypes for each participant.

This is not surprising given the high level of variation in body types and the fact that Sheldon and his colleagues did not employ the technology that is used today, such as caliper tests and submersion in water tanks, which provide the information for which he was searching. People may alter their weights, going from an ectomorphic or mesomorphic builds to more endomorphic forms, or vice versa. Presented with the argument that individuals often alter their body types via diet or exercise, Sheldon responded that he could tell what the "natural" body type of each individual was from the three pictures that were taken. Obviously, this position is not a strong one, which is backed up by the poor inter-rater reliability shown by his staff. Therefore, Sheldon's methodology is questionable, which casts doubt on the entire theoretical framework.

Despite the problematic issues regarding his methodology, Sheldon clearly showed that mesomorphs, or individuals who had muscular builds and tended to be more risk-taking, were more delinquent and criminal than individuals who had other body types or temperaments.[34] Furthermore, other researchers, even those who despised Sheldon's theory, found the same associations between mesomorphy and criminality, as well as related temperaments (i.e., somotonic) and criminality.[35] Subsequent studies showed that mesomorphic boys were far more likely to have personality traits that predicted criminality, such as aggression, short temper, self-centeredness, and impulsivity.

Recent theorists have also noted the link between an athletic, muscular build and the highly extroverted, aggressive personality that is often associated with this body type.[36] In fact, some recent theorists have gone so far as to claim that chronic offenders, both male and female, can be identified early in life by

---

[33]See discussion in Brown et al., op cit., *Criminology,* 246.

[34]Sheldon et al., *Varieties.*

[35]Sheldon Glueck and Eleanor Glueck, *Physique and Delinquency* (New York: Harper and Row, 1956); see also Emil Hartl, *Physique and Delinquent Behavior* (New York: Academic Press, 1982); Juan Cortes, *Delinquency and Crime* (New York: Seminar Press, 1972); for the most recent applications, see Hans J. Eysenck and G. H. Gudjonsson, *The Causes and Cures of Criminality* (New York: Plenum, 1989).

[36]James Q. Wilson and Richard Herrnstein, *Crime and Human Nature* (New York: Simon and Schuster, 1985).

their relatively v-shaped pelvic structure as opposed to a u-shaped pelvic structure.[37] The v-shaped pelvis is said to indicate relatively high level of androgens (male hormones, like testosterone) in the system, which predisposes individuals toward crime. On the other hand, a more u-shaped pelvis indicates relatively low levels of such androgens and therefore lower propensity toward aggression and criminality. Using this logic, it may be true that more hair on an individual's arms (whether they be male or female) is predictive of a high likelihood for committing crime. However, no research exists regarding this factor.

Regarding the use of body types and characteristics in explaining crime, many of the hard-line sociologists who have attempted to examine or replicate Sheldon's results have never been able to refute the association between mesomorphs and delinquency or criminality, nor the association between mesomorphy and the somotonic temperament of being risk taking or aggressive.[38] Thus, the association between being muscular or athletically built and engaging in criminal activities is now undisputed and assumed to be true. Still, sociologists have taken issue with what is causing this association.

Whereas Sheldon claimed it was due to inherited traits for a certain body type, sociologists argue that this association is due to societal expectations: Muscular male youth would be encouraged to engage in risk-taking and aggressive behavior. For example, a young male with an athletic build would be encouraged to join sports teams and engage in high-risk behaviors by peers. Who would gangs most desire as members? More muscular, athletic individuals would be better at fighting and performing acts that require a certain degree of physical strength and stamina.

Ultimately, it is now established that mesomorphs are more likely to commit crime.[39] Furthermore, the personality traits linked to having an athletic or muscular build are dispositions toward risk taking and aggressiveness, and few scientists dispute this correlation. No matter which theoretical model is adopted, whether it be the biopsychologists' or the sociologists', the fact of the matter is that mesomorphs are indeed more likely to be risk taking and aggressive and, thus, to commit more crime than individuals of other body types.

However, whether the cause is biological or sociological is a debate that shows the importance of theory in criminological research. After all, the link between mesomorphy and criminality is now undisputed; the explanation of why this link exists has become a theoretical debate. Is it the biological influence or the sociological? Readers may make their own determination, if not now, then later.

Our position is that both biology and social environment are likely to interact with one another in explaining this link. Thus, it is most likely that both nature and nurture are in play in this association between mesomorphy and crime, and both Sheldon and his critics may be correct. A middle ground can often be found in theorizing on criminality. It is important to keep in mind that theories in criminology, as a science, are always considered subject to falsification and criticism and can always be improved. Therefore, our stance on the validity and influence of this theory, as well as others, should not be surprising.

Policy Implications

Many policy implications can be derived from the theories presented in this chapter. First, one could propose more thorough medical screening at birth and in early childhood, especially regarding minor physical anomalies (MPAs). The studies reviewed in this chapter obviously implicate numerous MPAs in

---

[37]Eysenck and Gudjonsson, *The Causes and Cures.*

[38]Cortes, *Delinquency and Crime;* Glueck and Glueck, *Physique and Delinquency;* for a review, see Eysenck and Gudjonsson, *The Causes and Cures.*

[39]For a review, see Lee Ellis, *Criminology: A Global Perspective* (Minot: Pyramid Press, 2000), supplemental tables and references.

developmental problems (mostly in the womb). These MPAs are a red flag signaling problems, especially in cognitive abilities, which are likely to have a significant impact on criminal behavior.[40]

Other policy implications derived from the theories and findings of this chapter involve having same-sex classes for children in school because they focus on deficiencies that have been shown for both young boys and girls. Numerous school districts now have policies that specify same-sex math courses for female children. This same strategy might be considered for male children in English or literature courses because males have a biological disposition for a lower aptitude level than females regarding this area of study. Furthermore, far more screening should be done regarding IQ and aptitude levels of young children to identify which children require extra attention because studies show that such early intervention can make a big difference in improving their IQ and aptitude.

A recent report that reviewed the extant literature regarding what types of programs work best for reducing crime noted the importance of diagnosing early head trauma and further concluded that one of the most consistently supported programs for such at-risk children are those that involve weekly infant home visitation.[41] Another obvious policy implication derived from biosocial theory is mandatory health insurance for pregnant mothers and children, which is quite likely the most efficient way to reduce crime in the long term.[42] Finally, all youth should be screened for abnormal levels of hormones, neurotransmitters, and toxins (especially lead).[43] These and other policy strategies will be discussed in the last chapter of the book.

Conclusion

In this chapter, we discussed the development of the early Positive School of criminology. The Positive School can be seen as the opposite of the Classical School perspective, which we covered in Chapters 2 and 3, because positivism assumes that individuals have virtually no free will; rather, criminal behavior is the result of determinism, which means that factors other than rational decision making, such as poverty, intelligence, bad parenting, and unemployment, influence us and determine our behavior.

The earliest positivist theories, such as craniometry and phrenology, were developed in the early 1800s but did not become popular outside of scientific circles, likely because they were presented prior to Charles Darwin's theory of evolution. In the 1860s, Darwin's theory became widely accepted, which set the stage for the father of criminology, Cesare Lombroso, to propose his theory of born criminals. Lombroso's theory was based on Darwin's theory of evolution and argued that the worst criminals are born that way, being biological throwbacks to an earlier stage of evolution. Unfortunately, Lombroso's theory led to numerous policies that fit the philosophy and politics of fascism, which found useful a theory that proposed certain people were inferior to others. However, Lombroso and many of his contemporaries became aware that the field should shift to a more multifactor approach, such as environment and social factors interacting with physiological influences.

We also discussed theories regarding low IQ scores, traditionally known as feeblemindedness. Although most recent studies show that there is a correlation between crime and low IQ, this association is not quite as strong as thought in the early 1900s. Modern studies show consistent evidence that low verbal IQ is

---

[40]D. Fishbein, *Biobehavioral Perspectives.*

[41]Sherman et al., *Preventing Crime* (see chap. 2, n. 47).

[42]John Wright, Stephen Tibbetts, and Leah Daigle, *Criminals in the Making: Criminality Across the Life Course* (Thousand Oaks: Sage, 2008).

[43]Ibid., 254–259.

related with criminality,[44] especially when coupled with sociological factors, such as weak family structure. This is the state of the criminological field today, and it will be discussed in the next chapter.

Finally, we explored the theories and evidence regarding body types in predisposing an individual toward criminality. Studies have shown that the more athletic or mesomorphic an individual is, the higher the probability that this individual will be involved in criminality. This relationship is likely based on hormonal levels, and this type of association will be explored in the next chapter.

Ultimately, we have examined a variety of physiological and psychological factors that predict criminal offending, according to empirical research. Still, the existence of such influence is largely conditional based on environmental and social factors.

Chapter Summary

- The Positive School of criminology assumes the opposite of the Classical School. Whereas the Classical School assumes that individuals commit crime because they freely choose to act after rationally considering the expected costs and benefits of the behavior, the Positive School assumes that individuals have virtually no free will or choice in the matter; rather, their behavior is determined by factors outside of free will, such as poverty, low intelligence, bad child rearing, and unemployment.
- The earliest positive theories, such as craniometry and phrenology, emphasized measuring the size and shape of the skull and brain. These perspectives did not become very popular because they preceded Charles Darwin's theory of evolution.
- Cesare Lombroso, the father of criminology, presented a theoretical model that assumed the worst criminals are born that way. Highly influenced by Darwin, Lombroso claimed that born criminals are evolutionary throwbacks who are not as highly developed as most people.
- Lombroso claimed these born criminals could be identified by physical features called stigmata. This led to numerous policy implications that fit with the societal beliefs at that time, such as fascism.
- In the early 1900s, the IQ test was invented in France and was quickly used by American researchers in their quest to identify the feebleminded. This led to massive numbers of deportation, sterilizations, and institutionalizations across the United States and elsewhere.
- Modern studies support a link between low verbal IQ and criminality, even within a given race, social class, or gender.
- Merging elements of the early physiological and psychological perspectives are body-type theories. The best-known of these is somatotyping, which was proposed by William Sheldon. Sheldon found that an athletic or muscular build (i.e., mesomorphy) is linked to an aggressive, risk-taking personality, which in turn is associated with higher levels of crime.
- Despite the methodological problems with Sheldon's body-type theory, many propositions and associations of the perspective hold true in modern studies.
- The early Positive School theories set the stage for most of the other theories we will be covering in this book because they place an emphasis on using the scientific method for studying and explaining criminal activity.

---

[44]Gibson et al., "The Contribution of Family Adversity."

KEY TERMS

Atavism	Endoderm	Phrenology
Atavistic	Endomorphic	Physiognomy
Cerebrotonic	Eugenics	Positive School
Craniometry	Feeblemindedness	Somotonic
Determinism	Mesoderm	Somatotyping
Ectoderm	Mesomorphic	Stigmata
Ectomorphic	Minor physical anomalies (MPAs)	Viscerotonic

DISCUSSION QUESTIONS

1. What characteristics distinguish the Positive School from the Classical School regarding criminal thought? Which of these schools do you lean toward in your own perspective of crime and why?

2. Name and describe the various early schools of positivistic theories that existed in the early to mid-1800s (pre-Darwin), as well as the influence that they had on later schools of thought regarding criminality. Do you see any validity in these approaches (because modern medical science does)? And articulate why.

3. What was the significant reason(s) that these early schools of positivistic theories did not gain much momentum in societal popularity? Does this lack of popularity relate to the neglect of biological perspectives of crime in modern times?

4. What portion of Lombroso's theory of criminality do you find least valid? Which do you find most valid?

5. Most readers have taken the equivalent of an IQ test (e.g., SAT or ACT tests). Do you believe that this score is a fair representation of your knowledge as compared to others and why? Do your feelings reflect the criticisms of experts regarding the use of IQ in identifying potential offenders, such as feeblemindedness theory?

6. In light of the scientific findings that show that verbal IQ is shown to be a consistent predictor of criminality among virtually all populations and samples, can you provide evidence from your personal experience for why this occurs?

7. Regarding Sheldon's body-type theory, what portion of this theory do you find most valid? What do you find least valid?

8. If you had to give yourself a somatotype (e.g., 3-6-2), what would it be? Explain why your score would be the one you provide, and note whether this would make you likely to be a criminal in Sheldon's model.

9. Provide a somatotype of five of your family members or best friends. Does the somatotype have any correlation with criminality according to Sheldon's predictions? Either way, describe your findings.

10. Ultimately, do you believe some of the positive theoretical perspectives presented in this section are valid, or do you think they should be entirely dismissed in terms of understanding or predicting crime? Either way, state your case.

11. What types of policies would you implement if you were in charge, given the theories and findings in this chapter?

WEB RESOURCES

Body-Type Theories/Somatotyping

http://www.innerexplorations.com/psytext/shel.htm

http://www.teachpe.com/multi/somatotyping_eating_disorders_performance.htm

IQ Testing/Feeblemindedness

http://www.vineland.org/history/trainingschool/history/eugenics.htm

Cesare Lombroso

http://www.museocriminologico.it/lombroso_1_uk.htm

Phrenology/Craniometry

http://www.phrenology.org/index.html

http://skepdic.com/cranial.html

Modern Biosocial Perspectives of Criminal Behavior

This chapter will discuss the more modern biological studies of the 20th century. We will begin with studies in the early 1900s, particularly those that sought to emphasize the influence of biological factors—such as families, twins, and children who were adopted in infancy—on criminality. Virtually all of these studies have shown a significant biological effect in the development of criminal propensities. Then, we will examine the influence of a variety of physiological factors, including chromosomal mutations, hormones, **neurotransmitters,** brain trauma, and other dispositional aspects in individuals' nervous systems A special emphasis will be placed on showing the consistent evidence that has been found for the interaction between physiological and environmental factors (i.e., biosocial factors).

This section will examine a variety of perspectives that deal with interactions between physiological and environmental factors, which is currently the dominant model explaining criminal behavior. First, we will discuss the early studies that attempted to place an emphasis on the biological aspects of offending: family, twin, and **adoption studies.** All of these studies show that biological influences are more important than social and environmental factors, and most also conclude that when both negative biological and disadvantaged environmental variables are combined, these individuals are by far the most likely to offend in the future, which fully supports the interaction between nature and nurture factors.

Later in this chapter, we will examine other physiological factors, such as hormones and neurotransmitters. We will see that chronic, violent offenders tend to have significantly different levels of hormones and other chemicals in their bodies than do normal individuals. Furthermore, we will examine brain trauma and activity among violent offenders, and we will see that habitual violent criminals tend to have lower levels of brain wave patterns and anxiety levels than normal persons. Ultimately, we will see that numerous physiological distinctions can be made between chronic violent offenders and others but that these differences are most evident when physical factors are combined with being raised in poor, disadvantaged environments.

⚑ Nature Versus Nurture: Studies Examining the Influence of Genetics and Environment

At the same time that Freud was developing his perspective of psychological deviance, other researchers were busy testing the influence of heredity versus environment to see which of these two components had the strongest effect on predicting criminality. This type of testing produced four waves of research: **family studies, twin studies,** adoption studies, and in recent years, studies of identical twins separated at birth. Each of these waves of research, some more than others, contributed to our understanding of how much criminality is inherited from our parents (or other ancestors) versus how much is due to cultural norms, such as family or community. Ultimately, all of the studies have shown that the interaction between these two aspects—genetics and environment—is what causes crime among individuals and groups in society.

Family Studies

The most notable family studies were done in the early 1900s by Dugdale, in his study of the Jukes family, and the previously discussed researcher Goddard, who studied the Kallikak family.[1] These studies were supposed to test the proposition that criminality is more likely to be found in certain families, which would indicate that crime is inherited. Due to the similarity of the results, we will focus here on Goddard's work on the Kallikak family.

This study showed that a much higher proportion of children from the Kallikak family became criminal. Furthermore, Goddard thought that many of the individuals (often children) from the Kallikak family actually looked like criminals, which fit Lombroso's theory of stigmata. In fact, Goddard had photographs made of many members of this family to back up these claims. However, follow-up investigations of Goddard's research show that many of these photographs were actually altered to make the subjects appear more sinister or evil (fitting Lombroso's stigmata) by altering their facial features, most notably their eyes.[2]

Despite the despicable methodological problems with Goddard's data and subsequent findings, two important conclusions can be made from the family studies that were done in the early 1900s. The first is that criminality is indeed more common in some families; in fact, no study has ever shown otherwise. However, this tendency cannot be shown to be a product of heredity or genetics. After all, individuals from the same family are also products of a similar environment—often a bad one—so this conclusion from the family studies does little in advancing knowledge regarding the relative influence of nature versus nurture in terms of predicting criminality.

The second conclusion of family studies was more insightful and interesting. Specifically, they showed that criminality by the mother (or head female caretaker) had a much stronger influence on the future criminality of the children than did the father's criminality. This is likely due to two factors. The first is that the father is often absent most of the time while the children are being raised. Perhaps more important is that it takes much more for a woman to transgress social norms and become a convicted offender, which indicates that the mother is highly antisocial; this gives some (albeit limited) credence to the argument that criminality is somewhat inherited. Despite this conclusion, it should be apparent from the weaknesses in the methodology of family studies that this finding did not hold much weight in the nature versus nurture

[1]Gould, *The Mismeasure* (see chap. 4, n. 2).

[2]For an excellent discussion of the alteration of Goddard's photographs, see Gould, *The Mismeasure*, 198–204.

debate. Thus, a new wave of research soon emerged that did a better job of measuring the influence of genetics versus environment, which was twin studies.

Twin Studies

After family studies, the next wave of tests done to determine the relative influence on criminality between nature and nurture involved twin studies, the examination of identical twin pairs versus fraternal twin pairs. Identical twins are also known as **monozygotic twins** because they come from a single (hence *mono*) egg *(zygote);* they are typically referred to in scientific literature as MZ (monozygotic) twins. Such twins share 100% of their genotype, meaning they are identical in terms of genetic makeup. Keep in mind that everyone shares approximately 99% of the human genetic makeup, leaving about 1% that can vary over the entire species. On the other hand, fraternal twins are typically referred to as **dizygotic twins** because they come from two (hence *di*) separate eggs *(zygotes);* they are known in the scientific literature as DZ (dizygotic) twins. Such DZ twins share 50% of genes that can vary, which is the same amount that any siblings from the same two parents share. DZ twins can be of different genders and may look and behave quite differently, as many readers have probably observed.

The goal of the twin studies was to examine the **concordance rates** between MZ twin pairs versus that of DZ twin pairs regarding delinquency. Concordance is a count based on whether two people (or a twin pair) share a certain trait (or lack of the trait); for our purposes, the trait is criminal offending. Regarding a count of concordance, if one twin is an offender, then we look to see if the other is also an offender. If that person is, then it would be concordant given the fact that the first twin was a criminal offender. Also, if neither of the two twins were offenders, then that also would be concordant because they both lack the trait. However, if one twin is a criminal offender and the other twin of the pair is not an offender, then this would be discordant, in the sense that one has a trait that the other does not.

▲ **Image 5.1**   Identical Twins

Thus, the twin studies focused on comparing the concordance rates of MZ twin pairs versus those of DZ twin pairs, with the assumption that any significant difference in concordance can be attributed to the similarity of the genetic makeup of the MZ twins (which is 100%) versus the DZ twins (which is significantly less, that is, 50%). If genetics plays a major role in determining the criminality of individuals, then it would be expected that MZ twins would have a significantly higher concordance rate for being criminal offenders than would DZ twins. It is assumed that both MZ and DZ twin pairs in these studies were raised in more or less the same environment if they were brought up in the same families at the same time.

A number of studies were performed in the early and mid-1900s that examined the concordance rates between MZ and DZ twin pairs. These studies clearly showed that identical twins had far higher concordance rates than did fraternal twins, with most studies showing twice as much concordance or more for MZ twins, even for serious criminality.[3]

---

[3]See review in Adrian Raine, *The Psychopathology of Crime* (San Diego: Academic Press, 1993); see also Juan B. Cortes, *Delinquency and Crime* (New York: Seminar Press, 1972); Karl O. Christiansen, "Seriousness of Criminality and Concordance Among Danish Twins," in *Crime, Criminology, and Public Policy*, ed. Roger Hood (New York: The Free Press, 1974).

However, the studies regarding the comparisons between the twins were strongly criticized for reasons that most readers see on an everyday basis. Specifically, identical twins, who look almost exactly alike, are typically dressed the same by their parents and treated the same by the public. In addition, they are generally expected to behave the same way. However, this is not true for fraternal twins, who often look very different and quite often are of different genders.

This produced the foundation for criticism of the twin studies, which was a very valid argument that the higher rate of concordance among MZ twins could have been due to the extremely similar way they were treated or expected to behave by society. Another criticism of the early twin studies had to do with the accuracy in determining whether twins were fraternal or identical, which was often done by sight in the early tests.[4] Although these criticisms were seemingly valid, the most recent meta-analysis, which examined virtually all of the twin studies done up to the 1990s, concluded that the twin studies showed evidence of a significant hereditary basis for criminality.[5] Still, the criticisms of such studies were quite valid; therefore, in the early to mid-1900s, researchers involved in the nature versus nurture debate attempted to address these valid criticisms by moving on to another methodological approach for examining this debate: adoption studies.

Adoption Studies

Due to the valid criticisms leveled at twin studies in determining the relative influence of nature (biological) or nurture (environmental), researchers in this area moved on to adoption studies, which examined the predictive influence of the biological parents versus that of the adoptive parents who raised the children from infancy to adulthood. It is important to realize that, in such studies, the adoptees were typically given up for adoption prior to 6 months of age, meaning that the biological parents had relatively no interaction with their natural children; rather, they were almost completely raised from infancy by the adoptive parents.

Perhaps the most notable of the adoption studies was done by Sarnoff Mednick and his colleagues, in which they examined male children born in Copenhagen between 1927 and 1941 who had been adopted early in life.[6] This study and virtually all others that have examined adoptees in this light found that, by far, the highest predictability for future criminality was found for adopted youth who had *both* biological parents and adoptive parents who were convicted criminals. However, the Mednick study also showed that the criminality of biological parent(s) had a far greater predictive effect on future criminality of offspring than did the criminality of adoptive parents. Still, the adopted children who were least likely to become criminal had no parent with a criminal background. In light of this last conclusion, these findings support the major contentions of this book's authors in the sense that they fully back up the nature *via* nurture argument as opposed to the nature *versus* nurture argument. They support the idea that both biological *and* environmental factors contribute to the future criminality of youth.

Unfortunately, the researchers who performed these studies focused on the other two groups of youth, those who had either only criminal biological parents or only criminal adoptive parents. Thus, these adoption studies found that the adoptees who had only biological parents who were criminal had a much higher likelihood of becoming criminal as compared to the youths who had only adoptive parents who were criminal. Obviously, this finding supports the genetic influence in predisposing people toward criminality. However, this methodology was subject to criticism.

---

[4]Raine, *The Psychopathology of Crime.*

[5]Glenn Walters, "A Meta-Analysis of the Gene-Crime Relationship," *Criminology 30* (1992): 595–613.

[6]Barry Hutchings and Sarnoff A. Mednick, "Criminality in Adoptees and Their Adoptive and Biological Parents: A Pilot Study," in *Biosocial Bases of Criminal Behavior*, eds. Sarnoff Mednick and Karl O. Christiansen (New York: Gardner Press, 1977): 127–41.

Perhaps the most notable criticism of adoption studies was that adoption agencies typically incorporated a policy of **selective placement** in which adoptees were placed with adoptive families similar in terms of demographics and background to their biological parents. Such selective placement could bias the results of adoption studies. However, recent analyses have examined the impact of such bias, concluding that, even when accounting for the influence of selective placement, the ultimate findings of the adoption studies are still somewhat valid.[7] Children's biological parents likely have more influence on their future criminality than the adoptive parents who raise them from infancy to adulthood. Still, the criticism of selective placement was strong enough to encourage a fourth wave of research in the "nature versus nurture" debate, which became studies on identical twins separated at birth.

Twins Separated at Birth

Until recently, studies of identical twins separated at birth were virtually impossible because it was so difficult to get a high number of identical twins who were indeed separated early. But, since the early 1990s, **twins separated at birth studies** have been possible. Readers should keep in mind that, in many of the identical twin pairs studied for these investigations, the individuals did not even know that they had a twin. Furthermore, the environments in which they were raised were often extremely different; one twin might be raised by a very poor family in an urban environment while the other twin was raised by a middle- to upper-class family in a rural environment.

These studies, the most notable being done at the University of Minnesota, found that the twin pairs often showed extremely similar tendencies for criminality, sometimes more than those seen in concordance rates for identical twins raised together.[8] This finding obviously supports the profound influence of genetics and heredity, which is not surprising to most well-read scientists, who now acknowledge the extreme importance of inheritance of physiological and psychological aspects regarding human behavior. Perhaps more surprising was why separated identical twins, who never knew that they had a twin and were often raised in extremely different circumstances, had just as similar or even more similar concordance rates than identical twins who were raised together.

The leading theory for this phenomenon is that identical twins who are raised together actually go out of their way to deviate from their natural tendencies to form an identity separate from their identical twin with whom they have spent their entire lives. As for criticisms of this methodology, none have been presented. Thus, it is somewhat undisputed at this point in the scientific literature that the identical twins separated at birth studies have shown that genetics has a significant impact on human behavior, especially regarding criminal activity.

Ultimately, taking all of the nature versus nurture methodological approaches and subsequent findings together, the best conclusion that can be made is that genetics and heredity both have significant impacts on criminality. Environment simply cannot account for all of the consistent results seen in the comparisons between identical twins and fraternal twins, those of identical twins separated at birth, and those of adoptees with criminal biological parents versus those who did not have such parents. Despite the taboo nature and controversial response from the findings of such studies, it is quite clear that when nature and nurture are compared, biological rather than environmental factors tend to have the most influence on the criminality of individuals. Still, the authors of this book hope that readers will emphasize the importance of the interaction

---

[7] See James Q. Wilson and Richard J. Herrnstein, *Crime and Human Nature* (New York: Simon and Schuster, 1985); Walters, "Gene-Crime Relationship."

[8] Thomas J. Bouchard, David T. Lykken, Matthew McGue, Nancy L. Segal, and Auke Tellegen, "Sources of Human Psychological Differences: The Minnesota Study of Twins Reared Apart," *Science 250* (1990): 223–28.

between nature and nurture (better stated as nature *via* nurture). Ultimately, we hope that we have shown quite convincingly through scientific study that it is the interplay between biology and the environment that is most important in determining human behavior.

Perhaps in response to this nature versus nurture emphasis debate, a new theoretical perspective was offered in the mid-1900s that merged biological and psychological factors in explaining criminality. Although leaning more toward the nature side of the debate, critics would use this same perspective to promote the nurture side, so this framework was useful in promoting the interaction between biology and sociological factors.

⊠ Cytogenetic Studies: The XYY Factor

Beyond the body-type theories, another theory was proposed in the early 1900s regarding biological conditions that predispose individuals toward crime: **cytogenetic studies.** Cytogenetic studies of crime focus on the genetic makeup of individuals, with a specific focus on abnormalities in chromosomal makeup, and specifically chromosomal abnormalities that occur randomly in the population. Many of the chromosomal mutations that have been studied (such as XYY) typically result not from heredity but from random mutations in chromosomal formation.

The normal chromosomal makeup for women is XX, which represents an X from the mother and an X from the father. The normal chromosomal makeup for men is XY, which represents an X from the mother and a Y from the father. However, as in many species of animals, there are often genetic mutations, which we see in human beings. Consistent with evolutionary theory, virtually all possible variations of chromosomes that are possible have been found in the human population, such as XXY, XYY, and many others. We will focus our discussion on the chromosomal mutations that have been most strongly linked to criminality.

One of the first chromosomal mutations recognized as a predictor of criminal activity was the mutation of XYY. In 1965, the first major study showed that this mutation was far more common in a Scottish male population of mental patients than in the general population.[9] Specifically, in the general population, XYY occurs in about 1 of every 1,000 males.

The first major study that examined the influence of XYY sampled about 200 men in the mental hospital; one occurrence would have been predicted, assuming what was known about the general population. The study, however, found 13 individuals who were XYY, which suggested that individuals who have mental disorders are more likely to have XYY than those who do not. Males who have XYY have at least 13 times (or 1300%) the chance of having behavioral disorders compared with those who do not have this chromosomal abnormality. Subsequent studies have not been able to dismiss the effect of XYY on criminality, but they have concluded that this mutation is more often linked with property crime than with violent crime.[10]

Would knowing this relationship help in policies toward crime? Probably not, considering the fact that there were still 90% of the male mental patients who were not XYY. Still, this study showed the importance of looking at chromosomal mutations as a predictor of criminal behavior.

Such mutations include numerous chromosomal abnormalities, such as XYY, which is a male who is given an extra Y chromosome, making him more "male-like." These individuals are often very tall but slow in terms of social and intelligence skills. Another type of mutation is XXY, which is otherwise known as Klinefelter's syndrome; it results in more feminine males (homosexuality has been linked to this mutation). Many other

---

[9]Patricia A. Jacobs, Muriel Brunton, Marie M. Melville, R. P. Brittian, and W. F. McClemmot, "Aggressive Behavior, Mental Sub-Normality, and the XYY Male," *Nature 208* (1965): 1351–1352.

[10]See review in Raymond Paternoster and Ronet Bachman, *Explaining Criminals and Crime* (Los Angeles: Roxbury, 2001), 53.

types of mutations have been observed, but it is the XYY mutation that has been the primary focus of studies, largely due to the higher levels of testosterone produced by this chromosomal mutation (see Figure 5.1).

One study examined the relative criminality and deviance of a group of individuals in each of these groups of chromosomal mutations (see Figure 5.1).[11] This study found the more male hormones produced by the chromosomal mutation, the more likely people with the mutation were to commit criminal and deviant acts. On the other hand, the more feminine hormones produced by the chromosomal mutation, the less likely the individuals were to commit criminal activity. Ultimately, all of these variations in chromosomes show that there is a continuum of degrees of femaleness and maleness, and that the more male-like the individual is in terms of chromosomes, the more likely he is to commit criminal behavior.

Ultimately, the cytogenetic studies showed that somewhat random abnormalities in an individual's genetic makeup can have a profound influence on his or her level of criminality. Whether or not this can or should be used in policy related to crime is another matter, but the point is that genetics does indeed contribute to dispositions toward whether someone is likely to commit criminal acts. The extent to which male hormones or androgens are increased by the mutation is an important predictor of criminalistic traits.

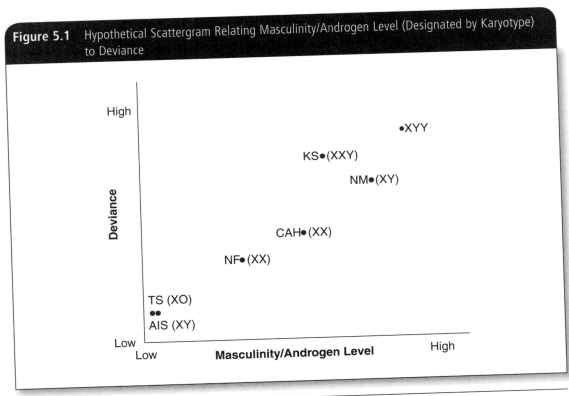

Figure 5.1   Hypothetical Scattergram Relating Masculinity/Androgen Level (Designated by Karyotype) to Deviance

NOTE: TS = Turner's syndrome; AIS = androgen insensitivity syndrome; NF = normal female; KS = Klinefelter's syndrome; CAH = congenital adrenal hyperplasia; Nm = normal male; XYY = Jacob's (Supermale) syndrome.

[11]Anthony Walsh, "Genetic and Cytogenetic Intersex Anomalies: Can They Help Us to Understand Gender Differences in Deviant Behavior?" *International Journal of Offender Therapy and Comparative Criminology* 39 (1995): 151–166.

⊠ Hormones and Neurotransmitters: Chemicals That Determine Criminal Behavior

Various chemicals in the brain and the rest of the body determine how we think, perceive, and react to various stimuli. Hormones, such as testosterone and estrogen, carry chemical signals to the body as they are released from certain glands and structures. Some studies have shown that a relatively excessive amount of testosterone in the body is consistently linked to criminal or aggressive behavior, with most studies showing a moderate relationship.[12] This relationship is seen even in the early years of life.[13] On the other side of the coin, studies have also shown that hormonal changes in females can cause criminal behavior. Specifically, studies have shown that a high proportion of the women in prison for violent crimes committed their crimes during their premenstrual cycle, at which time women experience a high level of hormones that make them more "male-like" during that time due to relatively low levels of estrogen as compared to progesterone.[14]

Anyone who doubts the impact of hormones on behavior should examine the scientific literature regarding performance on intelligence tests at different times of day. Virtually everyone performs better on spatial and mathematical tests early in the day, when people have relatively higher levels of testosterone and other male hormones in their bodies; on the other hand, virtually everyone performs better on verbal tasks in the afternoon or evening, when people have relatively higher levels of estrogen or other female hormones in their systems.[15] Furthermore, studies have shown that individuals who are given shots of androgens (male hormones) before math tests tend to do significantly better on spatial and mathematics tests than they would do otherwise. Scientific studies show the same is true for people who are given shots of female hormones prior to verbal or reading tests.

It is important to realize that this process of differential levels of hormones begins at a very early age, specifically at about the fifth week after conception. At that time, the Y chromosome of the male tells the developing fetus that it is a male and stimulates production of higher levels of testosterone. So, even during the first few months of gestation, the genes on the Y chromosome significantly alter the course of genital and thus hormonal development.[16]

This level of testosterone alters the genitals of the fetus through gestation, as well as later changes in the genital area, and produces profound increases in testosterone in the teenage and early adult years. This produces not only physical differences but also huge personality and behavioral alterations.[17] High levels of testosterone and other androgens tend to "masculinize" the brain toward risk-taking behavior, whereas the lower levels typically found in females tend to result in the default feminine model.[18] High levels of testosterone have

---

[12]Alan Booth and D. Wayne Osgood, "The Influence of Testosterone on Deviance in Adulthood: Assessing and Explaining the Relationship," *Criminology 31* (1993): 93–117; Hosanna Soler, P. Vinayak, and David Quadagno, "Biosocial Aspects of Domestic Violence," *Psychoneuroendocrinology 25* (2000): 721–773; for a review, see Lee Ellis and Anthony Walsh, *Criminology: A Global Perspective* (Boston: Allyn and Bacon, 2000).

[13]Jose R. Sanchez-Martin, Eduardo Fano, L. Ahedo, Jaione Cardas, Paul F. Brain, and Arantza Azpiroz, "Relating Testosterone Levels and Free Play Social Behavior in Male and Female Preschool Children," *Psychoneuroendocrinology 25* (2000): 773–783.

[14]Diane Halpern, *Sex Differences in Cognitive Abilities* (Mahwah: Lawrence Erlbaum, 2000).

[15]Ibid.

[16]Lee Ellis, "A Theory Explaining Biological Correlates of Criminality," *European Journal of Criminality 2* (2005): 287–315.

[17]See Ellis, "A Theory."

[18]Chandler Burr, "Homosexuality and Biology," *Atlantic Monthly 271* (1993): 47–65; Lee Ellis and M. Ashley Ames, "Neurohormonal Functioning and Sexual Orientation: A Theory of Homosexuality-Heterosexuality," *Psychological Bulletin 101* (1987): 233–258; Ellis, "A Theory."

numerous consequences, such as lowered sensitivity to pain, enhanced seeking of sensory stimulation, and a right-hemisphere shift in brain dominance, which has been linked to higher levels of spatial aptitude but lower levels of verbal reasoning and empathy. These consequences have profound implications for criminal activity and are more likely to occur in males than females.[19]

Ultimately, hormones have a profound effect on how individuals think and perceive their environments. All criminal behavior comes down to cognitive decisions in our 3-pound brains. So, it should not be surprising that hormones play a highly active role in this decision-making process. Nevertheless, hormones are probably secondary compared to levels of neurotransmitters, which are chemicals in the brain and body that help transmit electric signals from one neuron to another.

Neurotransmitters can be distinguished from hormones in the sense that hormones carry a signal that is not electric, whereas the signals that neurotransmitters carry are indeed electric. Neurotransmitters are chemicals that are released when a neuron, the basic unit of the nervous system, wants to send an electric message to a neighboring neuron(s). Sending such a message requires the creation of neural pathways, which means that neurotransmitters must be activated in processing the signal. At any given moment, healthy levels of various neurotransmitters are needed to pass messages from one neuron to the next across gaps between them, gaps called synapses.

Although there are many types of neurotransmitters, the most studied in relation to criminal activity are **dopamine** and **serotonin.** Dopamine is most commonly linked to feeling good. For example, dopamine is the chemical that tells us when we are experiencing good sensations, such as good food, sex, and so on. Most illicit drugs elicit a pleasurable sensation by enhancing the levels of dopamine in our systems. Cocaine and methamphetamine, for example, tell the body to produce more dopamine and inhibit the enzymes that typically mop up the dopamine in our systems after it is used.

Although a number of studies show that low levels of dopamine are linked to high rates of criminality, other studies show no association with—or even a positive link to—criminal behavior.[20] However, the relationship between dopamine and criminal behavior is probably curvilinear, such that both extremely high and extremely low levels of dopamine are associated with deviance. Unfortunately, no conclusion can be made at this point due to the lack of scientific evidence regarding this chemical.

On the other hand, a clear conclusion can be made about the other major neurotransmitter that has been implicated in criminal offending: serotonin. Studies have consistently shown that low levels of serotonin are linked with criminal offending.[21] Serotonin is important in virtually all information processing, whether it be learning or emotional; thus, it is vital in most aspects of interacting with the environment.

---

[19]Martin Reite, C. Munro Cullum, Jeanelle Stocker, Peter Teale, and Elizabeth Kozora, "Neuropsychological Test Performance and MEG-based Brain Lateralization: Sex Differences," *Brain Research Bulletin 32* (1993): 325–328; Jorge Moll, Ricardo de Oliveira-Souza, Paul J. Eslinger, Ivanei E. Bramanti, Janaina Mourao-Miranda, Pedro Angelo Andreiuolo, and Luiz Pessoa, "The Neural Correlates of Moral Sensitivity: A Functional Magnetic Resonance Imaging Investigation of Basic and Moral Emotions," *Journal of Neuroscience 22* (2002): 2730–2736; Kimberly Badger, Rebecca Simpson Craft, and Larry Jensen, "Age and Gender Differences in Value Orientation Among American Adolescents," *Adolescence 33* (1998): 591–596.

[20]For reviews see Raine, *The Psychopathology of Crime*, and Diana H. Fishbein, *Biobehavioral Perspectives in Criminology* (Belmont: Wadsworth/Thomson Learning, 2001).

[21]For a review, see Ellis, "A Theory"; see also Rachel Blumensohn, Gideon Ratzoni, Abraham Weizman, Malka Israeli, Nachman Greuner, Alan Apter, Sam Tyano, and Anat Biegon, "Reduction in Serotonin 5HT Receptor Binding on Platelets of Delinquent Adolescents," *Psychopharmacology 118* (1995): 354–356; Emil F. Coccaro, Richard J. Kavoussi, Thomas B. Cooper, and Richard L. Hauger, "Central Serotonin Activity and Aggression," *American Journal of Psychiatry 154* (1997): 1430–1435; Mairead Dolan, Bill J. F. Deakin, N. Roberts, and Ian Anderson, "Serotonergic and Cognitive Impairment in Impulsive Aggressive Personality Disordered Offenders: Are There Implications for Treatment?" *Psychological Medicine 32* (2002): 105–17; Richard J. Davidson, Katherine M. Putnam, and Christine L. Larson, "Dysfunction in the Neural Circuitry of Emotion Regulation: A Possible Prelude to Violence," *Science 289* (2000): 591–594.

Those who have low levels of serotonin are likely to have problems in everyday communication and life in general. Therefore, it is not surprising that low levels of serotonin are strongly linked to criminal activity.

Brain Injuries

Another area of physiological problems associated with criminal activity is that of trauma to the brain. As mentioned before, the brain is only 3 pounds, but it is responsible for every criminal act that an individual commits, so any problems related to this structure have profound implications regarding behavior, especially deviance and criminal activity.

Studies have consistently shown that damage to any part of the brain increases the risk of crime by individuals in the future. However, trauma to certain portions of the brain tends to have more serious consequences than injury to other areas. Specifically, damage to the frontal or **temporal lobes** (particularly those on the left side) appears to have the most consistent associations with criminal offending.[22] These findings make sense primarily because the **frontal lobes** (which include the prefrontal cortex) are the areas of the brain where the realm of higher-level problem solving and "executive" functioning takes place.[23] Thus, the frontal lobes, and especially the left side of the frontal lobes, process what we are thinking and inhibit us from doing what we are emotionally charged to do. Thus, any moral reasoning relies on this executive area of the brain because it is the region that considers long-term consequences.[24] If people suffer damage to their frontal lobes, they will be far more inclined to act on their emotional urges without any logical inhibitions imputed from this specialized region.

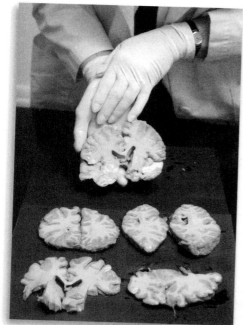

▲ **Image 5.2**   Harkening back to the 19th century, when postmortem examinations of the brains of criminals were a frequent phenomenon, researchers dissected the brain of serial killer John Wayne Gacy after his execution. The attempt to locate an organic explanation of his monstrous behavior was unsuccessful.

In a similar vein, the temporal lobe region is highly related to memory and emotion. To clarify, the temporal lobes cover and communicate almost directly with certain structures of our brain's limbic systems. Certain limbic structures govern our memories (the hippocampus) and emotions (the amygdala). Any damage to the temporal lobe, which is generally located above the ear, is likely to damage these structures or the effective communication of these structures to other

---

[22]For a review, see Raine, *The Psychopathology of Crime*, 129–54; see also Joseph M. Tonkonogy, "Violence and Temporal Lobe Lesion: Head CT and MRI Data," *Journal of Neuropsychiatry 3* (1991): 189–196; Percy Wright, Jose Nobrega, Ron Langevin, and George Wortzman, "Brain Density and Symmetry in Pedophilic and Sexually Aggressive Offenders," *Annals of Sex Research 3* (1990): 319–328.

[23]Ellis, "A Theory," 294; see also Diana Fishbein, "Neuropsychological and Emotional Regulatory Processes in Antisocial Behavior," in *Biosocial Criminology: Challenging Environmentalism's Supremacy*, eds. Anthony Walsh and Lee Ellis (New York: Nova Science, 2003): 185–208.

[24]Bobbi Jo Anderson, Malcolm D. Holmes, and Erik Ostresch, "Male and Female Delinquents' Attachments and Effects of Attachments on Severity of Self-Reported Delinquency," *Criminal Justice and Behavior 26* (1999): 435–452.

portions of the brain. Therefore, it is understandable why trauma to the temporal region of the brain is linked to future criminality.

Central and Autonomic Nervous System Activity

The brain is a key player in two different types of neurological systems that have been linked to criminal activity. The first is the **central nervous system (CNS),** which involves our brains and spinal columns and governs our voluntary motor activities. For example, the fact that you are reading this sentence means that you are in control of this brain-processing activity. Empirical studies of the influence of CNS functioning on criminality have traditionally focused on brain wave patterns, with most using electroencephalograms (EEGs). Although EEGs do not do a good job in describing which areas of the brain are active or inactive, they do reveal how much the brain as an entire organ is performing at certain times.

Studies have compared brain wave patterns of known chronic offenders (i.e., psychopaths, repeat violent offenders) to those of "normal" people (i.e., those who have never been charged with a crime).[25] These studies consistently show that the brain wave patterns of chronic offenders are abnormal as compared to the normal population, with most studies showing slower brain wave patterns in psychopaths.[26] Four types of brain wave patterns are found, from slowest to fastest: delta, theta, alpha, and beta.[27] Delta waves are often seen when people sleep, whereas theta is typically observed in lower levels of being awake, such as drowsiness. Relatively, alpha waves (which tend to be divided into slow alpha and fast alpha wave patterns, as are beta waves) tend to be related to a more relaxed wakefulness, and beta waves are observed with high levels of wakefulness, such as in times of extreme alertness and particularly in times of excited activity.

The studies that have compared brain wave patterns among chronic offenders and normals have shown significant differences. Psychopaths tend to have more activity in the theta (or sometimes slow alpha) patterns, whereas normals tend to show more activity in the fast alpha or beta types of waves. These consistent findings reveal that the cortical arousal of chronic offenders tends to be significantly slower than that of people who do not typically commit crimes. Thus, it is likely that chronic offenders typically do not have the mental functioning that would be disposed toward accurate assessments regarding the consequences of committing criminal behavior. Consistent with these relatively low levels of cortical activity in the CNS is that of the findings of studies examining the **autonomic nervous systems (ANSs)** of individuals.

The second area of the nervous system that has been most linked to criminal behavior is the ANS, which is primarily responsible for involuntary motor activities, such as heart rate, dilation of pupils, and electric conductivity in the skin. This is the type of physiological activity that is measured by polygraph

---

[25]Adrian Raine and Peter H. Venables, "Enhanced P3 Evoked Potentials and Longer P3 Recovery Times in Psychopaths," *Psychophysiology 25* (1988): 30–38; Lance O. Bauer and Victor M. Hesselbrock, "Frontal P300 Decrements, Childhood Conduct Disorder, Family History, and the Prediction of Relapse Among Abstinent Cocaine Abusers," *Drug and Alcohol Dependence 44* (1997): 1–10; for a review, see Lee Ellis, "A Theory."

[26]For a review, see Adrian Raine, *The Psychopathology of Crime,* 174–178; also see Robert D. Hare, *Psychopathy: Theory and Practice* (New York: Wiley, 1970); Jan Volavka, "Electroencephalogram Among Criminals," in *The Causes of Crime: New Biological Approaches,* eds. Sarnoff A. Mednick, Terrie E. Moffitt, and Susan Stack (Cambridge: Cambridge University Press, 1987): 137–145; Peter H. Venables, "Psychophysiology and Crime: Theory and Data," in *Biological Contributions to Crime Causation,* eds. Terrie E. Moffitt and Sarnoff A. Mednick (Dordrecht: Martinus Nijhoff, 1988); Peter H. Venables and Adrian Raine, "Biological Theory," in *Applying Psychology to Imprisonment: Theory and Practice,* eds. Barry McGurk, David Thornton, and Mark Williams (London: Her Majesty's Stationery Office, 1987): 3–28.

[27]For further discussion and explanation, see Raine, *The Psychopathology of Crime,* 174–77.

measures, or lie detector tests. Such measures capitalize on the inability of individuals to control physiological responses to anxiety, which occurs in most normal persons when they lie, especially regarding illegal behavior. However, such measures are not infallible because the individuals who are most at risk of being serious, violent offenders are the most likely to pass such tests even though they are lying.

Consistent with the findings regarding CNS arousal levels, studies have consistently shown that individuals who have significantly low levels of ANS functioning are far more likely to commit criminal acts.[28] For example, studies consistently show that chronic violent offenders tend to have much slower resting heartbeats than normal people, with a number of studies estimating this difference as much as 10 heartbeats per minute slower for the offenders.[29] This is a highly significant gap that cannot be explained away by alternative theories—for example, the explanation that offenders are just less excited in laboratory tests.

Furthermore, people who have such low levels of ANS arousal tend to experience what is known in the psychological literature as *stimulus hunger*. Stimulus hunger means that individuals with such a low level of ANS arousal may constantly seek out experiences and stimuli that are risky and thus often illegal. Readers may recall children they have known who can never seem to get enough attention, with some even seeming to enjoy being spanked or other forms of harsh punishment. In addition, people with a low level of ANS arousal may feel no anxiety about punishment, even corporal punishment, and thus they do not adequately learn right from wrong through normal forms of discipline. This is perhaps one of the reasons why children who are diagnosed with Attention Deficit-Hyperactivity Disorder (ADHD) have a higher likelihood of becoming criminal than their peers.

Because people who are accurately diagnosed with ADHD have a neurological abnormality—a significantly low-functioning ANS level of arousal—doctors prescribe stimulants (e.g., Ritalin) for such youth. It may seem counterintuitive to prescribe a hyperactive person a stimulant; however, the medication boosts the individual's ANS functioning to a normal level of arousal. This makes such individuals experience a healthy level of anxiety, which they would not normally experience from wrongdoing. Assuming that the medication is properly prescribed and at the correct dosage, children who are treated tend to become more attuned to the discipline that they face if they engage in rule violation.

Children who do not fear punishment at all—in fact, some of them do not feel anxiety even when being physically punished (e.g., spanking)—are likely to have lower-than-average levels of ANS functioning. Such individuals are likely to become chronic offenders if this disorder is not addressed because they will not respond to discipline or consider the long-term consequences of their risky behavior. If people didn't fear punishment or negative consequences from their behavior, they might be more likely to engage in selfish, greedy behavior. Thus, it is important to address this issue when it becomes evident. On the other hand, children will be children, and ADHD and other forms of disorders have been overly diagnosed in recent

---

[28]For a review, see Adrian Raine, *The Psychopathology of Crime,* 159–73; also see the following reviews and studies: Peter H. Venables, "Childhood Markers for Adult Disorders," *Journal of Child Psychology and Psychiatry and Allied Disciplines 30* (1989): 347–364; Adrian Raine and Peter H. Venables, "Skin Conductance Responsivity in Psychopaths to Orienting, Defensive, and Consonant-Vowel Stimuli," *Journal of Psychophysiology 2* (1988): 221–225; Traci Bice, "Cognitive and Psychophysiological Differences in Proactive and Reactive Aggressive Boys," Doctoral dissertation, Department of Psychology, University of Southern California (Ann Arbor: UMI, 1993).

[29]See Enrico Mezzacappa, Richard E. Tremblay, Daniel Kindlon, J. Philip Saul, Louise Arseneault, Jean Seguin, Robert O. Pihl, and Felton Earls, "Anxiety, Antisocial Behavior, and Heart Rate Regulation in Adolescent Males," *Journal of Psychiatry 38* (1997): 457–469; Graham A. Rogeness, Claudio Cepeda, Carlos A. Macedo, Charles Fischer, and William R. Harris, "Differences in Heart Rate and Blood Pressure in Children with Conduct Disorder, Major Depression, and Separation Anxiety," *Psychiatry Research 33* (1990): 199–206; Daniel J. Kindlon, Richard E. Tremblay, Enrico Mezzacappa, Felton Earls, Denis Laurent, and Benoist Schaal, "Longitudinal Patterns of Heart Rate and Fighting Behavior in 9- Through 12-year-old Boys," *Journal of the American Academy of Child and Adolescent Psychiatry 34* (1995): 371–377.

years. A well-trained physician should decide whether an individual has such a low level of ANS functioning that medication or therapy is required to curb deviant behavior.

Individuals who have significantly lower ANS arousal are likely to pass lie detector tests because they feel virtually no or little anxiety when they lie; many of them lie all the time. Thus, it is ironic, but the very people who lie-detecting measures are meant to capture are the most likely to pass such tests, which is probably why they are typically not allowed to be used in court. Only through medication or cognitive-behavioral therapy can such individuals learn to consider the long-term consequences of the decisions that they make regarding their behavior.

Individuals with low levels of ANS functioning are not always destined to become chronic offenders. Some evidence has shown that people with low ANS arousal often become successful corporate executives, decorated military soldiers, world-champion athletes, and high-level politicians. Most of these occupations require people who constantly seek out exciting, risky behaviors, and others require constant and convincing forms of lying to others. So, there are many legal outlets and productive ways to use the natural tendencies of individuals with low levels of ANS functioning. These individuals could perhaps be steered toward such occupations and opportunities when they present themselves. It is clearly a better option than committing antisocial acts against others in society.

Ultimately, low levels of cortical arousal in both the CNS and ANS are clearly linked to a predisposition toward criminal activity. However, modern medical research and societal opportunities exist to help such individuals divert their tendencies toward more prosocial outlets.

Biosocial Approaches Toward Explaining Criminal Behavior

Perhaps the most important and most recent perspective of how criminality is formed is that of biosocial approaches toward explaining crime. Specifically, if there is any conclusion that can be made regarding the previous theories and research in this chapter, it is that both genetics and environment influence behavior, particularly the interaction between the two. Even the most fundamental aspects of life can be explained by these two groups of factors.

For example, if we look at the height of individuals, we can predict with a great amount of accuracy how tall a person will be by looking at the individual's parents and other ancestors because much of height is determined by a person's genotype. However, even for something as physiological as height, the environment plays a large role. As many readers will observe, individuals who are raised in poor, underdeveloped areas (e.g., Mexico, Asia) are smaller in height than children raised in the United States. However, individuals who descend from parents and relatives in these underdeveloped areas but are raised in the United States tend to be just as tall (if not taller) than children born here. This is largely due to diet, which obviously is an environmental factor.

In other words, our genotype provides a certain range or window that determines the height of an individual, which is based on ancestral factors. But, the extent to which individuals grow to the maximum or minimum, or somewhere in between, is largely dependent on what occurs in the environment as they develop. This is why biologists make a distinction between genotype, which is directly due to genetics, and **phenotype,** which is a manifestation of genetics interacting with the environment. The same type of biosocial effect is seen on criminal behavior.

Furthermore, over the last decades, a number of empirical investigations have examined the extent to which physiological variables interact with environmental variables, and the findings of these studies have shown consistent predictions regarding criminality. Such studies have been more accurate than those that rely on either physiological and genetic variables or environmental factors separately. For example, findings from a cohort study in Philadelphia showed that individuals who had low birth weights were more likely to commit crime, but that was true primarily if they were raised in a lower income family or a family with weak social structure.[30] Those who were raised in a relatively high-income household or a strong family structure were unlikely to become criminals. It was the coupling of both a physiological deficiency (i.e., low birth weight) and an environmental deficit (i.e., weak family structure or income) that had a profound effect on criminal behavior.

In addition, recent studies have shown that when incarcerated juveniles were assigned to diets with limited levels of simple carbohydrates (e.g., sugars), their reported levels of violations during incarceration declined by almost half (45%).[31] Furthermore, other studies have reported that various food additives and dyes, such as those commonly found in processed foods, can also have a significant influence on criminal behavior. Thus, the old saying "you are what you eat" appears to have some scientific weight behind it, at least regarding criminal behavior. Additional studies have found that high levels of certain toxins, particularly lead and manganese, can have profound effects on behavior, including criminality. Recent studies have found a consistent, strong effect of high levels of lead in predicting criminal behavior. Unfortunately, medical studies have also found many subtle sources of high lead levels, such as the fake jewelry that many children wear as toys. Also unfortunate is that the individuals who are most vulnerable to high levels of lead, as with virtually every toxin, are children, and yet they are the most likely to be exposed. Even more unfortunate is that the populations (e.g., poor, urban, etc.) most susceptible to biosocial interactions are most likely to experience high levels of lead, largely due to old paint, which often contains lead, and other household products that contain dangerous toxins.[32]

Consistently, other studies have shown that pre- and perinatal problems alone do not predict violence very well. However, when such perinatal problems are considered along with environmental deficits, such as weak family structure, this biosocial relationship often predicts violent rather than property crime.[33] Other studies have shown the effects on criminality of a biosocial interaction between the impact of physiological factors within the first minute of life, called Apgar scores, and environmental factors, including exposure to nicotine.[34] Additional studies have also found that the interaction of maternal cigarette smoking and the father's absence from the household was associated with criminal behavior, especially early in life, which is one of the biggest predictors of chronic offending in the future.[35] One of the most revealing studies showed that, although only 4% of a sample of 4,269 individuals had both birth complications and maternal rejection, this relatively small group of people accounted for more than 18% of the total violent crimes committed by the whole sample.[36] So, studies have clearly shown that the interaction between biological factors and environmental deficiencies has the most consistent effect on predicting criminality.

---

[30]Stephen Tibbetts and Alex Piquero, "The Influence of Gender, Low Birth Weight, and Disadvantaged Environment in Predicting Early Onset of Offending: A Test of Moffitt's Interactional Hypothesis," *Criminology* 37 (1999): 843–878.

[31]Stephen J. Shoenthaler, *Improve Your Child's IQ and Behavior* (London: BBC Books, 1991).

[32]John Wright, Stephen Tibbett, and Leah Daigle, *Criminals in the Making* (Thousand Oaks: Sage, 2008).

[33]Alex Piquero and Stephen Tibbetts, "The Impact of Pre/Perinatal Disturbances and Disadvantaged Familial Environment in Predicting Criminal Offending," *Studies on Crime and Crime Prevention* 8 (1999): 52–71.

[34]Chris Gibson and Stephen Tibbetts, "Interaction Between Maternal Cigarette Smoking and Apgar Scores in Predicting Offending," *Psychological Reports* 83 (1998): 579–586.

[35]Chris Gibson and Stephen Tibbetts, "A Biosocial Interaction in Predicting Early Onset of Offending," *Psychological Reports* 86 (2000): 509–518.

[36]Adrian Raine, Pauline A. Brennan, and Sarnoff A. Mednick, "Birth Complications Combined with Early Maternal Rejection at Age 1 Year Predispose to Violent Crime at Age 18 Years," *Archives of General Psychiatry* 51 (1994): 984–988.

Policy Implications

The theories in this section have plenty of policy implications; a few of the primary interventions are discussed here. First, there should be universal, funded preschool for all children. This early life stage is important not only for developing academic skills, but also for fostering healthy social and disciplinary skills, which children who do not attend preschool often fail to develop.[37] In addition, there should be funded mental health and drug counseling for all young children and adolescents who exhibit symptoms of mental disorders or drug problems.[38] There should also be universal funding for health care for all expectant mothers, especially those who have high risk factors (e.g., poor, inner city, etc.). Also, and perhaps most important, there should be far more thorough examinations of children's physiological makeups in terms of hormones, neurotransmitters, brain formation and functioning, and genetic design so that earlier interventions can take place. It has been shown empirically that the earlier interventions take place, the better the outcomes.[39] These and other policy implications based on modern biosocial theories will be discussed in the final chapter of the book.

Conclusion

This chapter has examined a large range of explanations of criminal behavior that place most of the blame on biological and psychological factors, which are typically intertwined. These types of explanations were primarily popular in the early years of the development of criminology as a science, but they have also been shown in recent years to be quite valid as significant factors in individual decisions to commit crime. This chapter examined the influence of genetics and environment in family studies, twin studies, adoption studies, and studies of identical twins separated at birth. These studies ultimately showed the consistent influence of inheritance and genetics in predisposing individuals toward criminal activity. This was supported by the influence of hormones (e.g., testosterone) in human behavior, as well as the influence of variations in chromosomal mutations (e.g., XYY). Recent research has supported both of these theories in showing that people with high levels of male androgens are far more likely to commit crime than those who do not have high levels of these hormones. The link between brain trauma and crime was also discussed, with an emphasis on the consistent association between damage to the left or frontal parts of the brain. We also examined theories regarding variations in levels of CNS and ANS functioning; all empirical studies have shown that low levels of functioning of these systems have links to criminality. Finally, we explored the extent to which the interaction between physiological factors and environmental variables contribute to the most consistent prediction of criminal offending. Ultimately, it is interesting that the very theories that were key in the early years of the development of criminology as a science are now showing strong support in studies as being primary influences on criminal behavior.

Chapter Summary

- Early studies that examined the influence of biology focused on case studies of certain families. These studies showed that criminality was indeed clustered among certain families, but such studies did not separate biology from environment.

---

[37]Wright et al., *Criminals in the Making.*

[38]Ibid.

[39]Samuel S. Wu, Chang-Xing Ma, Randy L. Carter, Mario Ariet, Edward A. Feaver, Michael B. Resnick, and Jeffrey Roth. "Risk Factors for Infant Maltreatment: A Population-Based Study," *Child Abuse & Neglect 28* (2004): 1253–1264.

- The next stage of studies examined the concordance rates of identical twins versus nonidentical twins. These studies led to a conclusion that genetic makeup was very important, but critics called these conclusions into question.
- The following stage of research examined adoptees, namely the predictive nature of which parents (biological or adoptive) had more influence in their future criminal behavior. These studies revealed that biological parents (whom they never knew) had far more influence than the adoptive parents who raised them. However, there were criticisms of these studies, so the findings were questioned.
- The final stage of the biology versus environment debate was that of identical twins separated at birth as compared to identical twins raised together. These studies showed that the twins who were separated at birth were just as similar, if not more so, than the twins who were reared together. To date, there are no criticisms of this method of study. Thus, it appears that all four waves of study are consistent in showing that biological influences are vitally important in explaining the criminality of individuals.
- This chapter also examined chromosomal mutations, such as the XYY mutation, which has consistently shown associations with criminality. Much, if not most, of this link is believed to be due to the increased male androgens (e.g., testosterone) produced by individuals who have the XYY chromosomal mutation.
- Studies have consistently shown that individuals with higher levels of testosterone and other androgens are more disposed toward criminality; for example, normal males are far more likely than normal females to engage in violent crimes.
- This chapter reviewed findings from studies that show that people with abnormal levels of certain neurotransmitters, such as serotonin, are far more likely to engage in crime than those who have normal levels of these chemicals in their brains or bodies.
- Studies show criminality is more likely among individuals who have experienced brain trauma or have lower levels of brain functioning, especially in certain regions of the brain, such as the frontal and temporal lobes, which are the regions that largely govern higher level, problem-solving functions.
- This chapter also reviewed the disposition of individuals regarding two aspects of the nervous system, specifically the CNS and ANS. Those who have significantly slower brain waves and lower anxiety levels are far more likely to commit crimes.

KEY TERMS

Adoption studies	Dizygotic twins	Phenotype
Autonomic nervous system (ANS)	Dopamine	Selective placement
	Family studies	Serotonin
Central nervous system (CNS)	Frontal lobes	Temporal lobes
Concordance rates	Monozygotic twins	Twins separated at birth studies
Cytogenetic studies	Neurotransmitters	Twin studies

DISCUSSION QUESTIONS

1. Is there any validity to family studies in determining the role of genetics in criminal behavior? Explain why or why not.

2. Explain the rationale of studies that compare the concordance rates of identical twins and fraternal twins who are raised together. What do most of these studies show regarding the influence of genetics on criminal behavior? What are the criticisms of these studies?

3. Explain the rationale of studies that examine the biological and adoptive parents of adopted children. What do most of these studies show regarding the influence of genetics on criminal behavior? What are the criticisms of these studies?

4. What are the general findings of identical twins who are separated at birth? What implications do these findings have for the importance of genetics or heritability regarding criminal behavior? Can you find a criticism for such findings?

5. Explain what cytogenetic disorders are and describe the related disorder that is most linked to criminal behavior. What characteristics of this type of disorder seem to be driving the higher propensity toward crime?

6. What types of hormones have been shown by scientific studies to be linked to criminal activity? Give specific examples that show this link to be true.

7. Explain what neurotransmitters are, and describe which neurotransmitters are key in predicting criminal offending. Provide support from previous scientific studies.

8. Which areas of the brain, given trauma, have shown the greatest vulnerability regarding criminal offending? Does the lack of healthy functioning in these areas make sense? Why?

9. How do brain wave patterns differ among chronic, violent criminals versus normal people? Does this make sense in biosocial models of criminality?

10. How does the ANS differ among chronic, violent criminals versus normal people? Does this make sense in biosocial models of criminality?

11. What types of policy implications would you support based on the information provided by empirical studies reviewed in this chapter?

WEB RESOURCES

ANS/CNS

http://www.mentalhelp.net/poc/view_doc.php?type=doc&id=263

http://www.crimetimes.org/

Brain Trauma and Crime

> http://www.crimetimes.org/

Cytogenetic Studies

> http://en.wikipedia.org/wiki/cytogenetics

Family/Twin/Adoption Studies

> http://www.criminology.fsu.edu/crimtheory/dugdale.htm
> http://www.bookrags.com/research/twin-studies-wog/

Hormones and Neurotransmitters

> http://www.gender.org.uk/about/06encrn/63faggrs.htm
> http://serendip.brynmawr.edu/biology/b103/f02/web1/kamlin.html

6

Early Social Structure and Strain Theories of Crime

This chapter will review the development of **anomie** or **strain theory,** starting with early social structure theorists, such as Durkheim, then Merton, and on to the most modern versions (e.g., **general strain theory**). We will also examine the empirical research findings on this perspective, which remains one of the dominant theoretical explanations of criminal behavior today. We will finish by discussing the policy implications of this research.

We'll begin with a review of explanations of criminal conduct that emphasize the differences among varying groups in societies, particularly in the United States. Such differences are easy to see in everyday life, and many theoretical models place the blame for crime on observed inequalities and cultural differences between groups. In contrast to the theories presented in previous chapters, social structure theories disregard any biological or psychological variations across individuals. Instead, social structure theories assume that crime is caused by the way societies are structurally organized.

These social structure theories vary in many ways, most notably in what they propose as the primary constructs and processes in causing criminal activity. For example, some structural models place an emphasis on variations in economic or academic success, whereas others focus on differences in cultural norms and values. Still others concentrate on the actual breakdown of the social structure in certain neighborhoods and the resulting social disorganization that occurs from this process, a topic we will reserve for Chapter 7. Regardless of their differences, all of the theories examined in this section emphasize a common theme: Certain groups of individuals are more likely to break the law because of disadvantages or cultural differences resulting from the way a society is structured.

The most important distinction between these theories and those discussed in previous chapters is that they emphasize group differences instead of individual differences. Structural models tend to focus on the macrolevel of analysis as opposed to the microlevel. Therefore, it is not surprising that social structure theories are commonly used to explain the propensity of certain racial or ethnic groups for committing crime, as well as the overrepresentation of the lower class in criminal activity.

As you will see, these theoretical frameworks were presented as early as the 1800s and reached prominence in the early to mid-1900s, when the political, cultural, and economic climate of society was most conducive to such explanations. Although social structural models of crime have diminished in popularity in recent decades,[1] there is much validity to their propositions in numerous applications to contemporary society.

Early Theories of Social Structure: Early to Mid-1800s

Most criminological and sociological historians trace the origin of social structure theories to the research done in the early to mid-1800s by a number of European researchers, with the most important including Auguste Comte, André-Michel Guerry, and Adolphe Quetelet.[2] It is important to understand why structural theories developed in 19th-century Europe. The Industrial Revolution, defined by most historians as beginning in the mid-1700s and ending in the mid-1800s, was in full swing at the turn of the century, so societies were quickly transitioning from primarily agriculturally based economies to industrial-based. This transition inevitably brought people from rural farmlands to dense urban cities, with a resulting enormous increase in social problems. These social problems ranged from failure to properly dispose of waste and garbage, to constantly losing children and not being able to find them, to much higher rates of crime (which urban areas continue to show today, as compared to suburban and rural areas).

The problems associated with such fast urbanization, as well as the shift in economics, led to drastic changes in the basic social structures in Europe, as well as the United States. At the same time, other types of revolutions were also having an effect. Both the American (1776) and French Revolutions (1789) occurred in the last quarter of the 18th century. These two revolutions, inspired by the Enlightenment movement (see Chapter 2), shared an ideology that rejected tyranny and insisted that people should have a voice in how they were governed. Along with the Industrial Revolution, these political revolutions affected intellectual theorizing on social structures, as well on crime, throughout the 1800s.

Auguste Comte

One of the first important theorists in the area of social structure theory was Auguste Comte (1798–1857), who is widely credited with coining the term *sociology*.[3] Comte distinguished the concepts of **social statics** and **social dynamics.** Social statics are aspects of society that relate to stability and social order; they allow societies to continue and endure. Social dynamics are aspects of social life that alter how societies are structured and pattern the development of societal institutions. Although such conceptualization seems elementary by today's standards, it had a significant influence on sociological thinking at the time. Furthermore, the distinction between static and dynamic societal factors was incorporated in several criminological theories in decades to come.

---

[1] Anthony Walsh and Lee Ellis, "Political Ideology and American Criminologists' Explanations for Criminal Behavior," *The Criminologist* 24 (1999): 1, 4; Lee Ellis and Anthony Walsh, "Criminologists' Opinions About Causes and Theories of Crime and Delinquency," *The Criminologist* 24 (1999): 1–4.

[2] Much of the discussion of the development of structural theories of the 19th century is drawn from James W. Vander Zanden, *Sociology: The Core,* 2nd ed. (New York: McGraw-Hill, 1990), 8–14.

[3] Vander Zanden, *Sociology: The Core,* 8–9.

Between 1851 and 1854, Comte published a four-volume work titled *A System of Positive Polity* that encouraged the use of scientific methods to observe and measure societal factors.[4] Although we tend to take this for granted in modern times, the idea of applying such methods to help explain social processes was rather profound at the time; probably for this reason, he is generally considered the founder or father of sociology. Comte's work set the stage for the positivistic perspective, which emphasized social determinism and rejected the notion of free will and individual choice that was common up until that time.

André-Michel Guerry and Adolphe Quetelet

After the first modern national crime statistics were published in France in the early 1800s, a French lawyer named André-Michel Guerry (1802–1866) published a report that examined these statistics and concluded that property crimes were higher in wealthy areas, but violent crime was much higher in poor areas.[5] Some experts have claimed that this report likely represents the first study of scientific criminology[6]; it was later expanded and published as a book, titled *Essay on the Moral Statistics of France* in 1833. Ultimately, Guerry concluded that the explanation was opportunity: The wealthy had more to steal, and that is the primary cause of property crime. Interestingly, this conclusion is supported by recent U.S. DOJ statistics, which show that, compared to lower class households, property crime is just as common—and maybe more so—in middle- to upper-class households, but violent crime is not.[7] As Guerry stated centuries ago, there is more to steal in wealthier areas, and poor individuals take the opportunity to steal goods and currency from these households and establishments.

Adolphe Quetelet (1796–1874) was a Belgian researcher who, like Guerry, examined French statistics in the mid-1800s. Besides showing relative stability in the trends of crime rates in France, such as in age distribution and female-to-male ratios of offending, Quetelet also showed that certain types of individuals were more likely to commit crime.[8] Specifically, young, male, poor, uneducated, and unemployed individuals were more likely to commit crime than their counterparts,[9] a finding also supported by modern research. Like Guerry, Quetelet concluded that opportunities had a lot to do with where crime was concentrated in addition to the demographic characteristics.

However, Quetelet added a special component: Greater inequality or gaps between wealth and poverty in the same place tends to excite temptations and passions. This is a concept referred to as **relative deprivation,** a quite distinct condition from simple poverty.

For example, a number of deprived areas in the United States do not have high rates of crime, likely because virtually everyone is poor, so people are generally content with their lives relative to their neighbors. However, in areas of the country where very poor people live in close proximity to very wealthy people, this

---

[4]Auguste Comte, *A System of Positive Polity,* trans. John Henry Bridges (New York: Franklin, 1875).

[5]For more thorough discussions of Guerry and Quetelet, see the sources from which I have drawn the information presented here: Piers Beirne, *Inventing Criminology* (Albany: SUNY Press, 1993); Vold et al., *Theoretical Criminology* (see chap. 2, n. 40).

[6]Terrence Morris, *The Criminal Area* (New York: Routledge, 1957), 42–53, as cited in Vold et al., *Theoretical Criminology,* 22.

[7]U.S. Department of Justice, Bureau of Justice Statistics, *Sourcebook of Criminal Justice Statistics, 2000, NCJ-190251* (Washington, DC: USGPO, 2001), Table 3.26, 202; U.S. Department of Justice, Bureau of Justice Statistics, *Sourcebook of Criminal Justice Statistics, 2000, NCJ-190251* (Washington, DC: USGPO, 2001), Table 3.13, 194.

[8]Beirne, *Inventing Criminology,* 78–81.

[9]Vold et al., *Theoretical Criminology,* 23–26.

causes animosity and feelings of being deprived. Studies have supported this hypothesis.[10] It may well explain why Washington, DC, perhaps the most powerful city in the world and one with many neighborhoods that are severely run-down and poor, has a higher crime rate than any other jurisdiction in the country.[11] Modern studies have also showed a clear linear association between higher crime rates and relative deprivation. For example, David Sang-Yoon Lee found that crime rates were far higher in cities that had wider gaps in income: The larger the gap between the 10th and 90th percentiles, the greater the crime levels.[12]

In addition to the concept of relative deprivation, Quetelet also showed that areas with the most rapidly changing economic conditions also showed high crime rates. He is perhaps best known for his comment, "The crimes . . . committed seem to be a necessary result of our social organization. . . . Society prepares the crime, and the guilty are only the instruments by which it is executed."[13] This statement makes it clear that crime is a result of societal structure and not the result of individual propensities or personal decision making. Thus, it is not surprising that Quetelet's position was controversial at the time in which he wrote, and he was rigorously attacked by critics for removing all decision-making capabilities from his model of behavior. In response, Quetelet argued that his model could help lower crime rates by leading to social reforms that address the inequalities due to the social structure.[14]

One of the essential points of Guerry's and Quetelet's work is the positivistic nature of their conclusions, that the distribution of crime is not random but rather the result of certain types of people committing certain types of crimes in particular places largely due to the way society is structured and the way it distributes resources. This perspective of criminality strongly supports the tendency of crime to be clustered in certain neighborhoods, as well as among certain people. Such findings support a structural, positivistic perspective of criminality, in which criminality is seen as being deterministic and, thus, caused by factors outside of an individual's control. In some ways, early structural theories were a response to the failure of the classical approach to crime control. As the 19th century drew to a close, classical and deterrence-based perspectives of crime fell out of favor, while social structure theories and other positivist theories of crime, such as the structural models by Guerry and Quetelet, attracted far more attention.

Durkheim and the Concept of Anomie

Although influenced by earlier theorists (e.g., Comte, Guerry, and Quetelet), Emile Durkheim (1858–1916) was perhaps the most influential theorist in the state of modern structural perspectives on criminality.[15] Like most other social theorists of the 19th century, he was strongly affected by the

---

[10]Velmer Burton and Frank Cullen, "The Empirical Status of Strain Theory," *Journal of Crime and Justice 15* (1992): 1–30; Nikos Passas, "Continuities in the Anomie Tradition," in *Advances in Criminological Theory, Vol. 6: The Legacy of Anomie Theory,* eds. Freda Adler and William S. Laufer (New Brunswick: Transaction, 1995); Nikos Passas, "Anomie, Reference Groups, and Relative Deprivation," in *The Future of Anomie Theory,* eds. Nikos Passas and Robert Agnew (Boston: Northeastern University Press, 1997).

[11]U.S. Department of Justice, Bureau of Justice Statistics, *Sourcebook of Criminal Justice Statistics, 2000, NCJ-190251* (Washington, DC: USGPO, 2001), Table 3.124, 290.

[12]David Sang-Yoon Lee, "An Empirical Investigation of the Economic Incentives for Criminal Behavior," B.A. thesis in economics (Boston: Harvard University, 1993), as cited in Richard B. Freeman, "The Labor Market," in *Crime,* eds. James Q. Wilson and Joan Petersilia (San Francisco: ICS Press, 1995), 171–192.

[13]Beirne, *Inventing Criminology,* 88, as cited in Vold et al., *Theoretical Criminology,* 25.

[14]Vold et al., *Theoretical Criminology,* 25–26.

[15]Much of this discussion of Durkheim is taken from Vold et al., *Theoretical Criminology,* 100–116, as well as Vander Zanden, *Sociology: The Core,* 11–13.

American and French Revolutions and the Industrial Revolution. In his doctoral dissertation (1893) at the University of Paris, the first sociological dissertation at that institution, Durkheim developed a general model of societal development largely based on economic and labor distribution, in which societies are seen as evolving from a simplistic mechanical society toward a multilayered, organic society (see Figure 6.1).

As outlined in this dissertation, titled *The Division of Labor in Society*, in primitive **mechanical societies,** all members essentially perform the same functions, such as hunting (typically males) and gathering (typically females). Although there are a few anomalies (e.g., medicine man), virtually everyone essentially experiences the same daily routine. Such similarities in work, as well as constant interaction with like members of the society, lead to a strong uniformity in values, which Durkheim called the **collective conscience.** The collective conscience is the degree to which individuals of a society think alike, or as Durkheim put it, the totality of social likenesses. The similar norms and values among people in these primitive, mechanical societies create *mechanical solidarity,* a very simple-layered social structure with a very strong collective conscience. In mechanical societies, law functions to enforce the **conformity** of the group.

However, as societies progress toward more modern, **organic societies** in the industrial age, the distribution of labor becomes more highly specified. An *organic solidarity* arises in which people tend to depend on other groups because of the highly specified division of labor, and laws have the primary function of regulating interactions and maintaining solidarity among the groups.

For example, modern researchers at universities in the United States tend to be trained in extremely narrow topics—one might be an expert on something as specific as the antennae of certain species of ants. On the other hand, some individuals are still gathering trash from the cans on the same streets every single day. The antennae experts probably have little interaction and not much in common with the garbage collectors other than to pay them. According to Durkheim, to move from the universally shared roles of mechanical societies to the extremely specific roles of organic societal organization results in huge cultural

---

Figure 6.1   Durkheim's Continuum of Development From Mechanical to Organic Societies

Mechanical Societies	Organic Societies
⟶ Industrialization ⟶	
Primitive	Modern
Rural	Urban
Agricultural-based economy	Industrial-based economy
Simple division of labor (few divisions)	Complex division of labor (many specialized divisions)
Law used to enforce conformity	Law used to regulate interactions among divisions
Typically stronger collective conscience	Typically weaker collective conscience

differences and giant contrasts in normative values and attitudes across groups. Thus, the collective conscience in such societies is weak, largely because there is little agreement on moral beliefs or opinions. The preexisting solidarity among members breaks down, and the bonds are weakened, which creates a climate for antisocial behavior.

Durkheim was clear in stating that crime is not only normal but necessary in all societies. As a result, his theory is often considered a good representation of structural functionalism. He claimed that all social behaviors, especially crime, provide essential functions in a society. Durkheim thought crime serves several functions. First, it defines the moral boundaries of societies. Few people know or realize what is against societal laws until they see someone punished for a violation. This reinforces their understanding of both what the rules are and what it means to break them. Furthermore, the identification of rule breakers creates a bond among the other members of the society, perhaps through a common sense of self-righteousness or superiority.

In later works, Durkheim said the resultant bonding is what makes crime so necessary in a society. Given a community that has no law violators, society will change the legal definitions of what constitutes a crime to define some of its members as criminals, he thought. Examples of this are prevalent, but perhaps the most convincing is that of the Salem witch trials, in which hundreds of individuals were accused and tried for an almost laughable offense, and more than a dozen were executed. Durkheim would say this was inevitable because crime was so low in the Massachusetts Bay Colony, as historical records confirm, that society had to come up with a fabricated criterion for defining certain members of the population as offenders.

Other examples are common in everyday life. The fastest way to have a group of strangers bond is to give them a common enemy, which often means forming into cliques and ganging up on others in the group. In a group of three or more college roommates, for example, two or more of the individuals will quickly join together and complain about the others. This is an inevitable phenomenon of human interaction and group dynamics, which has always existed throughout the world across time and place. As Durkheim said, even in "a society of saints . . . crimes . . . will there be unknown; but faults which appear venial to the layman will create there the same scandal that the ordinary offense does in ordinary consciousnesses. . . . This society . . . will define these acts as criminal and will treat them as such."[16]

Law enforcement should always be cautious in cracking down on gangs, especially during relatively inactive periods, because it may make the gang stronger. Like all societal groups, when a common enemy appears, gang members—even those who do not typically get along—will come together and "circle the wagons" to protect themselves via strength in numbers. This very powerful bonding effect is one that many sociologists, and especially gang researchers, have consistently observed.[17]

Traditional, mostly mechanical societies could count on relative consensus about moral values and norms, and this sometimes led to too much control and a stagnation of creative thought. Durkheim thought that progress in society typically depends on deviating from established moral boundaries, especially if the society is in the mechanical stage. Some of the many examples include virtually all religious icons. Jesus, Buddha, and Mohammed were persecuted as criminals for deviating from societal norms in the times they preached. Political heroes have also been prosecuted and incarcerated as criminals, such as Gandhi in India, Mandela in South Africa, and Dr. King in the United States. In one of the most compelling cases, scientist and astronomer Galileo proposed a theory that Earth was not the center of the universe. Even though he was right, he was persecuted for his belief in a society that strictly adhered to its beliefs many centuries ago.

---

[16]Emile Durkheim, *The Rules of the Sociological Method,* trans. Sarah A. Solovay and John H. Mueller, ed. George E. G. Catlin (New York: Free Press, 1965), as cited in Vold et al., *Theoretical Criminology.*

[17]See discussion in Malcolm Klein, "Street Gang Cycles," in *Crime,* eds. James Q. Wilson and Joan Petersilia (San Francisco: ICS Press, 1995), 217–236.

Durkheim was clearly accurate in saying that the normative structure in some societies is so strong that it hinders progress and that crime is the price societies pay for progress.

In contrast to mechanical societies, modern societies do not have such extreme restraint against deviations from established norms. Rather, almost the opposite is true; there are too many differences across groups because the division of labor is highly specialized. Thus, the varying roles in society, such as farmers versus scientific researchers, lead to extreme differences in the cultural values and norms of each group. There is a breakdown in the collective conscience because there is really no longer a collective nature in society. Therefore, law focuses not on defining the norms of society but on governing the interactions that take place among the different classes. According to Durkheim, law provides a service in regulating such interactions as societies progress toward more organic (more industrial) forms.

Durkheim emphasized that human beings, unlike other animal species who live according to their spontaneous needs, have no internal mechanism to signal when their needs and desires are satiated. Therefore, the selfish desires of humankind are limitless; the more an individual has, the more he or she wants. People are greedy by nature, and without something to tell them what they need and deserve, they will never feel content.[18] According to Durkheim, society provides the mechanism for limiting this insatiable appetite, having the sole power to create laws that set tangible limits.

Durkheim also noted that in times of rapid change, society fails in this role of regulating desires and expectations. Rapid change can be due to numerous factors, such as war or social movements (like the changes seen in the United States in the 1960s). The transitions Durkheim likely had in mind when he wrote were the American and French Revolutions and the Industrial Revolution. When society's ability to serve as a regulatory mechanism breaks down, the selfish, greedy tendencies of individuals are uncontrolled, causing a state Durkheim called *anomie,* or *normlessness.* Societies in such anomic states experience increases in many social problems, particularly criminal activity.

Durkheim was clear that, whether the rapid change was for good or bad, it would have negative effects on society. For example, whether the U.S. economy was improving, as it did during the late 1960s, or quickly tanking, as it did during the Depression of the 1930s, criminal activity would increase due to the lack of stability in regulating human expectations and desires. Interestingly, these two periods of time experienced the greatest crime waves of the 20th century, particularly for murder.[19] Another fact that supports Durkheim's predictions is that middle- and upper-class individuals have higher suicide rates than those from lower classes. This is consistent with the idea that it is better to have stability, even if it means always being poor, than it is to have instability at higher levels of income. In his best known work, *Suicide,* Durkheim took an act that would seem to be the ultimate form of free choice or free will and showed that the decision to take one's own life was largely determined by external social factors. He argued that suicide was a *social fact,* a product of meanings and structural aspects that result from interactions among people.

Durkheim showed that the rate of suicide was significantly lower among individuals who were married, younger, and practiced religions that were more interactive and communal (e.g., Jewish). All of these characteristics boil down to one aspect: The more social interaction and bonding with the community, the less the suicide. So, Durkheim concluded that variations in suicide rates are due to differences in social solidarity or bonding to society. Examples of this are still seen today, as in recent reports of high rates of suicide among

---

[18]For more details on these issues, see Emile Durkheim's works: *The Division of Labor in Society* (New York: Free Press, 1893/1965) and *Suicide* (New York: Free Press, 1897/1951).

[19]U.S. Department of Justice, Bureau of Justice Statistics, *Violent Crime in the United States* (Washington, DC: Bureau of Justice Statistics, 1996), as illustrated in Joseph P. Senna and Larry G. Siegel, *Introduction to Criminal Justice,* 8th ed. (Belmont: West/Wadsworth, 1999).

people who live in remote areas, like Alaska (which has the highest rate of juvenile suicide), Wyoming, Montana, and the northern portions of Nevada. Another way of looking at the implications of Durkheim's conclusions is that social relationships are what make people happy and fulfilled. If they are isolated or have weak bonds to society, they will likely be depressed and discontented with their lives.

Another reason that Durkheim's examination of suicide was important was that he showed that suicide rates increased in times of rapid economic growth or rapid decline. Although researchers later argued that crime rates did not always follow this pattern,[20] Durkheim used quantified measures to test his propositions as the positivistic approach recommended. At the least, Durkheim created a prototype of how theory and empirical research could be combined in testing differences across social groups. This theoretical framework was drawn on heavily for one of the most influential and accepted criminological theories of the 20th century: strain theory.

✉ Strain Theories

All forms of strain theory share an emphasis on frustration as a factor in crime causation, hence the name *strain* theories. Although the theories differ regarding what exactly is causing the frustration—and the way individuals cope (or don't cope) with stress and anger—they all hold that strain is the primary causal factor in the development of criminality. Strain theories all trace their origin to Robert K. Merton's seminal theoretical framework.

Merton's Strain Theory

Working in the 1930s, Merton drew heavily on Durkheim's idea of anomie in developing his own theory of structural strain.[21] Although Merton altered the definition of anomie, Durkheim's theoretical framework was a vital influence in the evolution of strain theory. Combining Durkheimian concepts and propositions with an emphasis on American culture, Merton's structural model became one of the most popular perspectives in criminological thought in the early 1900s and remains one of the most cited theories of crime in criminological literature.

Cultural Context and Assumptions of Strain Theory

Some have claimed that Merton's seminal piece in 1938 was the most influential theoretical formulation in criminological literature and one of the most frequently cited papers in sociology.[22] Although its popularity is partially due to its strong foundation in previous structural theories, Merton's strain theory also benefited

---

[20]William J. Chambliss, "Functional and Conflict Theories of Crime," in *Whose Law? What Order?* eds. William J. Chambliss and Milton Mankoff (New York: John Wiley, 1976), 11–16.

[21]For reviews of Merton's theory, see both the original and more recent works by Merton himself: Robert K. Merton, "Social Structure and Anomie," *American Sociological Review 3* (1938): 672–682; Robert K. Merton, *Social Theory and Social Structure* (New York: Free Press, 1968); Robert K. Merton, "Opportunity Structure: The Emergence, Diffusion, and Differentiation as Sociological Concept, 1930s–1950s," in *Advances in Criminological Theory: The Legacy of Anomie Theory*, Vol. 6, eds. Freda Adler and William Laufer (New Brunswick: Transaction Press, 1995); for reviews by others, see the following: Ronald L. Akers and Christine S. Sellers, *Criminological Theories: Introduction, Evaluation, and Application*, 4th ed. (Los Angeles: Roxbury, 2004), 164–168; Stephen E. Brown, Finn-Aage Esbensen, and Gilbert Geis, *Criminology: Explaining Crime and Its Context*, 5th ed. (Cincinnati: Anderson, 2004), 297–307; Marshall B. Clinard, "The Theoretical Implications of Anomie and Deviant Behavior," in *Anomie and Deviant Behavior*, ed. Marshall B. Clinard (New York: Free Press, 1964), 1–56; Thomas J. Bernard, "Testing Structural Strain Theories," *Journal of Research in Crime and Delinquency 24* (1987): 262–280.

[22]Clinard, "The Theoretical Implications"; also see discussion in Brown et al., *Criminology: Explaining Crime*, 297.

from the timing of its publication. Virtually every theory discussed in this book became popular because it was well suited to the political and social climate of their times, fitting perspectives of how the world worked. Perhaps no other theory better represents this phenomenon than strain theory.

Most historians would agree that, in the United States, the most significant social issue of the 1930s was the economy. The influence of the Great Depression, largely a result of a stock market crash in 1929, affected virtually every aspect of life in the United States. Unemployment and extreme poverty soared, along with suicide rates and crime rates, particularly murder rates.[23] American society was fertile ground for a theory of crime that placed virtually all of the blame on the U.S. economic structure.

On the other side of the coin, Merton was highly influenced by what he saw happening to the country during the Depression, how much the economic institution impacted almost all other social factors, particularly crime. He watched how the breakdown of the economic structure drove people to kill themselves or others, not to mention the rise in property crimes, such as theft. Many once-successful individuals were now poor, and some felt driven to crime for survival. Notably, Durkheim's hypotheses regarding crime and suicide were supported during this time of rapid change, and Merton apparently realized that the framework simply had to be brought up to date and supplemented.

One of the key assumptions that distinguishes strain theory from Durkheim's perspective is that Merton altered his version of what anomie means. Merton focused on the nearly universal socialization of the American Dream in U.S. society, the idea that, as long as people work very hard and pay their dues, they will achieve their goals in the end. According to Merton, the socialized image of the goal is material wealth, whereas the socialized concept of the means of achieving the goal is hard work (e.g., education, labor). The conventional model of the American Dream was consistent with the Protestant work ethic, which called for working hard for a long time, knowing that you will be paid off in the distant future.

Furthermore, Merton thought that nearly everyone was socialized to believe in the American Dream, no matter what the economic class of their childhood. There is some empirical support for this belief, and it makes sense. Virtually all parents, even if they are poor, want to give their children hope for the future if they are willing to work hard in school and at a job. Parents and society usually use celebrities as examples of this process, individuals who started off poor and rose to become wealthy. Modern examples include former Secretary of State Colin Powell, owner of the NBA Dallas Mavericks Mark Cuban, Oscar-winning actress Hillary Swank, and Hollywood director and screenwriter Quentin Tarantino, not to mention Arnold Schwarzenegger's amazing rise from teenage immigrant to Mr. Olympia and California governor.

These stories epitomize the American Dream, but parents and society do not also teach the reality of the situation. As Merton points out, a small percentage of people rise from the lower class to become materially successful, but the vast majority of poor children don't have much chance of ever obtaining such wealth. This near-universal socialization of the American Dream—which turns out not to be true for most people—causes most of the strain and frustration in American society, Merton said. Furthermore, he thought that most of the strain and frustration was due not to the failure to achieve conventional goals (i.e., wealth) but rather to the differential emphasis placed on material goals and the de-emphasis on the importance of conventional means.

Merton's Concept of Anomie and Strain

Merton claimed that, in an ideal society, there would be an equal emphasis on the conventional goals and means. However, in many societies, one is emphasized more than the other. Merton thought the United

---

[23]Joseph J. Senna and Larry J. Siegel, *Introduction to Criminal Justice*, 8th ed. (Belmont: Wadsworth, 1999), 44.

States epitomized the type of society that emphasized the goals far more than the means. According to Merton, the disequilibrium in emphasis between the goals and means of societies is what he called anomie. So, like Durkheim, Merton's anomie was a negative state for society; however, the two men had very different ideas of how this state of society was caused. Whereas Durkheim believed that anomie was primarily caused by a society transitioning too fast to maintain its regulatory control over its members, for Merton, anomie represented too much focus on the goals of wealth in the United States at the expense of conventional means.

Some hypothetical situations will illustrate. Which of the following two men would be more respected by youth (or even adults) in our society: (1) John, who has his PhD in physics and lives in a one-bedroom apartment because his job as a postdoctoral student pays $25,000 a year; or (2) Joe, who is a relatively successful drug dealer and owns a four-bedroom home, drives a Hummer, dropped out of school in the 10th grade, and makes about $90,000 a year? In years of asking such a question to our classes, the answer is usually Joe, the drug dealer. After all, he appears to have obtained the American Dream, and little emphasis is placed on how he achieved it.

Still another way of supporting Merton's idea that America is too focused on the goal of material success is to ask you, the reader, to think about why you are taking the time to read this book or to attend college. Specifically, the question for you is this: If you knew that studying this book—or earning a college degree—would not lead to a better employment position, would you read it anyway just to increase your own knowledge? In over a decade of putting this question to about 10,000 students in various university courses, one of the authors of this book has found that only about 5% (usually fewer) of respondents said yes. Interestingly, when asked why they put all of this work into attending classes, many of them said it was for the partying or social life. Ultimately, it seems that most college students would not be engaging in the hard work it takes to educate themselves if it weren't for some payoff at the end of the task. In some ways, this supports Merton's claim that there is an emphasis on the goals with little or no intrinsic value being placed on the work itself (i.e., means). This phenomenon is not meant to be a disheartening or a negative statement; it is meant only to exhibit the reality of American culture. It is quite common in our society to place an emphasis on the goal of financial success as opposed to hard work or education.

Merton thought that individuals, particularly those in the lower class, eventually realize that the ideal of the American Dream is a lie or at least a false illusion for the vast majority. This revelation will likely take place when people are in their late teens to mid-20s, and according to Merton, this is when the frustration or strain is evident. That is consistent with the age-crime peak of offending at the approximate age of 17. Learning that hard work won't necessarily provide rewards, some individuals begin to innovate ways that they can achieve material success without the conventional means of getting it. Obviously, this is often through criminal activity. However, not all individuals deal with strain in this way; most people who are poor do not resort to crime. To Merton's credit, he explained that individuals deal with the limited economic structure of society in different ways. He referred to these variations as **adaptations to strain.**

Adaptations to Strain

There are five adaptations to strain according to Merton. The first of these is conformity, in which people buy into the conventional goals of society and the conventional means of working hard in school or labor.[24] Like conformists, most readers of this book would probably like to achieve material success and are willing to use conventional means of achieving success through educational efforts and doing a good job at work.

---

[24]Merton, *Social Theory*.

Another adaptation to strain is **ritualism.** Ritualists do not pursue the goal of material success, probably because they know they don't have a realistic chance of obtaining it. However, they do buy into the conventional means in the sense that they like to do their jobs or are happy with just making ends meet through their current positions. For example, studies have shown that some of the most contented and happy people in society are those who don't hope to become rich, are quite content with their blue-collar jobs, and often have a strong sense of pride in their work, even if it is sometimes quite menial. To these people, work is a type of ritual, performed without a goal in mind; rather, it is a form of intrinsic goal in and of itself. Ultimately, conformists and ritualists tend to be low risks for offending, which is in contrast to the other adaptations to strain.

The other three adaptations to strain are far more likely to engage in criminal offending: **innovation, retreatism,** and **rebellion.** Perhaps most likely to become predatory street criminals are the innovators, who Merton claimed greatly desire the conventional goals of material success but are not willing to engage in conventional means. Obviously, drug dealers and professional auto thieves, as well as many other variations of chronic property criminals (e.g., bank robbers, etc.), would fit this adaptation. They are innovating ways to achieve the goals without the hard work that is usually required. However, innovators are not always criminals. In fact, many of them are the most respected individuals in our society. For example, some entrepreneurs have used the capitalistic system of our society to produce useful products and services (e.g., the men who designed Google for the Internet) and have made a fortune at a very young age without advanced college education or years of work at a company. Other examples are successful athletes who sign multimillion-dollar contracts at age 18. So, it should be clear that not all innovators are criminals.

The fourth adaptation to strain is retreatism. Such individuals do not seek to achieve the goals of society, and they also do not buy into the idea of conventional hard work. There are many varieties of this adaptation: people who become homeless by choice or who isolate themselves in desolate places without human contact. A good example of a retreatist is Ted Kaczinski, the Unabomber, who left a good position as a professor at University of California at Berkeley to live in an isolated cabin in Montana that had no running water or electricity; he did not interact with humans for many months at a time. Other types of retreatists, perhaps the most likely to be criminal, are those who are heavy drug users who actively disengage from social life and try to escape via psychologically altering drugs. All of these forms of retreatists seek to drop out of society altogether, thus not buying into its means or goals.

Finally, the last adaptation to strain, according to Merton, is rebellion, which is the most complex of the five adaptations. Interestingly, rebels buy into the idea of societal goals and means, but they do not buy into those currently in place. Most true rebels are criminals by definition, largely because they are trying to overthrow the current societal structure. For example, the founders of the United States were all rebels because they actively fought the governing state—English rule—and clearly committed treason in the process. Had they lost or been caught during the American Revolution, they would have been executed as the criminals they were by law. However, because they won the war, they became heroes and presidents. Another example is Karl Marx. He bought into the idea of goals and means of society, just not those of capitalistic societies. Rather, he proposed socialism or communism as a means to the goal of utopia. So, there are many contexts in which rebels can be criminals, but sometimes they end up being heroes.

Merton also noted that one individual can represent more than one adaptation. Perhaps the best example is that of the Unabomber, who obviously started out as a conformist in that he was a respected professor at University of California, Berkeley, who was well on his way to tenure and promotion. He then seemed to shift to a retreatist in that he isolated himself from society. Later, he became a rebel who bombed innocent people in his quest to implement his own goals and means, which he described in his manifesto, which he coerced several national newspapers to publish. This subsequently resulted in his apprehension when his brother read it and informed authorities he thought his brother was the author.

Finally, some have applied an athletic analogy to these adaptations.[25] Assuming a basketball game is taking place, conformists will play to win, and they will always play by the rules and won't cheat. Ritualists will play the game just because they like to play; they don't care about winning. Innovators will play to win, and they will break any rules they can to triumph in the game. Retreatists don't like to play and obviously don't care about winning. Finally, rebels will not like the rules on the court, so they will try to steal the ball and go play by their own rules on another court. Although this is a somewhat simplistic analogy, it is likely to help readers remember the adaptations and perhaps enable them to apply these ways of dealing with strain to everyday situations, such as resorting to criminal activity.

Evidence and Criticisms of Merton's Strain Theory

Merton's framework, which emphasized the importance of economic structure, appeared to have a high degree of face validity during the Great Depression; however, many later scientific studies showed mixed support for strain theory. Although research that examined the effects of poverty on violence and official rates of various crimes has found relatively consistent support (albeit with weaker effects than strain theory implies), a series of studies of self-reported delinquent behavior found little or no relationship between social class and criminality.[26] Furthermore, the idea that unemployment drives people to commit crime has received little support.[27]

On the other hand, some experts have argued that Merton's strain theory is primarily a structural model of crime that is more a theory of societal groups, not individual motivations.[28] Therefore, some modern studies have used aggregated group rates (i.e., macrolevel measures) to test the effects of deprivation, as opposed to using individual (microlevel) rates of inequality and crime. Most of these studies provide some support for the hypothesis that social groups and regions with higher rates of deprivation and inequality have higher rates of criminal activity.[29] In sum, there appears to be some support for Merton's strain theory when the level of analysis is macrolevel and official measures are being used to indicate criminality.

However, many critics have claimed that these studies do not directly measure perceptions or feelings of strain, so they are only indirect examinations of Merton's theory. In light of these criticisms, some

---

[25]Vold et al., *Theoretical Criminology,* 140.

[26]For examples and reviews of this research, see F. Ivan Nye, *Family Relationships and Delinquent Behavior* (New York: Wiley, 1958); Ronald L. Akers, "Socio-economic Status and Delinquent Behavior: A Retest," *Journal of Research in Crime and Delinquency 1* (1964): 38–46; Charles R. Tittle and Wayne J. Villemez, "Social Class and Criminality," *Social Forces 56* (1977): 474–503; Charles R. Tittle, Wayne J. Villemez, and Douglas A. Smith, "The Myth of Social Class and Criminality: An Empirical Assessment of the Empirical Evidence," *American Sociological Review 43* (1978): 643–656; Michael J. Hindelang, Travis Hirschi, and Joseph C. Weis, "Correlates of Delinquency: The Illusion of Discrepancy Between Self-Report and Official Measures," *American Sociological Review 44* (1979): 995–1014; Michael J. Hindelang, *Measuring Delinquency* (Beverly Hills: Sage, 1980); Terence P. Thornberry and Margaret Farnworth, "Social Correlates of Criminal Involvement," *American Sociological Review 47* (1982): 505–517; Gregory R. Dunaway, Francis T. Cullen, Velmer S. Burton, and T. David Evans, "The Myth of Social Class and Crime Revisited: An Examination of Class and Adult Criminality," *Criminology 38* (2000): 589–632; for one of the most thorough reviews, see Charles R. Tittle and Robert F. Meier, "Specifying the SES/Delinquency Relationship," *Criminology 28* (1990): 271–299.

[27]Gary Kleck and Ted Chiricos, "Unemployment and Property Crime: A Target-Specific Assessment of Opportunity and Motivation as Mediating Factors," *Criminology 40* (2000): 649–680.

[28]Bernard, "Testing Structural Strain Theories"; Steven F. Messner, "Merton's 'Social Structure and Anomie': The Road Not Taken," *Deviant Behavior 9* (1988): 33–53; also see discussion in Burton and Cullen, "The Empirical Status."

[29]For a review of these studies, see Kenneth C. Land, Patricia L. McCall, and Lawrence E. Cohen, "Structural Covariates of Homicide Rates: Are There Any Invariances Across Time and Social Space?" *American Journal of Sociology 95* (1990): 922–963.

researchers have focused on the disparity in what individuals aspire to in various aspects of life (e.g., school, occupation, social life) versus what they realistically expect to achieve.[30] The rationale of these studies is that, if an individual has high aspirations (i.e., goals) but low expectations of actually achieving the goals due to structural barriers, then that individual is more likely to experience feelings of frustration and strain. Furthermore, it was predicted that the larger the gap between aspirations and expectations, the stronger the sense of strain. Of the studies that examined discrepancies between aspirations and expectations, most did not find evidence linking a large gap between these two levels with criminal activity. In fact, several studies found that, for most antisocial respondents, there was virtually no gap between aspirations and expectations. Rather, most of the subjects (typically young males) who reported the highest levels of criminal activity tended to report low levels of both aspirations and expectations.

Surprisingly, when aspirations were high, it seemed to inhibit offending, even when expectations to achieve those goals were low. One interpretation of these findings is that individuals who have high goals will not jeopardize their chances even if they are slim. On the other hand, individuals who don't have high goals are likely to be indifferent to their future and, in a sense, have nothing to lose. So, without a stake in conventional society, this predisposes them to crime. While this conclusion supports social control theories, it does not provide support for strain theory.

Some critics have argued that most studies on the discrepancies between aspirations and expectations have not been done correctly. For example, Farnworth and Leiber claimed that it was a mistake to examine differences between educational goals and expectations or differences between occupational goals and expectations, which is what most of these studies did.[31] Rather, they proposed testing the gap between economic aspirations (i.e., goals) and educational expectations (i.e., means of achieving the goals). This makes sense, and Farnworth and Leiber found support for a gap between these two factors and criminality. However, they also found that people who reported low economic aspirations were more likely to be delinquent, which supports the previous studies that they criticized. Another criticism of this type of strain theory study is that simply reporting a gap between expectations and aspirations may not mean that the individuals actually feel strain; rather, researchers have simply, and perhaps wrongfully, assumed that a gap between the two measures indicates feelings of frustration.[32]

Other criticisms of Merton's strain theory include some historical evidence and its failure to explain the age-crime curve. Regarding the historical evidence, it is hard to understand why some of the largest increases in crime took place during a period of relative economic prosperity, namely the late 1960s. Crime increased more than ever before (since recording crime rates began) between 1965 and 1973, which were generally good economic years in the United States. Therefore, if strain theory is presented as the primary explanation for criminal activity, it would probably have a hard time explaining this historical era. On the other hand, given the growth in the economy in the 1960s and early 1970s, this may have caused more disparity between rich and poor, thereby producing more relative deprivation.

---

[30]For examples and reviews of these types of studies, see Travis Hirschi, *Causes of Delinquency* (Berkeley: University of California Press, 1969), 4–10; Allen E. Liska, "Aspirations, Expectations, and Delinquency: Stress and Additive Models," *Sociological Quarterly 12* (1971): 99–107; Margaret Farnworth and Michael J. Leiber, "Strain Theory Revisited: Economic Goals, Educational Means, and Delinquency," *American Sociological Review 54* (1989): 263–274; Burton and Cullen, "The Empirical Status"; Velmer S. Burton, Francis T. Cullen, T. David Evans, and R. Gregory Dunaway, "Reconsidering Strain Theory: Operationalization, Rival Theories, and Adult Criminality," *Journal of Quantitative Criminology 10* (1994): 213–239; Robert F. Agnew, Francis T. Cullen, Velmer S. Burton, T. David Evans, and R. Gregory Dunaway, "A New Test of Classic Strain Theory," *Justice Quarterly 13* (1996): 681–704; also see discussion of this issue in Akers and Sellers, *Criminological Theories*, 173–175.

[31]Farnworth and Leiber, "Strain Theory."

[32]Agnew et al., "A New Test."

The other major criticism of strain theory is that it does not explain one of the most established facts in the field: the age-crime curve. In virtually every society in the world, across time and place, predatory street crimes (e.g., robbery, rape, murder, burglary, larceny, etc.) tend to peak sharply in the teenage years to early 20s and then drop off very quickly, certainly before age 30. However, most studies show that feelings of stress and frustration tend not to follow this pattern. For example, suicide rates tend to be just as high or higher as one gets older, with middle-aged and elderly people having much higher rates of suicide than those in their teens or early 20s.

On the other hand, it can be argued that crime rates go down even though strain can continue or even increase with age because individuals develop coping mechanisms for dealing with the frustrations they feel. But, even if this is true regarding criminal behavior, apparently this doesn't seem to prevent suicidal tendencies. General strain theory emphasized this concept. However, before we cover general strain theory, we will discuss two other variations of Merton's theory that were both developed between 1955 and 1960 to explain gang formation and behavior using a structural strain framework.

Variations of Merton's Strain Theory: Cohen and Cloward and Ohlin

Cohen's Theory of Lower-Class Status Frustration and Gang Formation

In 1955, Albert Cohen presented a theory of gang formation that used Merton's strain theory as a basis for why individuals resort to such group behavior.[33] In Cohen's model, young, lower-class males are at a disadvantage in school because they lack the normal interaction, socialization, and discipline instituted by educated parents of the middle class. This is in line with Merton's original framework of a predisposed disadvantage for underclass youth. According to Cohen, such youths are likely to experience failure in school because they are unprepared to conform with middle-class values and fail to meet the middle-class measuring rod, which emphasizes motivation, accountability, responsibility, deferred gratification, long-term planning, respect for authority and property, controlling emotions, and so on.

Like Merton, Cohen emphasized the youths' internalization of the American Dream and fair chances for success, so failure to be successful according to this middle-class standard is very frustrating for them. The strain that they feel as a result of failure in school performance and lack of respect among their peers is often referred to as *status frustration*. It leads them to develop a system of values opposed to middle-class standards and values. Some have claimed that this represents a Freudian defense mechanism known as **reaction formation,** which involves individuals adopting attitudes or committing behaviors that are opposite of what is expected as a form of defiance so that they will feel less guilt for not living up to the standards they cannot achieve. Specifically, instead of abiding by middle-class norms of obedience to authority, school achievement, and respect for authority, these youths change their normative beliefs to value the opposite characteristics: malicious, negativistic, and nonutilitarian delinquent activity.

Delinquent youths will begin to value destruction of property and skipping school, not because these behaviors lead to a payoff or success in the conventional world, but simply because they defy the conventional order. In other words, they turn middle-class values upside down and consider activity that violates the conventional norms and laws as good, thereby psychologically and physically rejecting the cultural system that has been imposed on them without preparation and fair distribution of resources. Furthermore, Cohen claimed that while these behaviors do not appear to have much utility or value, they are quite valuable and important from the perspective of the strained youth. Specifically, they do these acts to gain respect from their peers.

---

[33]Albert Cohen, *Delinquent Boys: The Culture of the Gang* (New York: Free Press, 1955).

Cohen stated that he believed that this tendency to reject middle-class values was the primary cause of gangs, a classic example of birds of a feather flock together. Not all lower-class males resort to crime and join a gang in response to this structural disadvantage. Other variations, beyond that of the **delinquent boy** who is described above, include the **college boy** and the **corner boy.** The college boy responds to his disadvantaged situation by dedicating himself to overcoming the odds and competing in middle-class schools despite the unlikely chances for success. The corner boy responds to the situation by accepting his place as a lower-class individual who will somewhat passively make the best of life at the bottom of the social order.

As compared to Merton's original adaptations, Cohen's delinquent boy is probably best seen as similar to a rebel because he rejects the means and goals (middle-class values and success in school) of conventional society, substituting new means and goals (negativistic behaviors and peer respect in the gang). Some would argue that delinquent boys should be seen as innovators because their goals are ultimately the same: peer respect. However, the actual peers involved completely change, so we argue that, through the reaction formation process, the delinquent boy actually creates his own goals and means that go against the conventional, middle-class goals and means. Regarding the college boy, the adaptation that seems to fit the best is that of conformity because the college boy continues to believe in the conventional goals (i.e., financial success and achievement) and means (i.e., hard work via education or labor) of middle-class society. Finally, the corner boy probably best fits the adaptation of ritualism because he knows that he likely will never achieve the goals of society but resigns himself to not obtaining financial success; at the same time, he does not resort to predatory street crime but, rather, holds a stable, blue-collar job or makes ends meet in other typically legal ways. Some corner boys end up simply collecting welfare and give up working altogether; they may actually become more like the adaptation of retreatism because they have virtually given up on the conventional means (hard work) of society, as well as the goals.

At the time that Cohen developed his theory, official statistics showed that virtually all gang violence, and most violence for that matter, was concentrated among lower-class male youth. However, with the development of self-report studies in the 1960s, Cohen's theory was shown to be somewhat overstated: Middle-class youth were well represented among those who commit delinquent acts.[34] Other studies have also been critical of Cohen's theory, particularly the portions that deal with his proposition that crime rates would increase after youths drop out of school and join gangs. Although the findings are mixed, many studies have found that delinquency is often higher before youths drop out of school and may actually decline once they drop out and become employed.[35] Some critics have pointed out that such findings discredit Cohen's theory, but this is not necessarily true. After all, delinquency may be peaking right before the youths drop out because they feel the most frustrated and strained, whereas delinquency may be decreasing after they drop out because some are raising their self-esteem by earning wages and taking pride in having jobs.

Still, studies have clearly shown that lower-class youths are far more likely to have problems in school and that school failure is consistently linked to criminality.[36] Furthermore, there is little dispute that much

---

[34]Tittle et al., "The Myth of Social Class"; Hindelang et al., "Correlates of Delinquency."

[35]See Bernard, "Testing Structural Strain Theories"; Merton, "Opportunity Structure"; Delbert Elliott and Harwin Voss, *Delinquency and Dropout* (Lexington: D. C. Heath, 1974); Terence P. Thornberry, Melanie Moore, and R. L. Christenson, "The Effect of Dropping Out of High School on Subsequent Criminal Behavior," *Criminology 23* (1985): 3–18; G. Roger Jarjoura, "Does Dropping Out of School Enhance Delinquent Involvement? Results From a Large-Scale National Probability Sample," *Criminology 31* (1993): 149–172; G. Roger Jarjoura, "The Conditional Effect of Social Class on the Dropout Delinquency Relationship," *Journal of Research in Crime and Delinquency 33* (1996): 232–255; see discussion in Donald J. Shoemaker, *Theories of Delinquency,* 5th ed. (New York: Oxford University Press, 2005).

[36]Alexander Liazos, "School, Alienation, and Delinquency," *Crime and Delinquency* 24 (1978): 355–370; Joseph W. Rogers and G. Larry Mays, *Juvenile Delinquency and Juvenile Justice* (New York: Wiley, 1987); Clarence E. Tygart, "Strain Theory and Public School Vandalism: Academic Tracking, School Social Status, and Students' Academic Achievement," *Youth and Society 20* (1988): 106–118; Hirschi, *Causes of Delinquency.*

of delinquency represents malicious, negativistic, and nonutilitarian activity. For example, what do individuals have to gain from destroying mailboxes or spraying graffiti on walls? These acts will never earn much money or any payoff other than peer respect. So, ultimately, it appears that there is some face validity to what Cohen proposed in the sense that some youths engage in behavior that has no value other than earning peer respect, even though that behavior is negativistic and nonutilitarian, according to the values of conventional society. Regardless of some criticisms of Cohen's model, he provided an important structural strain theory of the development of gangs and lower-class delinquency.

Cloward and Ohlin's Theory of Differential Opportunity

Five years after Cohen published his theory, Cloward and Ohlin presented yet another structural strain theory of gang formation and behavior.[37] Like Merton and Cohen, Cloward and Ohlin assumed in their model that all youth, including those in the lower class, are socialized to believe in the American Dream, and when individuals realize that they are blocked from conventional opportunities, they become frustrated and strained. What distinguishes Cloward and Ohlin's theory from that of the previous strain theories is that they identified three different types of gangs based on the characteristics of each neighborhood's social structure. They thought the nature of gangs varied according to the availability of illegal opportunities in the social structure. So, whereas previous strain theories focused only on lack of legal opportunities, Cloward and Ohlin's model emphasized both legal and illegal opportunities; the availability or lack of these opportunities largely determined what type of gang would form in that neighborhood, hence the name *differential opportunity theory*. Furthermore, the authors acknowledged Edwin Sutherland's influence on their theory (see Chapter 8), and it is evident in their focus on the associations made in the neighborhood. According to differential opportunity theory, the three types of gangs that form are **criminal gangs, conflict gangs,** and **retreatist gangs.**

Criminal gangs form in lower-class neighborhoods that have an organized structure of adult criminal behavior. Such neighborhoods are so organized and stable that criminal networks are often known and accepted by the conventional individuals in the area. In these neighborhoods, the adult gangsters mentor the youth and take them under their wings. This can pay off for the adult criminals, too, because youths can often be used to do the dirty work for the criminal enterprises in the neighborhood without risk of serious punishment if they are caught. The successful adult offenders supply the youths with the motives and techniques for committing crime. So, while members of criminal gangs are blocked from legal opportunities, they are offered ample opportunities in the illegal realm.

Criminal gangs tend to reflect the strong organization and stability of such neighborhoods. Therefore, criminal gangs primarily commit property or economic crimes with the goal of making a profit through illegal behavior. These crimes can range from running numbers as local bookies to fencing stolen goods to running businesses that are fronts for vice crimes (e.g., prostitution, drug trading). All of these businesses involve making a profit illegally, and there is often a system or structure in which the criminal activity takes place. Furthermore, these criminal gangs are most like the Merton adaptation of innovation because the members still want to achieve the goals of conventional society (financial success). Because of their strong organizational structure, these gangs favor members who have self-control and are good at planning over individuals who are highly impulsive or uncontrolled.

Examples of criminal gangs are seen in movies depicting highly organized neighborhoods often consisting of primarily one ethnicity, such as *The Godfather, The Godfather II, A Bronx Tale, State of Grace,*

---

[37]Richard A. Cloward and Lloyd E. Ohlin, *Delinquency and Opportunity: A Theory of Delinquent Gangs* (New York: The Free Press, 1960). For discussion and theoretical critique of the model, see Bernard, "Testing Structural Strain Theories" and Merton, "Opportunity Structure."

Sleepers, New Jack City, Clockers, Goodfellas, Better Luck Tomorrow, and many others that were partially based on real events. All of these depictions involve a highly structured hierarchy of a criminal enterprise, which is largely a manifestation of the organization of the neighborhood. Hollywood motion pictures also produce stories about older criminals taking younger males from the neighborhood under their wings and training them in the ways of criminal networks. Furthermore, virtually all ethnic groups offer examples of this type of gang or neighborhood; the list of movies above includes Italian American, Irish American, African American, and Asian American examples. Thus, criminal gangs can be found across the racial and ethnic spectrum, largely because all groups have certain neighborhoods that exhibit strong organization and stability.

Conflict gangs were another type of gang that Cloward and Ohlin identified. Conflict gangs tend to develop in neighborhoods that have weak stability and little or no organization. In fact, the neighborhood often seems to be in a state of flux with people constantly moving in and out. Because the youths in the neighborhood do not have a solid crime network or adult criminal mentors, they tend to form as relatively disorganized gangs, and they typically lack the skills and knowledge to make a profit through criminal activity. Therefore, the primary illegal activity of conflict gangs is violence, which is used to gain prominence and respect among themselves and the neighborhood. Due to the disorganized nature of the neighborhoods, as well as the gangs themselves, conflict gangs never quite achieve the respect and stability that criminal gangs typically achieve. The members of conflict gangs tend to be more impulsive and lack self-control as compared to members of criminal gangs, largely because there are no adult criminal mentors to control them.

According to Cloward and Ohlin, conflict gangs are blocked from both legitimate and illegitimate opportunities. If applying Merton's adaptations, conflict gangs would probably fit the category of rebellion, largely because none of the other categories fit well, but it can be argued that conflict gangs have rejected the goals and means of conventional society and implemented their own values, which emphasize violence. Examples of motion pictures that depict this type of breakdown in community structure and a mostly violent gang culture are *Menace to Society, Boyz in the Hood, A Clockwork Orange, Colors, The Outsiders,* and others, which all emphasize the chaos and violence that results when neighborhood and family organization is weak.

Finally, if individuals are double failures in both legitimate and illegitimate worlds, meaning that they can't achieve success in school or status in their local gangs, they join together to form retreatist gangs. Because members of retreatist gangs are no good at making a profit from crime or using violence to achieve status, the primary form of offending is usually drug usage. Like Merton's retreatist adaptation to strain, the members of retreatist gangs often want simply to escape from reality. Therefore, the primary activity of the gang when they get together is usually just to get high, which is well represented by Hollywood in such movies as *Trainspotting, Drug-Store Cowboy,* and *Panic in Needle Park.* In all of these movies, the only true goal of the gangs was getting stoned to escape from the worlds where they had failed.

There are a number of empirical studies and critiques of Cloward and Ohlin's theory, with the criticisms being similar to those of Merton's strain theory. Specifically, the critics argue that there is little evidence that gaps between what lower-class youths aspire to and what they expect to achieve produce frustration and strain, nor do such gaps appear predictive of gang membership or criminality.[38] Another criticism of Cloward

---

[38]James F. Short, "Gang Delinquency and Anomie," in *Anomie and Deviant Behavior,* ed. Marshall B. Clinard (New York: Free Press, 1964), 98–127; James F. Short, Ramon Rivera, and Ray A. Tennyson, "Perceived Opportunities, Gang Membership, and Delinquency," *American Sociological Review* 30 (1965): 56–67; James F. Short and Fred L. Strodtbeck, *Group Processes and Gang Delinquency* (Chicago: University of Chicago Press, 1965); Liska, "Aspirations"; Hirschi, *Causes of Delinquency;* see discussion in Shoemaker, *Theories of Delinquency,* 121–130.

and Ohlin's theory is the inability to find empirical evidence that supports their model of the formation of three types of gangs and their specializations in offending. While some research supports the existence of gangs that appear to specialize in certain forms of offending, many studies find that the observed specializations of gangs are not exactly the way that Cloward and Ohlin proposed.[39] Additional studies have shown that many gangs tend not to specialize but, rather, engage in a wider variety of offending behaviors.

Despite the criticisms of Cloward and Ohlin's model of gang formation, their theoretical framework inspired policy largely due to the influence their work had on Attorney General Robert Kennedy, who had read their book. In fact, Kennedy asked Lloyd Ohlin to assist in developing federal policies regarding delinquency,[40] which resulted in the Juvenile Delinquency Prevention and Control Act of 1961. Cloward and Ohlin's theory was a major influence on the Mobilization for Youth project in New York City, which along with the federal legislation, stressed creating education and work opportunities for youths. Although evaluations of this program showed little effect on reducing delinquency, it was impressive that such theorizing about lower-class male youths could have such a large impact on policy interventions.

Ultimately, the variations presented by Cohen, as well as Cloward and Ohlin, provided additional revisions that seemed at the time to advance the validity of strain theory. However, most of these revisions were based on official statistics that showed lower-class male youth committed most crime, which was later shown by self-reports to be exaggerated.[41] Once scholars realized that most of the earlier models were not empirically valid for most criminal activity, strain theory became unpopular for several decades. But during the 1980s, another version was devised by Robert Agnew, who rejuvenated interest in strain theory by devising a way to make it more general and applicable to a larger variety of crimes and forms of deviance.

General Strain Theory

In the 1980s, Robert Agnew proposed *general strain theory,* which covers a much larger range of behavior by not concentrating on simply the lower class and which provides a more applicable model for the frustrations that all individuals feel in everyday life.[42] Unlike other strain theories, general strain theory does not rely on assumptions about the frustration arising when people realize that the American Dream is a false promise to those of the lower classes. Rather, this theoretical framework assumes that people of all social classes and economic positions deal with frustrations in routine daily life.

Previous strain theories, such as the models proposed by Merton, Cohen, and Cloward and Ohlin focused on individuals' failure to achieve positively valued goals that they had been socialized to work to obtain. General strain theory also focuses on this source of strain; however, it identifies two additional categories of strain: *presentation of noxious stimuli* and *removal of positively valued stimuli.* In addition to the

---

[39]Irving Spergel, *Racketville, Slumtown, and Haulburg: An Exploratory Study of Delinquent Subcultures* (Chicago: University of Chicago Press, 1964); Short and Strodtbeck, *Group Processes;* Paul E. Tracy, Marvin E. Wolfgang, and Robert M. Figlio, *Delinquency Careers in Two Birth Cohorts* (New York: Plenum, 1990).

[40]Lamar T. Empey, *American Delinquency,* 4th ed. (Homewood: Dorsey, 1991).

[41]Tittle et al., "The Myth of Social Class"; Hindelang et al., "Correlates of Delinquency."

[42]For Agnew's works regarding this theory, see Robert Agnew, "A Revised Strain Theory of Delinquency," *Social Forces 64* (1985): 151–167; Robert Agnew, "Foundation for a General Strain Theory of Crime and Delinquency," *Criminology 30* (1992): 47–88; Robert Agnew and Helene Raskin White, "An Empirical Test of General Strain Theory," *Criminology 30* (1992): 475–500; Robert Agnew, "Controlling Delinquency: Recommendations From General Strain Theory," in *Crime and Public Policy: Putting Theory to Work,* ed. Hugh Barlow (Boulder: Westview Press, 1995), 43–70; Agnew et al., "A New Test," 681–704; Robert Agnew, "Building on the Foundation of General Strain Theory: Specifying the Types of Strain Most Likely to Lead to Crime and Delinquency," *Journal of Research in Crime and Delinquency 38* (2001): 319–361.

failure to achieve one's goals, Agnew claimed that the presentation of noxious stimuli (i.e., bad things) in one's life could cause major stress and frustration. Examples of noxious stimuli would include things like an abusive parent, a critical teacher, or an overly demanding boss. These are just some of the many negative factors that can exist in one's life—the number of examples is endless.

The other strain category Agnew identified was the removal of positive stimuli, which is likely the largest cause of frustration. Examples of removal of positively valued stimuli include the loss of a good job, loss of the use of a car for a period of time, or the loss of a loved one(s). As with the other two sources of strain, a number of examples may have varying degrees of influence depending on the individual. One person may not feel much frustration in losing a job or divorcing a spouse, whereas another person may experience severe anxiety or depression from such events.

Ultimately, general strain theory proposes that these three categories of strain (failure to achieve goals, noxious stimuli, and removal of positive stimuli) will lead to stress and that this results in a propensity to feel anger. Anger can be seen as a primary mediating factor in the causal model of the general strain framework. It is predicted that, to the extent that the three sources of strain cause feelings of anger in an individual, he or she will be predisposed to commit crime and deviance. However, Agnew was clear in stating that, if an individual can somehow cope with this anger in a positive way, then such feelings do not necessarily have to result in criminal activity. These coping mechanisms vary widely across individuals, with different strategies working for some people better than others. For example, some people relieve stress by working out or running, whereas others watch television or a movie. One type of activity that has shown relatively consistent success in relieving stress is laughter, which psychologists are now prescribing as a release of stress. Another is yoga, which includes simple breathing techniques, such as taking several deep breaths, which are shown to physiologically enhance release of stress.

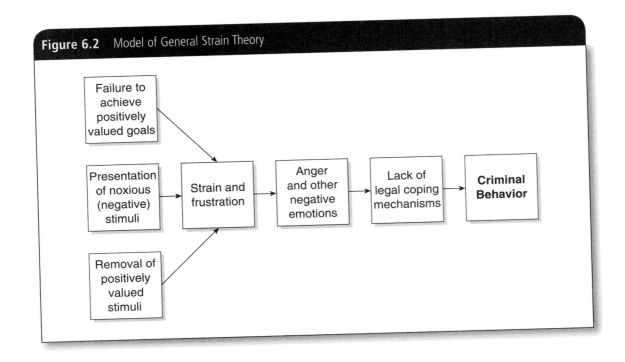

Figure 6.2 Model of General Strain Theory

Although he did not originally provide details on how coping mechanisms work or explore the extant psychological research on these strategies, Agnew specifically pointed to such mechanisms in dealing with anger in prosocial ways. The primary prediction regarding coping mechanisms is that individuals who find ways to deal with their stress and anger positively will no longer be predisposed to commit crime, whereas individuals who do not find healthy, positive outlets for their anger and frustrations will be far more likely to commit crime. Obviously, the goal is to reduce the use of antisocial and negative coping to strain, such as drug usage or aggression, which are either criminal in themselves or increase the likelihood of offending.

Recent research and theoretical development has more fully examined various coping mechanisms and their effectiveness in reducing anger and, thus, preventing criminal activity. Obviously, in focusing on individuals' perceptions of stress and anger, as well as their personal abilities to cope with such feelings, general strain theory places more emphasis on the microlevel of analysis. Still, due to its origins deriving from structural strain theory, it is included in this section and is typically classified as belonging to the category of strain theories that includes the earlier, more macrolevel-oriented theories. In addition, recent studies and revisions of the theory have attempted to examine the validity of general strain theory propositions at the macro-, structural-level.[43]

Since it was first proposed in the mid-1980s, there has been a vast amount of research examining various aspects of the general strain theory.[44] For the most part, studies have generally supported the model. Most studies find a link between the three categories of strain and higher rates of criminality, as well as a link between the sources of strain and feelings of anger or other negative emotions (e.g., anxiety, depression).[45] However, the theory and especially the way it has been tested have also been criticized.

It is important for strain research to measure subjects' perceptions and feelings of frustration, not simply the occurrence of certain events themselves. Unfortunately, some studies have only looked at the latter, and the validity of such findings is questionable.[46] Other studies, however, have directly measured subjective perceptions of frustration, as well as personal feelings of anger.[47]

---

[43]Timothy Brezina, Alex Piquero, and Paul Mazerolle, "Student Anger and Aggressive Behavior in School: An Initial Test of Agnew's Macro-Level Strain Theory," *Journal of Research in Crime and Delinquency* 38 (2001): 362–386.

[44]Agnew and White, "An Empirical Test"; Agnew et al., "A New Test"; Raymond Paternoster and Paul Mazerolle, "General Strain Theory and Delinquency: A Replication and Extension," *Journal of Research in Crime and Delinquency* 31 (1994): 235–263; Timothy Brezina, "Adapting to Strain: An Examination of Delinquent Coping Responses," *Criminology* 34 (1996): 39–60; John P. Hoffman and Alan S. Miller, "A Latent Variable Analysis of General Strain Theory," *Journal of Quantitative Criminology* 14 (1998): 83–110; John P. Hoffman and Felicia Gray Cerbone, "Stressful Life Events and Delinquency Escalation in Early Adolescence," *Criminology* 37 (1999): 343–374; Paul Mazerolle and Alex Piquero, "Linking Exposure to Strain With Anger: An Investigation of Deviant Adaptations," *Journal of Criminal Justice* 26 (1998): 195–211; Paul Mazerolle, "Gender, General Strain, and Delinquency: An Empirical Examination," *Justice Quarterly* 15 (1998): 65–91; John P. Hoffman and S. Susan Su, "Stressful Life Events and Adolescent Substance Use and Depression: Conditional and Gender Differentiated Effects," *Substance Use and Misuse* 33 (1998): 2219–2262; Lisa M. Broidy, "A Test of General Strain Theory," *Criminology* 39 (2001): 9–36; Nicole Leeper Piquero and Miriam Sealock, "Generalizing General Strain Theory: An Examination of an Offending Population," *Justice Quarterly* 17 (2000): 449–484; Paul Mazerolle and Jeff Maahs, "General Strain and Delinquency: An Alternative Examination of Conditioning Influences," *Justice Quarterly* 17 (2000): 753–778; Paul Mazerolle, Velmer Burton, Francis Cullen, T. David Evans, and Gary Payne, "Strain, Anger, and Delinquent Adaptations: Specifying General Strain Theory," *Journal of Criminal Justice* 28 (2000): 89–101; Stephen W. Baron and Timothy F. Hartnagel, "Street Youth and Labor Market Strain," *Journal of Criminal Justice* 30 (2002): 519–533; Carter Hay, "Family Strain, Gender, and Delinquency," *Sociological Perspectives* 46 (2003): 107–135; Sung Joon Jang and Byron R. Johnson, "Strain, Negative Emotions, and Deviant Coping Among African Americans: A Test of General Strain Theory," *Journal of Quantitative Criminology* 19 (2003): 79–105; Paul Mazerolle, Alex Piquero, and George E. Capowich, "Examining the Links Between Strain, Situational and Dispositional Anger, and Crime: Further Specifying and Testing General Strain Theory," *Youth and Society* 35 (2003): 131–157; Stephen W. Baron, "General Strain, Street Youth and Crime: A Test of Agnew's Revised Theory," *Criminology* 42 (2004): 457–484.

[45]A recent review of this research can be found in Baron, "General Strain, Street Youth" 457–467.

[46]For examples, see Paternoster and Mazerolle, "General Strain Theory"; Hoffman and Cerbone, "Stressful Life Events and Delinquency"; Hoffman and Su, "Stressful Life Events and Adolescent."

[47]Broidy, "A Test"; Baron, "General Strain, Street Youth."

Such studies have found mixed support for the hypothesis that certain events lead to anger[48] but less support for the prediction that anger leads to criminality, and this link is particularly weak for nonviolent offending.[49] On the other hand, the most recent studies have found support for the links between strain and anger, as well as anger and criminal behavior, particularly when coping variables are considered.[50] Still, many of the studies that examine the effects of anger use time-stable, trait measures as opposed to incident-specific, state measures that would be more consistent with the situation-specific emphasis of general strain theory.[51] This is similar to the methodological criticism that has been leveled against studies of self-conscious emotions, particularly shame and guilt; when it comes to measuring emotions such as anger and shame, criminologists should choose their measures carefully and make sure the instruments are consistent with the theory they are testing. Thus, future research on general strain theory should employ more effective, subjective measures of straining events and situational states of anger.

Regardless of the criticisms of general strain theory, it is hard to deny its face validity. After all, virtually everyone can relate to reacting differently to similar situations based on what type of day they are having. For example, we all have days in which everything seems to be going great—it's a Friday, you receive accolades at work, and you are looking forward to a nice weekend with your friends or family. If someone says something derogatory to you or cuts you off in traffic, being in such a good mood, you are probably inclined to let it go. On the other hand, we also all have days in which everything seems to be going horribly—it's Monday, you get blamed for mishaps at work, and you have a fight with your spouse or significant other. At this time, if someone yells at you or cuts you off in traffic, you may be more inclined to respond aggressively in some way. Or perhaps more commonly, you will overreact and snap at a loved one or friend when they really didn't do much to deserve it; this is often a form of displacement in which a cumulative buildup of stressors results in taking it out on another individual(s). In many ways, this supports general strain theory, and therefore, it is very prevalent and easy to see in everyday life.

Summary of Strain Theories

The common assumption that is found across all variations of strain theory is that crime is far more common among individuals who are under a great degree of stress and frustration, especially those who can't cope with stress in a positive way. The origin of most variations of strain theory can be traced to Durkheim's and Merton's concept of anomie, which essentially means a state of chaos, or normlessness, in society due to a breakdown in the ability of societal institutions to regulate human desires, thereby resulting in feelings of strain.

Although different types of strain theories were proposed and gained popularity at various periods of time throughout the 20th century, they all became accepted during eras that were politically and culturally conducive to such perspectives, especially regarding the differences across the strain models. For example, Merton's formulation of strain in the 1930s emphasized the importance of the economic institution, which was developed and became very popular during the Great Depression. Then, in the late 1950s, two strain theories that focused on gang formation were developed, by Cohen and by Cloward and Ohlin; they became popular among politicians and society due to the focus on official statistics suggesting that most crime at that time was being committed by lower-class, inner-city male youths, many of whom were gang members.

---

[48]See Brezina, "Adapting to Strain"; Broidy, "A Test"; Mazerolle and Piquero, "Linking Exposure."

[49]Mazerolle and Piquero, "Linking Exposure"; Piquero and Sealock, "Generalizing General Strain Theory"; Mazerolle et al., "Strain, Anger."

[50]Baron, "General Strain, Street Youth"; Mazerolle et al., "Examining the Links."

[51]For examples, see Mazerolle et al., "Strain, Anger"; Baron, "General Strain, Street Youth"; see discussion in Mazerolle et al., "Examining the Links"; Akers and Sellers, *Criminological Theories*, 180–182.

Finally, Agnew developed his general strain model in the mid- to late 1980s, during a period in which a number of general theories of crime were being developed (e.g., Gottfredson and Hirschi's low self-control theory and Sampson and Laub's developmental theory), so such models were popular at that time, particularly those that emphasized personality traits (such as anger) and experiences of individuals. So, all of the variations of strain, like all of the theories discussed in this book, were manifestations of the period in which they were developed and became widely accepted by academics, politicians, and society.

Policy Implications

Although this chapter dealt with a wide range of theories regarding social structure, the most applicable policy implications are those suggested by the most recent theoretical models of this genre. Thus, we will focus on the key policy factors in the most modern versions of this perspective. The factors that are most vital for policy implications regarding social structure theories are those regarding educational and vocational opportunities and programs that develop healthy coping mechanisms to deal with stress.

Empirical studies have shown that intervention programs are needed for high-risk youths that focus on educational or vocational training and opportunities because developing motivation for such endeavors can have a significant impact on reducing their offending rates.[52] Providing an individual with a job, or the preparation for one, is key to building a more stable life, even if it is not a high-paying position. As a result, the individual is less likely to feel stressed or strained. In modern times, a person is lucky to have a stable job, and this must be communicated to our youths, and hopefully, they will find some intrinsic value in the work they do.

Another key area of recommendations from this perspective involves developing healthy coping mechanisms. Everyone deals with stress virtually every day. The key is not avoiding stress or strain because that is inevitable. Rather, the key is to develop healthy, legal ways to cope with such strain. Many programs have been created to train individuals on how to handle stress without resorting to antisocial behavior. There has been some success in anger management programs, particularly the ones that take a cognitive-behavioral approach and often involve role-playing to teach individuals to think before they act.[53]

Conclusion

This chapter examined the theories that emphasize inequitable social structure as the primary cause of crime. We examined early perspectives, which established that societies vary in the extent to which they are stratified, and looked at the consequences of inequalities and complexities of such structures. Our examination of strain theories explored theoretical models proposing that individuals and groups who are not offered equal opportunities for achieving success can develop stress and frustration and, in turn, dispositions for committing criminal behavior. Finally, we examined the policy recommendations suggested by this theoretical model, which included the need to provide individuals with educational and job opportunities and help them develop healthy coping mechanisms.

---

[52]Wright et al., *Criminals in the Making* (see chap. 5, n. 32), 204–207.

[53]Patricia Van Voorhis, Michael Braswell, and David Lester, *Correctional Counseling and Rehabilitation,* 3rd ed. (Cincinnati: Anderson, 1997).

Chapter Summary

- First, we discussed the primary distinction between social structure theories and other types of explanations (e.g., biological or social process theories).
- We examined the importance of the early sociological positivists, particularly Guerry, Quetelet, and Durkheim, and their contributions to the study of deviance and crime.
- We explored reasons why strain theory was developed and became popular in its time and discussed the primary assumptions and propositions of Merton's strain theory.
- We identified, defined, and examined examples of all five adaptations to strain.
- We discussed the variations of strain theory that were presented by Cohen, Cloward and Ohlin, and Agnew, as well as the empirical support that has been found regarding each.

KEY TERMS

Adaptations to strain

Anomie

Collective conscience

College boy

Conflict gangs

Conformity

Corner boy

Criminal gangs

Delinquent boy

General strain theory

Innovation

Mechanical societies

Organic societies

Reaction formation

Rebellion

Relative deprivation

Retreatism

Retreatist gangs

Ritualism

Social dynamics

Social statics

Strain theory

DISCUSSION QUESTIONS

1. How does sociological positivism differ from biological or psychological positivism?

2. Which of the early sociological positivism theorists do you think contributed the most to the evolution of social structure theories of crime? Why? Do you think their ideas still hold up today?

3. Can you think of modern examples of Durkheim's image of mechanical societies? Do you think such societies have more or less crime than modern organic societies?

4. What type of adaptation to strain do you think you fit the best? The least? What adaptation do you think best fits your professor? Your postal delivery worker? Your garbage collector?

5. Do you know school friends who fit Cohen's model of status frustration? What did they do in response to the feelings of strain?

6. How would you describe the neighborhood where you or others you know grew up in terms of Cloward and Ohlin's model of organization or disorganization? Can you relate to the types of gangs they discussed?

WEB RESOURCES

Emile Durkheim

http://www.emile-durkheim.com

Strain Theory

http://www.umsl.edu/~keelr/200/strain.html

http://www.criminology.fsu.edu/crimtheory/agnew.htm

http://www.apsu.edu/oconnort/crim/crimtheory11.htm

CHAPTER

7

The Chicago School and Cultural and Subcultural Theories of Crime

This chapter will examine the origin and evolution of the Chicago or **Ecological School** theory, otherwise known as the ecological perspective or the theory of **social disorganization.** We will also discuss modern research on this theory, which assumes that the environments people live in determine their behavior. Finally, the assumptions and dynamics of **cultural and subcultural theory** in society will be discussed, with an emphasis on differences in certain models emphasizing inner-city subcultures and other modern examples (e.g., street gangs). We will finish by reviewing the policy implications that have been suggested by this perspective of crime.

The Chicago School evolved there because the city, during the late 19th and early 20th centuries, desperately needed answers for its exponentially growing problem of delinquency and crime. Thus, this became a primary focus in the city of Chicago, where total chaos prevailed at the time.

A significant portion of the Chicago perspective involved the transmission of cultural values to other peers, and even across generations, as the older youths relayed their antisocial values and techniques to the younger children. Thus, the cultural and subcultural perspective is also a key area of this theoretical model. This cultural aspect of the Chicago model is also examined in this chapter, as well as other subculture frameworks of offending behaviors.

The Ecological School and the Chicago School of Criminology

Despite its name specifying one city, the **Chicago School of criminology** represents one of the most valid and generalizable theories we will discuss in this book in the sense that many of its propositions can be readily applied to the growth and evolution of virtually all cities around the world. The Chicago School, which is often referred to as the Ecological School or the theory of social disorganization, also represents one of the earliest examples of balancing theorizing with scientific analysis and at the same time guiding important programs and policy implementations that still thrive today. Perhaps most important, the Chicago School of criminology was the epitome of using theoretical development and scientific testing to help improve conditions in society when it was most needed, which can be appreciated only by understanding the degree of chaos and crime that existed in Chicago in the late 1800s and early 1900s.

Cultural Context: Chicago in the 1800s and Early 1900s

Experts have determined that 19th-century Chicago was the fastest-growing city in U.S. history.[1] Census data show the population went from about 5,000 in the early 1800s to more than 2 million by 1900; put another way, the population more than doubled every decade during the 19th century.[2] This massive rate of growth— much faster than that seen in other large U.S. cities such as Boston, Baltimore, New York, Philadelphia, and San Francisco—was due to Chicago's central geographic position. It was in many ways land-locked because, although it sits on Lake Michigan, there was no water route to the city from the Atlantic Ocean until the Erie Canal opened in 1825, which provided access to the Great Lakes region for shipping and migration of people. Three years later came the first U.S. passenger train, the Baltimore & Ohio railroad, with a route from a mid-Atlantic city to central areas. These two transportation advancements created a continuous stream of migration to the Chicago area, which increased again when the transcontinental railroad was completed in 1869, linking both coasts with the U.S. Midwest.[3]

It is important to keep in mind that, in the early to mid-1800s, many large U.S. cities had virtually no formal social agencies to handle the problems of urbanization—no social workers, building inspectors, garbage collectors, or even police officers. Once police agencies were introduced, their duties often included finding lost children and collecting the garbage, primarily because there weren't other agencies to perform these tasks. Therefore, communities were largely responsible for solving their own problems, including crime and delinquency. By the late 1800s, however, Chicago was largely made up of citizens who did not speak a common language and did not share each other's cultural values. This phenomenon is consistent with Census Bureau data from that era, which show that 70% of Chicago residents were foreign born and another 20% were first-generation Americans. Thus, it was almost impossible for these citizens to organize themselves to solve community problems because, in most cases, they could not even understand each other.

This resulted in the type of chaos and normlessness that Durkheim predicted would occur when urbanization and industrialization occurred too rapidly; in fact, Chicago represented the archetypal example of a society in an anomic state, with almost a complete breakdown in control. One of the most notable manifestations of this breakdown in social control was that children were running wild on the streets in gangs, with adults making little attempt to intervene. So, delinquency was soaring, and it appeared that the gangs controlled the streets as much as any other group.

The leaders and people of Chicago needed theoretical guidance to develop solutions to their problems, particularly regarding the high rates of delinquency. This was a key factor in why the Department of

---

[1]For an excellent discussion of the early history of Chicago, see Thomas J. Bernard, *The Cycle of Juvenile Justice* (New York: Oxford University Press, 1992).

[2]See discussion in Vold et al., *Theoretical Criminology,* 117–22 (see chap. 1, n. 4).

[3]These dates were taken from *The World Almanac, 2000,* Millennium Collector's Edition (Mahwah: Primedia Reference, 2000).

Sociology at the University of Chicago became so important and dominant in the early 1900s. Essentially, modern sociology developed in Chicago because the city needed it the most to solve its social problems. Thus, Chicago became a type of laboratory for sociological researchers, and they developed a number of theoretical models of crime and other social ills that are still shown to be empirically valid today.

Ecological Principles in City Growth and Concentric Circles

In the 1920s and 1930s, several new perspectives of human behavior and city growth were offered by sociologists at the University of Chicago. The first relevant model was proposed by Robert E. Park, who claimed that much of human behavior, especially the way cities grow, follows the basic principles of ecology that had been documented and applied to wildlife for many years at that point.[4] Ecology is essentially the study of the dynamics and processes through which plants and animals interact with the environment. In an application of Darwinian theory, Park proposed that the growth of cities follows a natural pattern and evolution.

Specifically, Park claimed that cities represent a type of complex organism with a sense of unity composed of the interrelations among its citizens and groups. Park applied the ecological principle of symbiosis to explain the dependency of various citizens and units on each other: Everyone is better off working together as a whole. Furthermore, Park claimed that all cities would contain identifiable clusters, which he called **natural areas,** where each cluster had taken on a life or organic unity by itself. To clarify, many cities have neighborhoods that are made up of primarily one ethnic group or are distinguished by certain features. For example, New York City's Hell's Kitchen, Times Square, and Harlem represent areas of one city that have each taken on unique identities; however, each of them contributes to the whole makeup and identity of the city. The same can be seen in other cities, such as Baltimore, which in a two-mile area has the Inner Harbor, Little Italy, and Fell's Point, with each area complementing the other zones. From Miami to San Francisco to New Orleans, all cities across America, and throughout the world for that matter, contain these identifiable natural areas.

Applying several other ecological principles, Park also noted that some areas (or species) may invade and dominate adjacent areas (species). The dominated area or species can recede, migrate to another location, or die off. In wildlife, an example is the incredible proliferation of a plant called kudzu, which grows at an amazing pace and has large leaves. It grows on top of other plants, trees, fields, and even houses, seeking to cover everything in its path and steal all of the sunlight needed by other plants. Introduced to the United States in the 1800s at a world exposition, this plant was originally used to stop erosion but got out of control, especially in the southeastern region of the United States. Now, kudzu costs the government more than $350 million each year because of its destruction of crops and other flora. This is a good example of a species that invades, dominates, and causes the recession of other species in the area.

A similar example can be found in the introduction of bison on Santa Catalina Island off the Southern California coast in the 1930s. About three dozen buffalo were originally imported to the island for a movie shoot, and the producers decided not to spend the money to remove them after the project, so they have remained and multiplied. Had this occurred in other parts of the United States, it would not have caused a problem. However, the largest mammal native to the island before the bison was a 4-pound fox. So, the buffalo—now numbering in the hundreds, to the point where several hundred were recently shipped to their native Western habitat—have destroyed much of the environment, driving to extinction some plants

---

[4]Robert E. Park, "Human Ecology," *American Journal of Sociology* 42 (1936): 158–164; Robert E. Park, *Human Communities* (Glencoe: The Free Press, 1952).

and animals unique to Catalina Island. Like the kudzu, the bison came to dominate the environment; in this case, other species couldn't move off the island and died off.

Park claimed that a similar process occurs in human cities as some areas invade other zones or areas, and the previously dominant area must relocate or die off. This is easy to see in modern times with the growth of what is known as *urban sprawl*. Geographers and urban planners have long acknowledged the detriment to traditionally stable residential areas when businesses move in. Some of the most recent examples involve the battles between long-time homeowners against the introduction of malls, businesses, and other industrial centers in previously zoned residential districts. The media have documented such fights, especially with the proliferation of such establishments as Wal-Mart or Super K-Marts, which residents perceive, and perhaps rightfully so, as an invasion. Such an invasion can create chaos in a previously stable residential community due to increased traffic, transient populations, and perhaps most important, crime. Furthermore, some cities are granting power to such development through eminent domain, in which the local government can take land from homeowners to rezone and import businesses.

When Park developed his theory of ecology, he observed the trend of businesses and factories invading the traditionally residential areas of Chicago, which caused major chaos and breakdown in the stability of those areas. Readers, especially those who were raised in suburban or rural areas, can likely relate to this; going back to where they grew up, they can often see fast growth. Such development can devastate the informal controls (such as neighborhood networks or family ties, etc.) as a result of invasion by a highly transient group of consumers and residents who do not have strong ties to the area.

This leads to a psychological indifference toward the neighborhood in which no one cares about protecting the community any longer. Those who can afford to leave the area do, and those who can't afford to get out remain until they can save enough money to do so. When Park presented his theory of ecology in the 1920s, factories moving into the neighborhood often meant having a lot of smoke billowing out of chimneys. No one wanted to live in such a place, particularly at a time when the effects of pollution were not understood and smokestacks had no filters. Certain parts of Chicago and other U.S. cities were perpetually covered by the smog these factories created. In highly industrial areas, the constant and vast coverage of smoke and pollutants made it seem to be snowing or overcast most of the time. It is easy to see how such invasions can completely disrupt the previously dominant and stable residential areas of a community.

Park's ideas became even more valid and influential with the complementary perspective offered by Ernest W. Burgess.[5] He proposed a theory of city growth in which cities were seen as growing not simply on the edges but from the inside outward. It is easy to observe cities growing on the edges, as in the example of urban sprawl, but Burgess claimed that the source of growth was in the cities' centers. Growth of the inner city puts pressure on adjacent zones, which in turn begin to grow into the adjacent zones, following the ecological principle of *succession* identified by Park. This type of development is referred to as radial growth, meaning beginning on the inside and rippling outward.[6]

An example of this can be seen by watching a drop of water fall into the center of a bucket filled with water. The waves from the impact will form circles that ripple outward. This is exactly how Burgess claimed that cities grow. Although the growth of cities is most visible on the edges, largely due to development of business and homes where only trees or barren land existed before, the reason for growth there is due to pressure forming at the very heart of the city. Another good analogy is the domino effect because pressure from the center leads to growing pressure on the next zone, which leads to pressure on the adjacent zones, and so forth.

---

[5]Ernest W. Burgess, "The Growth of the City," in *The City,* eds. Robert Park, Ernest W. Burgess, and Roderick D. McKenzie (Chicago: University of Chicago Press, 1928).

[6]See Vold et al., *Theoretical Criminology,* 118–21.

Burgess also specified the primary zones—five pseudodistinctive natural areas in a constant state of flux due to growth—that all cities appear to have. He depicted these zones as a set of concentric circles. The first innermost circle was called Zone I, or the central business district due to costs. This area of a city contains the large business buildings, modern skyscrapers that are home to banking, chambers of commerce, courthouses, and other essential business and political centers such as police headquarters, post offices, and so on.

Just outside the business district is the unnumbered factory zone. It is perhaps the most significant in terms of causing crime because it invaded the previously stable residential zones in Zone II, which was identified as the **zone in transition.** Zone II is appropriately named because it was truly in a state of transition from residential to industrial, primarily because this was the area of the city in which businesses and factories were invading residential areas. Zone II was the area that was most significantly subjected to the ecological principles Park suggested: invasion, domination, recession, and succession. Subsequent criminological theorists focused on this zone.

According to Burgess's theory of concentric circles, Zone III was the workingmen's homes, relatively modest houses and apartment buildings; Zone IV consisted of higher-priced family dwellings and more expensive apartments; and Zone V was the suburban or commuter zone. These outer three zones Burgess identified were of less importance in terms of crime, primarily because, as a general rule, the farther a family could move out of the city, the better the neighborhood was in terms of social organization and the lower the rate of social ills (e.g., poverty, delinquency). The important point of this theory of concentric circles is that the growth of each inner zone puts pressure on the next zone to grow and push into the next adjacent zone.

It is easy for readers to see examples of **concentric circles theory.** Wherever you live in the United States, any major city provides real-life evidence of the validity of this perspective. For example, whether people drive on Interstate 95 through Baltimore or Interstate 5 through Los Angeles, they will see the same pattern of city structure. As they approach each of the cities, they see suburban wealth in the homes and buildings, often hidden by trees off the highway. Closer to the cities, they see homes and buildings deteriorate in terms of value. Because parts of the highway systems near Baltimore and Los Angeles are somewhat elevated, drivers entering Zone II can easily see the prevalence of factories and the highly deteriorated nature of the areas. Today, many 20th-century factories have been abandoned or are limited in use; these factory zones consist of rusted-out or demolished buildings. Zone II is also often the location of subsidized or public housing. Only the people who can't afford to live anywhere else are forced to live in these neighborhoods. Finally, as drivers enter the inner city of skyscrapers, the conditions improve dramatically because the major businesses have invested their money there. Compared to Zone II, this innermost area is a utopia.

This theory applies around the world, and we challenge readers to find any major city throughout the world that did not develop this way. Nowadays, some attempts have been made to plan community development, and others have experienced the convergence of several patterns of concentric circles as central business districts (i.e., Zone Is) are developed in what was previously suburb (i.e., Zone Vs). However, for the most part, the theoretical framework of concentric circles still has a great deal of support. In fact, even cities found in Eastern cultures have evolved this way. Therefore, Park's application appears to be correct: Cities grow in a natural way across time and place, abiding by the natural principles of ecology.

Shaw and McKay's Theory of Social Disorganization

Clifford Shaw and Henry McKay drew heavily on their colleagues at the University of Chicago in devising their theory of social disorganization, which became known as the Chicago School theory of criminology.[7]

---

[7]Clifford Shaw and Henry D. McKay, *Juvenile Delinquency and Urban Areas* (Chicago: University of Chicago Press, 1942); Clifford Shaw and Henry D. McKay, *Juvenile Delinquency and Urban Areas,* rev. ed. (Chicago: University of Chicago Press, 1969).

Shaw had been producing excellent case studies for years on the individual (i.e., micro) level before he took on theorizing on the macro (i.e., structural) level of crime rates.[8] However, once he began working with McKay, he devised perhaps the most enduring and valid model of why certain neighborhoods have more social problems, such as delinquency, than others.

In this model, Shaw and McKay proposed a framework that began with the assumption that certain neighborhoods in all cities have more crime than other parts of the city, most of them located in Burgess's Zone II, which is the zone in transition from residential to industrial due to the invasion of factories. According to Shaw and McKay, the neighborhoods that have the highest rates of crime typically have at least three common problems (see Figure 7.1): physical dilapidation, poverty, and heterogeneity (which is a fancy way of saying a high cultural mix). There were other common characteristics that Shaw and McKay noted, such as a highly transient population, meaning that people constantly move in and out of the area, as well as unemployment among the residents of the neighborhood.

As noted in Figure 7.1, other social ills are included as antecedent factors in the theoretical model. The antecedent social ills tend to lead to a breakdown in social organization, which is why this model is referred to as the theory of social disorganization. Specifically, it is predicted that the antecedent factors of poverty, heterogeneity, and physical dilapidation lead to a state of social disorganization, which in turn leads to crime and delinquency. This means that in a neighborhood that fits the profile of having a high rate of poor, culturally mixed residents in a dilapidated area, they cannot come together to solve problems, such as delinquency among youth.

One of the most significant contributions of Shaw and McKay was that they demonstrated the prevalence and frequency of various social ills—such as poverty, disease, and low birth weights—tend to overlap with higher delinquency rates. Regardless of what social problem is measured, higher rates are almost always clustered in the zone in transition. Shaw and McKay believed there is a breakdown of informal social controls in these areas and that children begin to learn offending norms from their interactions with peers on the street, through what they call *play activities*.[9] Thus, the breakdown in the conditions of the neighborhood leads to social disorganization, which in turn leads to delinquents learning criminal activities from older youths. Ultimately, the failure of the neighborhood residents to organize themselves allows the older youths to govern the behavior of the younger children. Basically, the older youths in the area provide a system of organization where the neighborhood adults cannot, so younger children follow them.

One of the best things about Shaw and McKay's theoretical model is that the researchers support their propositions with data from U.S. census and city records showing that neighborhoods with high rates of poverty and physical dilapidation and high cultural mixes also had the highest rates of delinquency and crime. Furthermore, the high rates of delinquency and other social problems were consistent with Burgess's framework of concentric circles in that the highest rates were observed for the areas that were in Zone II, the zone in transition. There was one exception to the model: The Gold Coast area along the northern coast of Lake Michigan was notably absent from the high rates of social problems, particularly delinquency, even though it was geographically in Zone II according to the otherwise consistent model of concentric circles and neighborhood zones.

---

[8]Clifford Shaw, *Brothers in Crime* (Chicago: University of Chicago Press, 1938); Clifford Shaw, *The Jackroller* (Chicago: University of Chicago Press, 1930); Clifford Shaw, *The Natural History of a Delinquent Career* (Chicago: University of Chicago Press, 1931).

[9]Shaw, *Brothers in Crime*, 354–55.

Figure 7.1   Model of Shaw and McKay's Theory of Social Disorganization

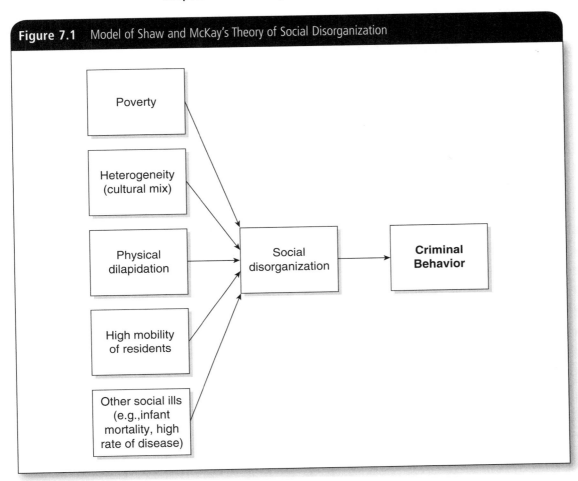

So, the findings of Shaw and McKay were as predicted in the sense that high delinquency rates occurred in areas where factories were invading residential districts. Furthermore, Shaw and McKay's longitudinal data showed that it did not matter which ethnic groups lived in Zone II (zone in transition); all groups (with the exception of Asians) that lived in that zone had high delinquency rates during their residency. On the other hand, once most of an ethnic group had moved out of Zone II, the delinquency rate among its youths decreased significantly.

This finding rejects the notion of **social Darwinism** because it is clearly not the culture that influences crime and delinquency but rather the criminogenic nature of the environment. If ethnicity or race made a difference, the delinquency rates in Zone II would fluctuate based on who lived there, but the rates continued to be high from one group to the next. Rather, the zone determined the rates of delinquency.

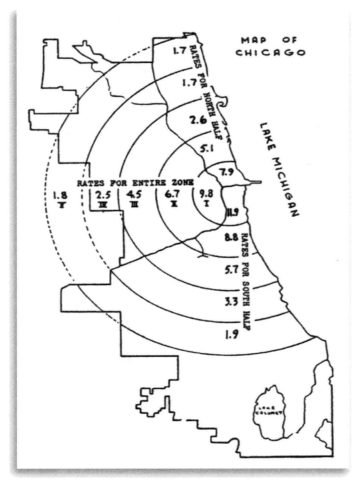

▲ **Image 7.1**   Zone Map of Male Delinquents in Chicago 1925–1933

Reaction and Research on Social Disorganization Theory

Over the last few decades, the Chicago School theoretical framework has received an enormous amount of attention from researchers.[10] Virtually all of the research has supported Shaw and McKay's version of social disorganization and the resulting high crime rates in neighborhoods that exhibit such deprived conditions. Modern research has supported the theoretical model proposed by Shaw and McKay, specifically in terms of the high crime rates in disorganized neighborhoods. Also, virtually every city that has an elevated highway (such as Richmond, Virginia; Baltimore, Maryland; Los Angeles, California) visually supports Shaw and McKay's model of crime in concentric circles. Drivers entering those cities can see the pattern of dilapidated structures in the zone of transition surrounding the inner-city area. Before and after this layer of dilapidated structures, drivers encounter a layer of houses and residential areas that seem to increase in quality as the driver gets farther away from the inner-city area.

Some critics, however, have raised some valid concerns regarding the original model, arguing that Shaw and McKay's original research did not actually measure their primary construct: social disorganization. Although this criticism is accurate, recent research has shown that the model is valid even when valid measures of social disorganization are included.[11] Such measures of social disorganization

---

[10]John Laub, "Urbanism, Race, and Crime," *Journal of Research in Crime and Delinquency 20* (1983): 283–298; Robert Sampson, "Structural Sources of Variation in Race-Age-Specific Rate of Offending Across Major U.S. Cities," *Criminology 23* (1985): 647–673; J. L. Heitgard and Robert J. Bursik, "Extracommunity Dynamics and the Ecology of Delinquency," *American Journal of Sociology 92* (1987): 775–787; Robert Bursik, "Social Disorganization and Theories of Crime and Delinquency: Problems and Prospects," *Criminology 26* (1988): 519–551; Ralph Taylor and Jeanette Covington, "Neighborhood Changes in Ecology and Violence," *Criminology 26* (1988): 553–589; Robert Bursik, "Ecological Stability and the Dynamics of Delinquency," in *Crime and Community,* eds. Albert J. Reiss and Morris H. Tonry (Chicago: University of Chicago Press, 1986), 35–66; Robert Sampson, "Transcending Tradition: New Directions in Community Research, Chicago Style—The American Society of Criminology 2001 Sutherland Address," *Criminology 40* (2002): 213–230; Robert Sampson, J. D. Morenoff, and T. Gannon-Rowley, "Assessing 'Neighborhood Effects': Social Processes and New Directions in Research," *Annual Review of Sociology 28* (2002): 443–478; P. O. Wikstrom and Rolf Loeber, "Do Disadvantaged Neighborhoods Cause Well-Adjusted Children to Become Adolescent Delinquents?" *Criminology 38* (2000): 1109–1142.

[11]Robert Sampson and W. Byron Grove, "Community Structure and Crime: Testing Social Disorganization Theory," *American Journal of Sociology 94* (1989): 774–802.

include simply asking members of the neighborhood how many neighbors they know by name or how often they observe unsupervised peer groups in the area.

Additional criticisms of Shaw and McKay's formulation of social disorganization focus on the emphasis that the theory places on the macro, or aggregate, level of analysis. Although their theory does a good job of predicting which neighborhoods have higher crime rates, the model does not even attempt to explain why most youths in the worst areas do not become offenders. Furthermore, their model does not attempt to explain why some youths—although a very small number—in the best neighborhoods (in Zone V) choose to commit crime. However, the previous case studies completed and published by Clifford Shaw, such as *The Jackroller* or *Brothers in Crime*, attempted to address the individual (micro) level of offending.

Also, there was one notable exception to Shaw and McKay's proposition that all ethnic and racial groups have high rates of delinquency and crime while they live in Zone II. Evidence showed that when Japanese made up a large portion of residents in this zone in transition, they had very low rates of delinquency. Thus, as in most theoretical models in social science, there was an exception to the rule.

Perhaps the biggest criticism of Shaw and McKay's theory, one that has yet to be adequately addressed, deals with their blatant neglect in proposing ways to ameliorate the most problematic source of criminality in Zone II, transitional zone neighborhoods. Although they clearly point to the invasion of factories and businesses into residential areas as a problem, they do not recommend how to slow such invasion. This is likely due to political and financial concerns: The owners of the factories and businesses partially financed their research and later funded their primary policy implementation. Furthermore, this neglect is represented in their failure to explain the exception of the Gold Coast in their results and conclusions.

Despite the criticisms and weaknesses of the Chicago School perspective of criminology, this theory resulted in one of the largest programs to date in attempting to reduce delinquency rates. Clifford Shaw was put in charge of establishing the Chicago Area Project (CAP), which created neighborhood centers in the most crime-ridden parts of Chicago. These centers offered activities for youths and tried to establish ties between parents and officials in the neighborhood. Although this program was never scientifically evaluated, it still exists, and many cities have implemented programs that are based on this model. For example, Boston implemented a very similar program, which was evaluated by Walter Miller.[12] This evaluation showed that, although the project was effective in establishing relationships and interactions between local gangs and community groups and in providing more educational and vocational opportunities, it seemed to fail in reducing delinquent and criminal behavior. Thus, the overall conclusion made by experts was that the Boston project and other similar programs, like the CAP, typically fail to prevent criminal behavior.[13]

Cultural and Subcultural Theories of Crime

Cultural and subcultural theories of crime assume that there are unique groups in society that socialize their children to believe that certain activities that violate conventional law are good and positive ways to behave. Although it is rather difficult to find large groups of people or classes who fit this definition, it may be that some subcultures or isolated groups of individuals buy into a different set of norms than the conventional, middle-class set of values.

---

[12]Walter B. Miller, "The Impact of a 'Total-Community' Delinquency Control Project," *Social Problems 10* (1962): 168–191.

[13]For a review, see Richard Lundman, *Prevention and Control of Juvenile Delinquency,* 2nd ed. (New York: Oxford University Press, 1993). Also see review by Vold et al., *Theoretical Criminology,* 125–126.

Early Theoretical Developments and Research in Cultural and Subcultural Theory

One of the key developments of cultural theory is the 1967 work of Ferracuti and Wolfgang, who examined the violent themes of a group of inner-city youth from Philadelphia.[14] Ferracuti and Wolfgang's primary conclusion was that violence is a culturally learned adaptation to deal with negative life circumstances and that learning such norms occurs in an environment that emphasizes violence over other options.[15] These researchers based their conclusion on an analysis of data that showed great differences in the rates of homicide across racial groups. However, Ferracuti and Wolfgang were clear that their theory was based on subcultural norms. Specifically, they proposed that no subculture can be totally different from or totally in conflict with the society of which it is a part.[16] This brings the distinction of culture and subculture to the forefront.

A culture represents a distinct set of norms and values among an identifiable group of people, values that are distinctly different from those of the mainstream culture. For example, communism is distinctly different from capitalism because it emphasizes equality over competition, and it values utopia (i.e., everyone gets to share all profits) over the idea that the best performer gets the most reward. So, it can be said that communists tend to have a different culture than capitalists. There is also a substantial difference between a culture and a subculture, which is typically only a pocket of individuals who may have a set of norms that deviate from conventional values. Therefore, what Ferracuti and Wolfgang developed is not so much a cultural theory as much as a subcultural theory.

This is also seen in the most prominent (sub)culture theory, which was presented by Walter Miller.[17] Miller's theoretical model proposed that the entire lower class had its own cultural value system. In this model, virtually everyone in the lower class believed in and socialized the values of six **focal concerns:** fate, autonomy, trouble, toughness, excitement, and smartness. Fate means luck, or whatever life dealt you; it disregards responsibility and accountability for one's actions. Autonomy is the value of independence from authority. Trouble means staying out of legal problems, as well as getting into and out of personal difficulties (e.g., pregnancy). Toughness is maintaining your reputation on the street in many ways. Excitement is engagement in activities, some illegal, that help liven up an otherwise mundane existence of being lower class. Smartness puts an emphasis on street smarts or the ability to con others. Miller thought that members of the lower class taught these six focal concerns as a culture or environment (or "milieu," as stated in the title of his work).

A more recent subculture model, proposed by Elijah Anderson, has received a lot of attention in the past few years.[18] This theory focuses on African Americans; because of the very deprived conditions in inner cities, Black Americans who live there feel a sense of hopelessness, isolation, and despair, Anderson asserts. He clearly notes that, although many African Americans believe in middle-class values, these values have no weight on the street, particularly among young urban males. According to Anderson, the *Code of the Street*, which was the appropriate title of his book, is to maintain one's reputation and demand respect. For example, to be treated with disrespect ("dissed") is considered grounds for a physical attack. Masculinity and control of one's immediate environment are treasured characteristics; this is perceived as the only thing people can control given the harsh conditions in which they live (e.g., unemployment, poverty, etc.).

---

[14]Vold et al., *Theoretical Criminology,* 165–169; Frank Schmalleger, *Criminology Today,* 4th ed. (Upper Saddle River: Prentice Hall, 2006).

[15]Franco Ferracuti and Marvin Wolfgang, *The Subculture of Violence: Toward an Integrated Theory of Criminology* (London: Tavistock, 1967).

[16]As quoted from Schmalleger, *Criminology Today,* 230–231.

[17]Walter B. Miller, "Lower Class Culture as a Generating Milieu of Gang Delinquency," *Journal of Social Issues 14* (1958): 5–19.

[18]Elijah Anderson, *Code of the Street* (New York: W. W. Norton, 1999).

Criticisms of Cultural Theories of Crime

Studies on cultural theories of crime, at least in the United States, find no large groups that blatantly deny the middle-class norms of society. Miller's model of lower class focal concerns is not consistent across the entire lower class. Studies show that most adults in the lower class attempt to socialize their children to believe in conventional values, such as respect for authority, hard work, and delayed gratification, and not the focal concerns that Miller specified in his model.[19] Ferracuti and Wolfgang admitted that their research findings led them to conclude that their model had more of a subcultural perspective than a distinctly different culture. There may be small groups or gangs who have subcultural normative values, but that doesn't constitute a completely separate culture in society. Perhaps the best subculture theories are those presented by Cohen or Cloward and Ohlin (see Chapter 6) in their variations of strain theory that emphasize the formations of gangs among lower-class male youths. If there are subcultural groups in U.S. society, they seem to make up a very small percentage of the population, which somewhat negates the cultural and subcultural perspective of criminality. But, this type of perspective may be important regarding the criminality of some youth offenders.

Policy Implications

Many of the policy implications that are suggested by the theoretical models proposed in this chapter are rather ironic. Regarding social disorganization, there is a paradox that exists in the sense that the very neighborhoods most desperately in need of becoming organized to fight crime are the same inner-city ghetto areas where it is, by far, the most difficult to cultivate such organization (e.g., neighborhood watch or block watch groups). Rather, the neighborhoods that have high levels of organization tend to be those that already have very low levels of crime because the residents naturally police their neighbors' well-being and property; they have a stake in the area remaining crime free. Although there are some anecdotal examples of success of neighborhood watch programs in high-crime neighborhoods, most of the empirical evidence shows that this approach is "almost uniformly unsupportive" in its ability to reduce crime there.[20] Furthermore, many studies of these neighborhood watch programs find that the groups actually increase the fear of crime in some places, perhaps due to the heightened awareness regarding the crime issues in these areas.[21]

Perhaps the most notable programs that resulted from the Chicago School or social disorganization model—the CAP and similar programs—have been dubbed failures in reducing crime rates among the participants. Still, there have been some advances in trying to get residents of high-crime areas to become organized in fighting crime. The more specific the goals regarding crime reduction—such as more careful monitoring of high-level offenders (more intensive supervised probation) or better lighting in dark places—the more effective the implementation will be.[22]

Regarding cultural and subcultural programs, some promising intervention and outreach programs have been suggested by such models. Many programs attempt to build prosocial attitudes among high-risk youths, often young children. For example, a recent evaluation showed that a program called Peace Builders, which focuses on children in early grades, was effective in producing gains in conflict resolution, development

[19]Vold et al., *Theoretical Criminology.*

[20]John Worrall, *Crime Control in America: An Assessment of the Evidence* (Boston: Allyn & Bacon, 2006), 95.

[21]Ibid.

[22]Lundman, *Prevention and Control;* Sampson and Grove, "Community Structure"; Vold et al., *Theoretical Criminology.*

of prosocial values, and reductions in aggression; a follow-up showed that these attributes were maintained for a long period of time.[23] Another recent antiaggression training program for foster-home boys showed positive effects in levels of empathy, self-efficacy, and attribution style among boys who had exhibited early-onset aggression.[24] Ultimately, there are effective programs out there that promote prosocial norms and culture. More efforts should be given to promoting such programs that will help negate the antisocial cultural norms of individuals, especially among high-risk youth.

✑ Conclusion

In this chapter, we examined theoretical perspectives proposing that the lack of social organization in broken-down and dilapidated neighborhoods leads to the inability to contain delinquency and crime. Furthermore, we discussed how this model of crime was linked to processes derived from ecological principles. This type of approach has been tested numerous times, and virtually all studies show that the distribution of delinquents and crime activity is consistent with this model.

We then discussed the ability of cultural and subcultural theories to explain criminal activity. Empirical evidence shows that cultural values make a contribution to criminal behavior, but that the existence of an actual alternative culture in our society has not been found. However, some subcultural pockets, particularly regarding inner-city youth gangs, certainly exist and provide some validity for this perspective of crime. Furthermore, the Chicago School perspective plays a role because these subcultural groups tend to be found in zones of transition.

Finally, we examined some policy implications that are suggested by these theoretical models. Regarding social disorganization, we noted that neighborhood crime-fighting groups are hardest to establish in high-crime neighborhoods and easiest to build in those neighborhoods with an already low rate of crime. Nevertheless, there have been some successes. We also looked at intervention and outreach programs based on cultural and subcultural perspectives.

✑ Chapter Summary

- We examined how principles of ecology were applied to the study of how cities grow, as well as to the study of crime, by researchers at the University of Chicago, which became known as the Chicago (or Ecological) School of criminology.
- We reviewed the various zones of concentric circles theory, also a key contribution of the Chicago School of criminology, and explored which zones are most crime prone.
- We examined why the findings from the Chicago School of criminology showed that social Darwinism was not accurate in attributing varying crime rates to ethnicity or race.
- We reviewed much of the empirical evidence regarding the theory of social disorganization and examined the strengths and weaknesses of this theoretical model.

---

[23]Daniel J. Flannery, Mark I. Singer, and Kelly L. Wester, "Violence, Coping, and Mental Health in a Community Sample of Adolescence," *Violence and Victims 18* (2003): 403–418; for a review of this research, see Stephen G. Tibbetts, "Perinatal and Developmental Determinants of Early Onset of Offending: A Biosocial Approach for Explaining the Two Peaks of Early Antisocial Behavior," in *The Development of Persistent Criminality*, ed. Joanne Savage (New York: Oxford University Press, 2009), 179–201.

[24]Karina Weichold, "Evaluation of an Anti-aggressiveness Training with Antisocial Youth," *Gruppendynamik 35* (2004): 83–105; for a review, see Tibbetts, "Perinatal and Developmental."

- We discussed the cultural and subcultural model presented by Ferracuti and Wolfgang, as well as the cultural model of inner-city urban youths presented by Anderson.
- We discussed Miller's theory of lower-class culture, particularly its six focal concerns.
- We reviewed the strengths and weaknesses of cultural and subcultural theories of crime, based on empirical evidence.

KEY TERMS

Chicago School of criminology	Ecological School	Social Darwinism
Concentric circles theory	Focal concerns	Social disorganization
Cultural and subcultural theories	Natural areas	Zone in transition

DISCUSSION QUESTIONS

1. Identify and discuss an example of the ecological principles of invasion, domination, and succession among animals or plants that was not discussed in this chapter.

2. Can you see examples of the various zones that Shaw and McKay described in the town or city where you live (or the one nearest you)? Try obtaining a map or sketching a plot of this city or town, and then draw the various concentric circles where you think the zones are located.

3. What forms of organization and disorganization have you observed in your own neighborhood? Try to supply examples of both, if possible.

4. Can you provide modern-day examples of different cultures and subcultures in the United States? What regions, or parts of the country, would you say have different cultures that are more conducive to crime?

5. Do you know people who believe most or all of Miller's focal concerns? What are their social classes? What are their other demographic features (age, gender, urban versus rural, etc.)?

6. Do you know individuals who seem to fit either Ferracuti and Wolfgang's cultural theory or Anderson's model of inner-city youths' street code? Why do you believe they fit such a model?

WEB RESOURCES

Chicago School of Criminology

   http://www.helium.com/items/865770-an-overview-of-the-chicago-school-theories-of-criminology

   http://www.associatedcontent.com/article/771561/an_overview_of_the_chicago_school_theories.html

Subcultural Theories

   http://www.umsl.edu/~keelr/200/subcult.html

8

Social Process and Learning Theories of Crime

This chapter will discuss Sutherland's development of **differential association theory** and how this evolved into Akers's work of differential reinforcement and other social **learning theories,** such as techniques of neutralization. Then, the modern state of research on these theories will be presented. We will also discuss the evolution of **control theories** of crime, with an emphasis on social bonding and the scientific evidence found regarding the key constructs in Hirschi's control theories, two of the most highly regarded perspectives according to criminological experts and their studies.[1]

Most of the social process theories assume that criminal behavior is learned behavior, which means that criminal activity is actually learned from others through social interaction, much like riding a bike or playing basketball. Namely, they learn this from significant others, such as from family, peers, or coworkers. However, other social process theories, namely control theories, assume that offending is the result of natural tendencies and thus must be controlled by social processes. Social process theories examine how individuals interact with other individuals and groups and how the learning that takes place in these inter-actions leads to a propensity toward criminal activity. This chapter will explore both of these theoretical frameworks and explain how social processes are vital to both perspectives in determining criminal behavior.

This chapter begins with social process theories known as learning theories. Such learning theories attempt to explain how and why individuals learn from significant others to engage in criminal rather than conventional behavior. Next, we discuss control theories, which emphasize personal or socialization factors that prevent individuals from engaging in selfish, antisocial behaviors.

---

[1]Walsh and Ellis, "Political Ideology," 1, 14 (see chap. 1, n. 13).

✖ Learning Theories

In this section, we review theories that emphasize how individuals learn criminal behavior through interacting with significant others, people with whom they typically associate. These learning theories assume that people are born with no tendency toward or away from committing crime. This concept is referred to as ***tabula rasa,*** or blank slate, meaning that all individuals are completely malleable and will believe what they are told by their significant others and act accordingly. Thus, such theories of learning tend to explain how criminal behavior is learned through cultural norms. One of the main concepts in learning theories is the influence of peers and significant others on an individual's behavior.

Here, three learning theories are discussed: (1) differential association theory, (2) **differential identification theory,** and (3) **differential reinforcement theory;** then we examine techniques of neutralization.

Differential Association Theory

Edwin Sutherland introduced his differential association theory in the late 1930s.[2] He proposed a theoretical framework that explained how criminal values could be culturally transmitted to individuals from their significant others. Sutherland proposed a theoretical model that included nine principles, but rather than list them all, we will summarize the main points of his theory.

To summarize, perhaps the most interesting principle is the first: Criminal behavior is learned. This was a radical departure from previous theories (e.g., Lombroso's born criminal theory, Goddard's feeble-mindedness theory, Sheldon's body-type theory). Sutherland was one of the first to state that criminal behavior is the result of normal social processes, resulting when individuals associate with the wrong type of people, often by no fault on their part. By associating with crime-oriented people, whether they are parents or peers, an individual will inevitably choose to engage in criminal behavior because that is what he or she has learned, Sutherland thought.

Perhaps the most important of Sutherland's principles, and certainly the most revealing one, was No. 6 in his framework: "A person becomes delinquent because of an excess of definitions favorable to violation of law over definitions unfavorable to violation of law."[3] Sutherland noted that this principle represented the essence of differential association theory. It suggests that people can have associations that favor both criminal and noncriminal behavior patterns. If an individual is receiving more information and values that are pro-crime than anti-crime, the individual will inevitably engage in criminal activity. Also, Sutherland claimed that such learning could take place only in interactions with significant others and not via television, movies, radio, or other media.

It is important to understand the cultural context at the time Sutherland was developing his theory. In the early 20th century, most academics, and society for that matter, believed that there was something abnormal or different about criminals. Sheldon's body-type theory was popular in the same time period, as was the use of IQ to pick out people who were of lower intelligence and supposedly predisposed to crime (both of these theories were covered in Chapters 4 and 5). So, the common assumption when Sutherland created the principles of differential association theory was that there was essentially something wrong with individuals who commit crime.

---

[2]Edwin H. Sutherland, *Principles of Criminology,* 3rd ed. (Philadelphia: Lippincott, 1939).

[3]Edwin H. Sutherland and Donald R. Cressey, *Principles of Criminology,* 5th ed. (Chicago: Lippincott, 1950), 78.

In light of this common assumption, Sutherland's proposal—that criminality is learned just like any conventional activity—was extremely profound. This suggests that any normal person, when exposed to attitudes favorable toward crime, will learn criminal activity and that the processes and mechanisms of learning are the same for crime as for most legal, everyday behaviors, namely, social interaction with family and friends and not reading books or watching movies. How many of us learned to play basketball or other sports by reading a book about it? Virtually no one learns how to play sports this way. Rather, we learn the techniques (e.g., how to do a jump shot) and motivations for playing sports (e.g., it is fun or you might be able to earn a scholarship) through our friends, relatives, coaches, and other people close to us. According to Sutherland, crime is learned the same way; our close associates teach us both the techniques (e.g., how to steal a car) and the motivations (e.g., it is fun; you might be able to sell it or its parts). While most criminologists tend to take it for granted that criminal behavior is learned, the idea was rather unique and bold when Sutherland presented his theory of differential association.

▲ **Image 8.1** Edwin H. Sutherland (1883–1950), University of Chicago and Indiana University, Author of Differential Association Theory

It is important to keep in mind that differential association theory is just as positivistic as earlier biological and psychological theories. Sutherland clearly believed that, if people were receiving more information that breaking the law was good, then they would inevitably commit crimes. There is virtually no allowance for free will and rational decision making in this model of offending. Rather, people's choices to commit crime are determined through their social interactions with those close to them; they do not actually make the decisions to engage (or not engage) in criminal activities. So, differential association can be seen as a highly positive, deterministic theory, much like Lombroso's born criminal and Goddard's feeblemindedness (see Chapter 4), except that, instead of biological or psychological traits causing crime, it is social interaction and learning. Furthermore, Sutherland claimed that individual differences in biological and psychological functioning have little to do with criminality; however, this idea has been discounted by modern research, which shows such variations do in fact affect criminal behavior, largely because such biopsychological factors influence the learning processes of individuals, thereby directly impacting the basic principles of Sutherland's theory (see Chapters 4 and 5).

Classical Conditioning: A Learning Theory With Limitations

Sutherland used the dominant psychological theory of learning in his era as the basis for his theory of differential association. This model was **classical conditioning,** which was primarily developed by Ivan Pavlov. Classical conditioning assumes that animals, as well as people, learn through associations between stimuli and responses.[4] Organisms, animals, or people are somewhat passive actors in this process, meaning

---

[4]For a discussion, see Vold et al., *Theoretical Criminology,* 156–157 (see chap. 2, n. 40).

that they simply receive and respond in natural ways to various forms of stimuli; over time, they learn to associate certain stimuli with certain responses.

For example, Pavlov showed that dogs, which naturally are afraid of loud noises such as bells, could be quickly conditioned not only to be less afraid of bells but to actually desire and salivate at their sound. A dog naturally salivates when presented with meat, so when this presentation of an unconditioned stimuli (meat) is given, a dog will always salivate (unconditioned response) in anticipation of eating. Pavlov demonstrated through a series of experiments that, if a bell (conditioned stimuli) is always rung at the same time the dog is presented with meat, then the dog will learn to associate what was previously a negative stimulus with a positive stimulus (food). Thus, the dog will very quickly begin salivating at the ringing of a bell, even when meat is not presented. When this occurs, it is called a conditioned response because it is not natural; however, it is a very powerful and effective means of learning, and it sometimes takes only a few occurrences of coupling the bell ringing with meat before the conditioned response takes place.

One modern use of this in humans is the administration of drugs that make people ill when they drink alcohol. Alcoholics are often prescribed drugs that make them very sick, often to the point of vomiting, if they ingest any alcohol. The idea is that they will learn to associate feelings of sickness with drinking and thus stop wanting to consume alcohol. One big problem with this strategy is that alcoholics often do not consistently take the drugs, so they quickly slip back into addiction. Still, if they were to maintain their regimen of drugs, it would likely work because people do tend to learn through association.

This type of learning model was used in the critically acclaimed 1964 novel (and later motion picture) *A Clockwork Orange*. In this novel, the author, Anthony Burgess, tells the story of a juvenile murderer who is "rehabilitated" by doctors who force him to watch hour after hour of violent images while simultaneously giving him drugs that make him sick. In the novel, the protagonist is "cured" after only 2 weeks of this treatment, having learned to consistently associate violence with sickness. However, once he is let out, he lacks the ability to choose violence and other antisocial behavior, which is seen as losing his humanity. Therefore, the ethicists order a reversal treatment and make him back into his former self, a violent predator. Although a fictional piece, *A Clockwork Orange* is probably one of the best illustrations of the use of classical conditioning in relation to criminal offending and rehabilitation.

Another example of classical conditioning is the association we make with certain smells and sounds. For example, all of us can relate good times to smells that were present during those occasions. If a loved one or someone we dated wore a certain perfume or cologne, the scent of that at a later time can bring back memories. When our partners go out of town, we can smell their pillows, and they remind us of them because we associate their smell with their being. Or perhaps the smell of a turkey cooking in an oven always reminds us of Thanksgiving or other holidays. Regarding associations of sounds, we all can remember one or more songs that remind us of both happy and sad times in our lives. Often these songs will replay on the radio, and they take us back to those occasions, whether they are good or bad. People with post-traumatic stress disorder (PTSD) also experience sound associations; war veterans, for example, may hit the deck when a car backfires. These are all clear examples of classical conditioning and associating stimuli with responses.

Since Sutherland's theory was published, many of the principles outlined in his model have come under scrutiny. Follow-up research has shown some flaws, as well as misinterpretations, of his work.[5] Specifically,

---

[5]Edwin H. Sutherland and Donald R. Cressey, *Criminology*, 9th ed. (Philadelphia: Lippincott, 1974).

Sutherland theorized that crime occurs when the ratio of associations favorable to violation of the law outweigh associations favorable to conforming to the law. However, measuring this type of ratio is all but impossible for social scientists.[6]

Another topic of criticism involves Sutherland's claim that all criminals learn the behavior from others before they engage in such activity. However, many theorists have noted that an individual may engage in criminal activity without being taught such behavior then seek out others with attitudes and behavior similar to their own.[7] So, do individuals learn to commit crime after they are taught by delinquent peers, or do they start associating with similar delinquents or criminals once they have initiated their offending career (i.e., birds of a feather flock together)? This exact debate was examined by researchers, and the most recent studies point to both causal processes occurring: Criminal associations cause more crime, and committing crime causes more criminal associations. Both are key in the causal process, so Sutherland was missing half this equation.[8]

Another key criticism is that, if each individual is born with a blank slate and all criminal behavior is learned, then who committed crime in the first place? Who could expose the first criminal to the definitions favorable to violation of law? Furthermore, what factor(s) caused that individual to do it first if it was not learned? Obviously, if it were due to any other factor(s) than learning—and it must have been because there was no one to teach it—then it obviously was not explained by learning theories. This criticism cannot be addressed, so it is somewhat ignored in the scientific literature.

Despite the criticisms and flaws, much research supports Sutherland's theory. For example, researchers have found that older criminals teach younger delinquents.[9] In addition, delinquents often associate with criminal peers prior to engaging in criminal activity.[10] Furthermore, research has shown that the nexus of criminal friends, attitudes, and activity are highly associated.[11] Still, Sutherland's principles are quite vague and elusive in terms of measurement, which renders them difficult in terms of measuring by social scientists.[12] Related to these issues, perhaps one of the biggest problems with Sutherland's formulation of differential association is that he used primarily one type of learning model—classical conditioning—to formulate most of his principles, and thus he neglects other important ways that we learn attitudes and behavior from others. Ultimately, Sutherland's principles are hard to test; more current models of his framework have incorporated other learning models and thus are easier to test so that empirical validity can be demonstrated.

---

[6]Ross L. Matsueda and Karen Heimer, "Race, Family Structure, and Delinquency: A Test of Differential Association and Social Control Theories," *American Sociological Review 47* (1987): 489–504; Charles R. Tittle, M. J. Burke, and E. F. Jackson, "Modeling Sutherland's Theory of Differential Association: Toward an Empirical Clarification," *Social Forces 65* (1986): 405–432.

[7]Tittle et al., "Modeling Sutherland's Theory."

[8]Terence Thornberry, "Toward an Interactional Theory of Delinquency," *Criminology 25* (1987): 863–887.

[9]Kenneth Tunnell, "Inside the Drug Trade: Trafficking from the Dealer's Perspective," *Qualitative Sociology 16* (1993): 361–381.

[10]Douglas Smith, Christy Visher, and G. Roger Jarjoura, "Dimensions of Delinquency: Exploring the Correlates of Participation, Frequency, and Persistence of Delinquent Behavior," *Journal of Research in Crime and Delinquency 28* (1991): 6–32.

[11]Matthew Ploeger, "Youth Employment and Delinquency: Reconsidering a Problematic Relationship," *Criminology 35* (1997): 659–675.

[12]Reed Adams, "The Adequacy of Differential Association Theory," *Journal of Research in Crime and Delinquency 1* (1974): 1–8; James F. Short, "Differential Association as a Hypothesis: Problems of Empirical Testing," *Social Problems 8* (1960): 14–25.

Glaser's Concept of Differential Identification

Another reaction to Sutherland's differential association dealt with the influence of movies and television, as well as other reference groups outside of one's significant others. As stated above, Sutherland claimed that learning of criminal definitions could take place only through social interactions with significant others as opposed to reading a book or watching movies. However, in 1956, Daniel Glaser proposed the idea of differential identification theory, which allows for learning to take place not only through people close to us but also through other reference groups, even distant ones, such as sports heroes or movie stars with whom the individual has never actually met or corresponded.[13] Glaser claimed that it did not matter much whether an individual had a personal relationship with a reference group(s); in fact, he argued that a group could be imaginary, such as fictitious characters in a movie or book. The important thing, according to Glaser, was that an individual must identify with a person or character and thus behave in ways that fit the norm set of this reference group or person.

Glaser's proposition has been virtually ignored, with the exception of Dawes's study of delinquency in 1973, which found that identification with people other than parents was strong when youths perceived a greater degree of rejection from their parents.[14] Given the profound influence of movies, music, and television on today's youth culture, it is obvious that differential identification was an important addition to Sutherland's framework, and more research should examine the validity of Glaser's theory in contemporary society. Although Glaser and others modified differential association, the most valid and respected variation is differential reinforcement theory.

Differential Reinforcement Theory

In 1965, C. R. Jeffery provided an extensive critique and reevaluation of Sutherland's differential association theory. He argued that the theory was incomplete without some attention to an updated social psychology of learning (e.g., **operant conditioning** and modeling theories of learning).[15] He wanted Sutherland to account for the fact that people can be conditioned into behaving certain ways, such as by being rewarded for conforming behavior. Then, in 1966, Robert Burgess and Ronald Akers criticized and responded to Jeffery's criticism by proposing a new theory that incorporated some of these learning models into Sutherland's basic framework.[16] The result was what is now known as differential reinforcement theory. Ultimately, Burgess and Akers argued that by integrating Sutherland's work with contributions from the field of social psychology, criminal behavior could be more clearly understood.

In some ways, differential reinforcement theory may appear to be no different than rational choice theory (see Chapter 3). To an extent, this is true because both models focus on reinforcements and punishments that occur after an individual offends. However, differential reinforcement theory can be distinguished from the rational choice perspective. The latter assumes that humans are born with the capacity for rational decision making, whereas the differential reinforcement perspective assumes people are born with a blank slate and, thus, must be socialized and taught how to behave through various forms of conditioning (e.g., operant and classical), as well as modeling.

---

[13]Daniel Glaser, "Criminality Theories and Behavioral Images," *American Journal of Sociology* 61 (1956): 433–444.

[14]Kenneth J. Dawes, "Family Relationships, Reference Others, Differential Identification and Their Joint Impact on Juvenile Delinquency," PhD dissertation (Ann Arbor: University Mircrofilms, 1973).

[15]C. Ray Jeffery, "Criminal Behavior and Learning Theory," *The Journal of Criminal Law, Criminology, and Police Science* 56 (1965): 294–300.

[16]Robert Burgess and Ronald Akers, "A Differential Association-Reinforcement Theory of Criminal Behavior," *Social Problems* 14 (1966): 131.

Burgess and Akers developed seven propositions to summarize differential reinforcement theory, which largely represent efficient modifications of Sutherland's original nine principles of differential association.[17] The strong influence of social psychologists is illustrated in their first statement, as well as throughout the seven principles. Although differential reinforcement incorporates the elements of modeling and classical conditioning learning models in its framework, the first statement clearly states that the essential learning mechanism in social behavior is operant conditioning, so it is important to understand what operant conditioning is and how it is evident throughout life.

Operant Conditioning

The idea of operant conditioning was primarily developed by B. F. Skinner,[18] who ironically was working just across campus from Edwin Sutherland when he was developing differential association theory at Indiana University. As in modern times, academia was too intradisciplinary and intradepartmental. Had Sutherland been aware of Skinner's studies and theoretical development, he likely would have included it in his original framework. In his defense, operant conditioning was not well known or researched at the time; as a result, Sutherland incorporated the then-dominant learning model, classical conditioning. Burgess and Akers went on to incorporate operant conditioning into Sutherland's framework.

Operant conditioning is concerned with how behavior is influenced by reinforcements and punishments. Furthermore, operant conditioning assumes that an animal or human being is a proactive player in seeking out rewards and not just a passive entity that receives stimuli, as classical conditioning assumes. Behavior is strengthened or encouraged through reward (**positive reinforcement**) and avoidance of punishment (**negative reinforcement**). For example, if someone is given a car for graduation from college, that would be a positive reinforcement. On the other hand, if a teenager who has been grounded is allowed to start going out again because he or she has better grades, this would be a negative reinforcement because he or she is now being rewarded via the avoidance of something negative. Like different types of reinforcement, punishment comes in two forms as well. Behavior is weakened, or discouraged, through adverse stimuli (**positive punishment**) or lack of reward (**negative punishment**). A good example of positive punishment would be a good, old-fashioned spanking, because it is certainly a negative stimulus; anything that directly presents negative sensations or feelings is a positive punishment. On the other hand, if parents take away car privileges from a teenager who broke curfew, that would be an example of negative punishment because the parents are removing a positive aspect or reward.

Some notable examples of operant conditioning include teaching a rat to successfully run a maze. When rats take the right path and finish the maze quickly, they are either positively reinforced (e.g., rewarded with a piece of cheese) or negatively reinforced (e.g., not zapped with electricity as they were when they chose the wrong path). On the other hand, when rats take wrong turns or do not complete the maze in adequate time, they are either positively punished (e.g., zapped with electricity) or negatively punished (e.g., not given the cheese they expect to receive). The rats, like humans, tend to learn the correct behavior very fast using such consistent implementation of reinforcements and punishments.

In humans, such principles of operant conditioning can be found even at very early ages. In fact, many of us have implemented such techniques (or been subjected to them) without really knowing they were

---

[17]For a more recent version of these principles, see Ronald Akers, *Deviant Behavior: A Social Learning Approach,* 3rd ed. (Belmont: Wadsworth, 1985); also see Ronald L. Akers and Christine S. Sellers, *Criminological Theories: Introduction, Evaluation, and Application,* 4th ed. (Los Angeles: Roxbury, 2004).

[18]B. F. Skinner, *Science and Human Behavior* (New York: Macmillan, 1953).

called operant conditioning. For example, during toilet training, children learn to use the bathroom to do their natural duty rather than doing it in their pants. To reinforce the act of going to the bathroom on a toilet, we encourage the correct behavior by presenting positive rewards, which can be as simple as applauding or giving a child a piece of candy for a successful job. While parents (we hope) rarely proactively use spanking in toilet training, there is an inherent positive punishment involved when children go in their pants; namely, they have to be in their dirty diaper for a while, not to mention the embarrassment that most children feel when they do this. Furthermore, negative punishments are present in such situations because they do not get the applause or candy, so the rewards have been removed.

Of course, this does not apply only to early behavior. An extensive amount of research has shown that humans learn attitudes and behavior best through a mix of reinforcements and punishments throughout life. In terms of criminal offending, studies have clearly shown that rehabilitative programs that appear to work most effectively in reducing recidivism in offenders are those that have opportunities for rewards as well as threats for punishments. Empirical research has combined the findings from hundreds of such studies of rehab programs, showing that the programs that are most successful in changing attitudes and behavior of previous offenders are those that offer at least four reward opportunities for every one punishment aspect of the program.[19] So, whether it is training children to go potty correctly or altering criminals' thinking and behavior, operant conditioning is a well-established form of learning that makes differential reinforcement theory a more valid and specified model of offending than differential association.

Whether deviant or conforming behavior occurs and continues "depends on the past and present rewards or punishment for the behavior, and the rewards and punishment attached to alternative behavior."[20] This is in contrast to Sutherland's differential association model, which looked only at what happened before an act (i.e., classical conditioning), not at what happens after the act is completed (i.e., operant conditioning); Burgess and Akers's model looks at both. Criminal behavior is likely to occur, Burgess and Akers theorized, when its rewards outweigh the punishments.

Bandura's Model of Imitation–Modeling

Another learning model that Burgess and Akers emphasized in their formulation of differential reinforcement theory was the element of **imitation–modeling.** Although Sutherland's original formulation of differential association theory was somewhat inspired by Gabriel Tarde's concept of imitation,[21] the nine principles did not adequately emphasize the importance of modeling in the process of learning behavior. Sutherland's failure was likely due to the fact that Albert Bandura's primary work in this area had not occurred when Sutherland was formulating differential association theory.[22]

Through a series of experiments and theoretical development, Bandura demonstrated that a significant amount of learning takes place without any form of conditioning. Specifically, he claimed that individuals can learn even if they are not rewarded or punished for behavior (i.e., operant conditioning) and even if they have not been exposed to associations between stimuli and responses (i.e., classical conditioning). Rather, Bandura proposed that people learn much of their attitudes and behavior from simply observing the behavior

---

[19]P. Van Voorhis, Michael Braswell, and David Lester. *Correctional Counseling*, 3rd ed. (Cincinnati: Anderson, 2000).

[20]Ronald L. Akers, *Deviant Behavior: A Social Learning Approach,* 2nd ed. (Belmont: Wadsworth Publishing, 1977), 57.

[21]Gabriel Tarde, *Penal Philosophy,* trans. Rapelje Howell (Boston: Little, Brown, 1912).

[22]See Albert Bandura, *Principles of Behavior Modification* (New York: Holt, Rinehart, and Winston, 1969); Albert Bandura, *Aggression: A Social Learning Analysis* (Englewood Cliffs: Prentice Hall, 1973); Albert Bandura, *Social Learning Theory* (Englewood Cliffs: Prentice Hall, 1977).

of others, namely through mimicking what others do. This is often referred to as *monkey see, monkey do,* but it is not just monkeys that do this. Like most animal species, humans are biologically hardwired to observe and learn the behavior of others, especially elders, to see what behavior is essential for survival and success.

Bandura showed that simply observing the behavior of others, especially adults, can have profound learning effects on the behavior of children. Specifically, he performed experiments in which a randomized experimental group of children watched a video of adults acting aggressively toward Bo-Bo dolls (which are blow-up plastic dolls) and a control group of children that did not watch such a video. Both groups of children were then sent into a room containing Bo-Bo dolls, and the experimental group, who had seen the adult behavior, mimicked their elders by acting far more aggressively toward the dolls than the children in the control group. The experimental group had no previous associations of more aggressive behavior toward the dolls and good feelings or motivations, let alone rewards, for such behavior. Rather, the children became more aggressive themselves simply because they were imitating what they had seen older people do.

Bandura's findings have important implications for the modeling behavior of adults (and peers) and for the influence of television, movies, video games, and other factors. Furthermore, the influences demonstrated by Bandura supported a phenomenon commonplace in everyday life. Mimicking is the source of fashion trends—wearing low-slung pants or baseball hats turned a certain way. Styles tend to ebb and flow based on how some respected person (often a celebrity) wears clothing. This can be seen very early in life; parents must be careful what they say and do because their children, as young as 2 years old, imitate what their parents do. This continues throughout life, especially in teenage years as young persons imitate the cool trends and styles, as well as behavior. Of course, sometimes this behavior is illegal, but individuals are often simply mimicking the way their friends or others are behaving with little regard for potential rewards or punishments. Ultimately, Bandura's theory of modeling and imitation adds a great deal of explanation in a model of learning, and differential reinforcement theory includes such influences, whereas Sutherland's model of differential association does not, largely because the psychological perspective had not yet been developed.

Burgess and Akers's theory of differential reinforcement has also been the target of criticism by theorists and researchers. Perhaps the most important criticism of differential reinforcement theory is that it appears *tautological,* which means that the variables and measures used to test its validity are true by definition. To clarify, studies testing this theory have been divided into four groups of variables or factors: associations, reinforcements, definitions, and modeling.

Some critics have noted that, if individuals who report that they associate with those who offend, are rewarded for offending, believe offending is good, and have seen many of their significant others offend, they will inevitably be more likely to offend. In other words, if your friends and family are doing it, there is little doubt that you will be doing it.[23] For example, critics would argue that a person who primarily hangs out with car thieves, knows he will be rewarded for stealing cars, believes stealing cars is good and not immoral, and has observed many respected others stealing cars, he will inevitably commit auto theft himself. However, it has been well argued that such criticisms of tautology are not valid because none of these factors necessarily make offending by the respondent true by definition.[24]

Differential reinforcement theory has also faced the same criticism that was addressed to Sutherland's theory, that delinquent associations may take place after criminal activity rather than before. However, Burgess and Akers's model clearly has this area of criticism covered in the sense that differential

[23]Mark Warr, "Parents, Peers, and Delinquency," *Social Forces 72* (1993): 247–264; Mark Warr and Mark Stafford, "The Influence of Delinquent Peers: What They Think or What They Do?" *Criminology 29* (1991): 851–866.

[24]Akers and Sellers, *Criminological Theories: Introduction,* 98–101.

reinforcement includes what comes after the activity, not just what happens before it. Specifically, it addresses the rewards or punishments that follow criminal activity, whether those rewards come from friends, parents, or other members or institutions of society.

It is arguable that differential reinforcement theory may have the most empirical validity of any contemporary (nonintegrated) model of criminal offending, especially considering that studies have examined a variety of behaviors, ranging from drug use to property crimes to violence. The theoretical model has also been tested in samples across the United States, as well as other cultures, such as South Korea, with the evidence being quite supportive of the framework. Furthermore, a variety of age groups have been examined, ranging from teenagers to middle-aged adults to the elderly, with all studies providing support for the model.[25]

Specifically, researchers have found that the major variables of the theory had a significant effect in explaining marijuana and alcohol use among adolescents.[26] The researchers concluded that the "study demonstrates that central learning concepts are amenable to meaningful questionnaire measurement and that social learning theory can be adequately tested with survey data."[27] Other studies have also supported the theory when attempting to understand delinquency, cigarette smoking, and drug use.[28] Therefore, the inclusion of three psychological learning models, namely classical conditioning, operant conditioning, and **modeling and imitation,** appears to have made differential reinforcement one of the most valid theories of human behavior, especially in regard to crime.

Neutralization Theory

Neutralization theory is associated with Gresham Sykes and David Matza's techniques of neutralization[29] and Matza's **drift theory.**[30] Like Sutherland, both Sykes and Matza thought that social learning influences delinquent behavior, but they also asserted that most criminals hold conventional beliefs and values. More specifically, Sykes and Matza argued that most criminals are still partially committed to the dominant social order. According to Sykes and Matza, youth are not immersed in a subculture that is committed to either extremes of complete conformity or complete nonconformity. Rather, these individuals vacillate, or drift, between these two extremes and are in a state of *transience.*[31]

---

[25]See studies including Ronald Akers and Gang Lee, "A Longitudinal Test of Social Learning Theory: Adolescent Smoking," *Journal of Drug Issues 26* (1996): 317–343; Ronald Akers and Gang Lee, "Age, Social Learning, and Social Bonding in Adolescent Substance Abuse," *Deviant Behavior 19* (1999): 1–25; Ronald Akers and Anthony J. La Greca, "Alcohol Use Among the Elderly: Social Learning, Community Context, and Life Events," in *Society, Culture, and Drinking Patterns Re-examined,* eds. David J. Pittman and Helene Raskin White (New Brunswick: Rutgers Center of Alcohol Studies, 1991), 242–262; Sunghyun Hwang, "Substance Use in a Sample of South Korean Adolescents: A Test of Alternative Theories," PhD dissertation (Ann Arbor: University Microfilms, 2000); Sunghyun Hwang and Ronald Akers, "Adolescent Substance Use in South Korea: A Cross-Cultural Test of Three Theories," in *Social Learning Theory and the Explanation of Crime: A Guide for the New Century,* eds. Ronald Akers and Gary F. Jensen (New Brunswick: Transaction Publishers, 2003).

[26]Ronald Akers, Marvin D. Krohn, Lonn Lanza-Kaduce, and Marcia Radosevich, "Social Learning and Deviant Behavior: A Specific Test of a General Theory," *American Sociological Review 44* (1979): 638.

[27]Akers et al., "Social Learning and Deviant Behavior," 651.

[28]Richard Lawrence, "School Performance, Peers and Delinquency: Implications for Juvenile Justice," *Juvenile and Family Court Journal 42* (1991): 59–69; Marvin Krohn, William Skinner, James Massey, and Ronald Akers, "Social Learning Theory and Adolescent Cigarette Smoking: A Longitudinal Study," *Social Problems 32* (1985): 455–471; L. Thomas Winfree, Christine Sellers, and Dennis Clason, "Social Learning and Adolescent Deviance Abstention: Toward Understanding the Reasons for Initiating, Quitting, and Avoiding Drugs," *Journal of Quantitative Criminology 9* (1993): 101–123; Ronald Akers and Gang Lee, "A Longitudinal Test of Social Learning Theory: Adolescent Smoking," *Journal of Drug Issues 26* (1996): 317–343.

[29]Gresham M. Sykes and David Matza, "Techniques of Neutralization: A Theory of Delinquency," *American Sociological Review 22* (1957): 664–670.

[30]David Matza, *Delinquency and Drift* (New York: John Wiley, 1964).

[31]Ibid., 28.

While remaining partially committed to conventional social order, youths can drift into criminal activity, Sykes and Matza claimed, and avoid feelings of guilt for these actions by justifying or rationalizing their behavior. This typically occurs in teenage years, when social controls (e.g., parents, family, etc.) are at their weakest point, and peer pressures and associations are at their highest level. Why is this called neutralization theory? The answer is that people justify and rationalize behavior through neutralizing it or making it appear not so serious. They make up situational excuses for behavior that they know is wrong to alleviate the guilt they feel for doing such immoral acts. In many ways, this resembles Freud's defense mechanisms, which allow us to forgive ourselves for the bad things we do, even though we know they are wrong. The specific techniques of neutralization outlined by Sykes and Matza in 1957 are much like excuses for inappropriate behavior.

Techniques of Neutralization

Sykes and Matza identified methods or techniques of neutralization[32] that people use to justify their criminal behavior. These techniques allow people to neutralize or rationalize their criminal and delinquent acts by making them look as though they are conforming to the rules of conventional society. If individuals can create such rationalizations, then they are free to engage in criminal activities without serious damage to their consciences or self-images. According to Sykes and Matza, there are five common techniques of neutralization:

1. *Denial of responsibility:* Individuals may claim they were influenced by forces outside themselves and that they are not responsible or accountable for their behavior. For example, many youths blame their peers for their own behavior.

2. *Denial of injury:* This is the rationalization that no one was actually hurt by their behavior. For instance, if someone steals from a store, they may rationalize this by saying that the store has insurance, so there is no direct victim.

3. *Denial of the victim:* Offenders see themselves as avengers and the victims as the wrongdoers. For example, some offenders believe that a person who disrespects or "disses" them deserves what he or she gets, even if it means serious injury.

4. *Condemnation of the condemners:* Offenders claim the condemners (usually the authorities who catch them) are hypocrites. For instance, one may claim that police speed on the highway all the time, so everyone else is entitled to drive higher than the speed limit.

5. *Appeal to higher loyalties:* Offenders often overlook the norms of conventional society in favor of the rules of a belief they have or a group to which they belong. For example, people who kill doctors who perform abortions tend to see their crimes as above the law because they are serving a higher power.

Although Sykes and Matza specifically labeled only five techniques of neutralization, it should be clear that there may be endless excuses people make up to rationalize behaviors they know are wrong. Techniques of neutralization have been applied to white-collar crime, for example. Several studies have examined the tendency to use such excuses to alleviate guilt for engaging in illegal corporate crime; they point out new types of excuses white-collar criminals use to justify their acts, techniques that were not discussed in Sykes and Matza's original formulation.[33]

---

[32]Sykes and Matza, "Techniques of Neutralization."

[33]For a review and a study on this topic, see Nicole Piquero, Stephen Tibbetts, and Michael Blankenship, "Examining the Role of Differential Association and Techniques of Neutralization in Explaining Corporate Crime," *Deviant Behavior* 26 (2005): 159–188.

Studies that have attempted to empirically test neutralization theory are, at best, inconsistent. For example, Agnew argued that there are essentially two general criticisms of studies that support neutralization theory.[34] First, theorists and researchers have noted that some neutralization techniques are very difficult to measure, as opposed to a commitment to unconventional attitudes or norms.[35] The second major criticism is the concern that criminals may not use techniques of neutralization prior to committing a criminal offense but rather only after committing a crime. As estimated by previous studies, this temporal ordering can be problematic in terms of causal implications when neutralization follows a criminal act.[36] This temporal ordering problem results from research conducted at a single point in time. Some would argue that the temporal ordering problem is not a major criticism because individuals may be predisposed to make up such rationalizations for their behavior regardless of whether they do it before or after the act of offending. Such a propensity may be related to **low self-control theory,** which we examine later in this chapter.

Summary of Learning Theories

Learning theories tend to emphasize the social processes of how and why individuals learn criminal behavior. These theories also focus on the impact of significant others involved in the socialization process, such as family, friends, and teachers. Ultimately, empirical research has shown that learning theories are key in our understanding of criminal behavior, particularly in terms of whether criminal behavior is rewarded or punished. In summary, if individuals are taught and rewarded for performing criminal acts by the people they interact with on a day-to-day basis, in all likelihood, they will engage in illegal activity.

⊠ Control Theories

The learning theories discussed in the previous section assume that individuals are born with a conforming disposition. By contrast, control theories assume that all people would naturally commit crimes if it weren't for restraints on their innate selfish tendencies. Social control perspectives of criminal behavior thus assume that there is some type of basic human nature and that all human beings exhibit antisocial tendencies. Such theories are concerned with why individuals don't commit crime or deviant behaviors. Control theorists ask questions like this: What is it about society and human interaction that causes people not to act on their impulses?

The assumption that people have innate antisocial tendencies is a controversial one because it is nearly impossible to test. Nevertheless, some recent evidence supports the idea that human beings are inherently selfish and antisocial by nature. Specifically, researchers have found that most individuals are oriented

---

[34]Robert Agnew, "The Techniques of Neutralization and Violence," *Criminology 32* (1994): 563–564.

[35]W. William Minor, "The Neutralization of Criminal Offense," *Criminology 18* (1980): 116; W. William Minor, "Neutralization as a Hardening Process: Considerations in the Modeling of Change," *Social Forces 62* (1984): 995–1019; see also Roy L. Austin, "Commitment, Neutralization, and Delinquency," in *Juvenile Delinquency: Little Brother Grows Up,* ed. Theodore N. Ferninand (Beverly Hills: Sage, 1977); Quint C. Thurman, "Deviance and the Neutralization of Moral Commitment: An Empirical Analysis," *Deviant Behavior 5* (1984): 291–304.

[36]Travis Hirschi, *Causes of Delinquency* (Berkeley: University of California Press, 1969), 207. See also Mark Pogrebin, Eric Poole, and Amos Martinez, "Accounts of Professional Misdeeds: The Sexual Exploitation of Clients by Psychotherapists," *Deviant Behavior 13* (1992): 229–252; John Hamlin, "The Misplaced Concept of Rational Choice in Neutralization Theory," *Criminology 26* (1988): 425–438.

toward selfish and aggressive behaviors at an early age, with such behaviors peaking at the end of the second year (see Figure 8.1).[37]

An example of antisocial dispositions appearing early in life was reported by Tremblay and LeMarquand, who found that most young children's (particularly boys') aggressive behaviors peaked at age 27 months. These behaviors included hitting, biting, and kicking others.[38] Their research is not isolated; virtually all developmental experts acknowledge that toddlers exhibit a tendency to show aggressive behaviors toward others. This line of research would seem to support the notion that people are predisposed toward antisocial, even criminal, behavior.

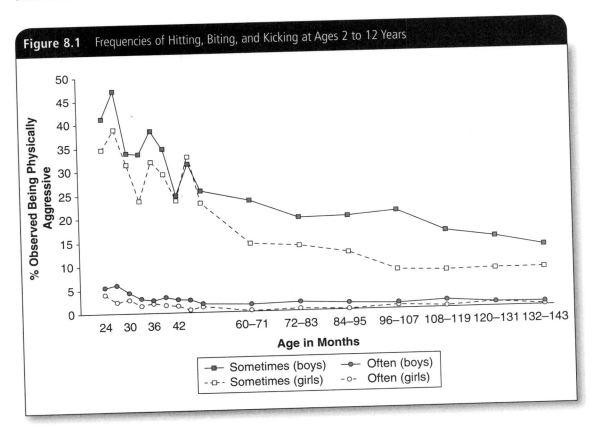

Figure 8.1 Frequencies of Hitting, Biting, and Kicking at Ages 2 to 12 Years

[37]Michael Lewis, Steven Alessandri, and Margaret Sullivan, "Expectancy, Loss of Control, and Anger in Young Infants," *Developmental Psychology 25* (1990): 745–751; Albert Restoin, Dayana Rodriguez, V. Ulmann, and Hubert Montagner, "New Data on the Development of Communication Behavior in the Young Child With His Peers," *Recherches de Psychologie Sociale 5* (1985): 31–56; Richard Tremblay, Christa Japel, Daniel Perusse, Pierre McDuff, Michel Boivin, Mark Zoccolillo, and Jacques Montplaisir, "The Search for the Age of 'Onset' of Physical Aggression: Rousseau and Bandura Revisited," *Criminal Behaviour and Mental Health 9* (1999): 8–23.

[38]Richard Tremblay and David LeMarquand, "Individual Risk and Protective Factors," in *Child Delinquents: Development, Intervention, and Service Needs,* eds. Rolf Loeber and David Farrington (London: Sage, 2001): 137–164.

Control theorists do not necessarily assume that people are predisposed toward crime in a way that remains constant throughout life. On the contrary, research shows that most individuals begin to desist from such behaviors starting at around age two. This trend continues until approximately age five, with only the most aggressive individuals (i.e., chronic offenders) continuing into higher ages.

It is important to note that, at the same time selfish and aggressive behaviors decline, self-consciousness is formed. In addition, social emotions—such as shame, guilt, empathy, and pride—begin to appear.[39] This observation is critical because it is what separates control theories from the classical school of criminology and the predispositional theories that we already discussed. According to control theories, without appropriate socialization, people act on their preprogrammed tendency toward crime and deviance.

In short, control theories claim that all individuals have natural tendencies to commit selfish, antisocial, and even criminal behavior. So, what is it that curbs this natural propensity? Many experts believe the best explanation is that individuals are socialized and controlled by social attachments and investments in conventional society. This assumption regarding the vital importance of early socialization is probably the primary reason why control theories are currently the most popular and accepted theories among criminologists.[40] We will now discuss several early examples of these control theories.

Early Control Theories of Human Behavior

Thomas Hobbes

Control theories are found in a variety of disciplines, including biology, psychology, and sociology. Perhaps the earliest significant use of social control in explaining deviant behavior is found in a perspective offered by the 17th-century Enlightenment philosopher, Thomas Hobbes (see Chapter 2). Hobbes claimed that the natural state of humanity was one of selfishness and self-centeredness to the point of constant chaos, characterized by a state of warfare between individuals. He stated that all individuals are inherently disposed to take advantage of others in order to improve their own personal well-being.[41]

However, Hobbes also claimed that the constant fear created by such selfishness resulted in humans rationally coming together to create binding contracts that would keep individuals from violating others' rights. Even with such controlling arrangements, however, Hobbes was clear that the selfish tendencies people exhibit could never be extinguished. In fact, they explained why punishments were necessary to maintain an established social contract among people.

Durkheim's Idea of Awakened Reflection and Collective Conscience

Consistent with Hobbes's view of individuals as naturally selfish, Durkheim later proposed a theory of social control in the late 1800s that suggested humans have no internal mechanism to let them know when they are fulfilled.[42] To this end, Durkheim coined the terms *automatic spontaneity* and *awakened reflection*. Automatic spontaneity can be understood with reference to animals' eating habits. Specifically, animals stop eating when they are full, and they are content until they are hungry again; they don't start hunting right

---

[39]For reviews of supporting studies, see June Price Tangney and K. W. Fischer, *Self-Conscious Emotions: The Psychology of Shame, Guilt, Embarrassment, and Pride* (New York: Guilford Press, 1995); Michael Lewis, *Shame: The Exposed Self* (New York: Macmillan, 1992).

[40]Walsh and Ellis, "Political Ideology."

[41]Thomas Hobbes, *Leviathan* (Cambridge: Cambridge University Press, 1651/1904).

[42]Emile Durkheim, *The Division of Labor in Society* (New York: Free Press, 1893/1965); Emile Durkheim, *Suicide* (see chap. 6, n. 18).

after they have filled their stomachs with food. In contrast, awakened reflection concerns the fact that humans do not have such an internal, regulatory mechanism. That is because people often acquire resources beyond what is immediately required. Durkheim went so far as to say that "our capacity for feeling is in itself an insatiable and bottomless abyss."[43] This is one of the reasons that Durkheim believed crime and deviance are quite normal, even essential, in any society.

Durkheim's awakened reflection has become commonly known as greed. People tend to favor better conditions and additional fulfillment because they apparently have no biological or psychological mechanism to limit such tendencies. As Durkheim noted, the selfish desires of humankind "are unlimited so far as they depend on the individual alone. . . . The more one has, the more one wants."[44] Thus, society must step in and provide the regulative force that keeps humans from acting too selfishly.

One of the primary elements of this regulative force is the *collective conscience*, which is the extent of similarities or likenesses that people share. For example, almost everyone can agree that homicide is a serious and harmful act that should be avoided in any civilized society. The notion of collective conscience can be seen as an early form of the idea of social bonding, which has become one of the dominant theories in criminology.[45]

According to Durkheim, the collective conscience serves many functions in society. One such function is the ability to establish rules that control individuals from following their natural tendencies toward selfish behavior. Durkheim also believed that crime allows people to unite together in opposition against deviants. In other words, crime and deviance allow conforming individuals to be bonded together in opposition against a common enemy, as can be seen in everyday life when groups come together to face opposition. This enemy consists of the deviants who have not internalized the code of the collective conscience.

Many of Durkheim's ideas hold true today. Just recall a traumatic incident you may have experienced with other strangers (e.g., being stuck in an elevator during a power outage, weathering a serious storm, or being involved in a traffic accident). Incidents such as these bring people together and permit a degree of bonding that would not take place in everyday life. Crime, Durkheim argued, serves a similar function.

How is all of this relevant today? Most control theorists claim that individuals commit crime and deviant acts not because they are lacking in any way but because certain controls have been weakened in their development. This assumption is consistent with Durkheim's theory, which we discussed previously (see Chapter 6).

Freud's Concepts of Id, Superego, and Ego

Although psychoanalytic theory would seem to have few similarities with sociological positivistic theory, in this case, it is extremely complementary. One of Freud's most essential propositions is that all individuals are born with a tendency toward inherent drives and selfishness due to the **id** domain of the psyche (see Figure 8.2).[46] According to Freud, all people are born with equal amounts of id drives (e.g., libido, food, etc.) and motivations toward selfishness and greed. Freud said this inherent selfish tendency must be countered by controls produced from the development of the **superego,** which is the subconscious domain of the

---

[43]Durkheim, *Suicide,* 246–247. Also, much of this discussion is adapted from Paternoster and Bachman, *Explaining Criminals and Crime* (see chap. 5, n. 10).

[44]Durkheim, *Suicide,* 254.

[45]A good discussion of Durkheim's concepts, particularly that of the collective conscience, can be found in Vold et al., *Theoretical Criminology,* 124–139 (see chap. 2, n. 40).

[46]Sigmund Freud, "The Ego and the Id," in *The Complete Psychological Works of Sigmund Freud,* Vol. 19, ed. James Strachey (London: Hogarth Press, 1923/1959).

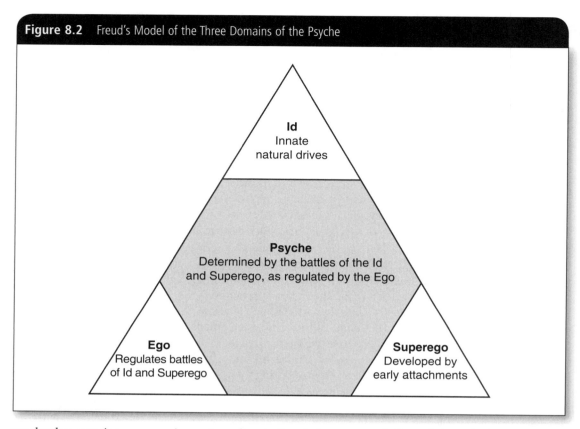

Figure 8.2 Freud's Model of the Three Domains of the Psyche

psyche that contains our conscience. According to Freud, the superego is formed through the interactions between a young infant or child and his or her significant others. As you can see, the control perspective has a long history in many philosophical and scientific disciplines.

These two drives of the subconscious domains of the id and superego are regulated, Freud thought, by the only conscious domain of the psyche: the **ego.** This ego mediates the battles between our innate drives (id) and our socialized constraints (superego); it represents our personality. There have been a number of applications of Freud's theoretical model to criminality, such as having a deficient superego (due to a lack of early attachments) or a weak ego (which fails to properly regulate the battle between the id and superego). The main point is that Freud was an early control theorist and that his theoretical model was highly influential among psychologists in the early 1900s as they tried to determine why certain individuals committed criminal offenses.[47]

Early Control Theories of Crime

Throughout the 1950s and 1960s, criminologists borrowed and built on some of the ideas just discussed. Until that time, most research in the criminological literature was dominated by the learning theories discussed earlier in this chapter or social structure theories, such as the Chicago School or Merton's strain

[47]Ibid.

theory (see Chapter 6). While early control theories may not be particularly popular in this day and age, they were vitally important in the sense that they laid the groundwork for future theoretical development.

Reiss's Control Theory

One of the first control theories of crime was proposed by Albert Reiss in 1951. Reiss claimed that delinquency was a consequence of weak ego or superego controls among juvenile probationers.[48] Reiss found no explicit motivation for delinquent activity. Rather, he thought it would occur in the absence of controls or restraints against such behavior.

Like Freud, Reiss believed that the family was the primary source through which deviant predispositions were discouraged. Furthermore, Reiss claimed that a sound family environment would provide for an individual's needs and the essential emotional bonds that are so important in socializing individuals. Another important factor in Reiss's model was close supervision, not only by the family but also by the community. He said that individuals must be closely monitored for delinquent behavior and adequately disciplined when they break the rules.

Personal factors, such as the ability to restrain one's impulses and delay gratification, were also important in Reiss's framework. These concepts are very similar to later, more modern concepts of control theory, which have been consistently supported by empirical research.[49] For this reason, Reiss was ahead of his time when he first proposed his control theory. Although the direct tests of Reiss's theory have provided only partial support for it, his influence is apparent in many contemporary criminological theories.[50]

Toby's Concept of Stake in Conformity

Soon after Reiss's theory was presented, a similar theory was developed. In 1957, Jackson Toby proposed a theory of delinquency and gangs.[51] He claimed that individuals were more inclined to act on their natural inclinations when the controls on them were weak. Like most other control theorists, Toby claimed that such inclinations toward deviance were distributed equally across all individuals. Furthermore, he emphasized the concept of a **stake in conformity** that supposedly prevents most people from committing crime. The stake in conformity Toby was referring to is the extent to which individuals have investments in conventional society. In other words, how much is a person willing to risk when he or she violates the law?

Studies have shown that stake in conformity is one of the most influential factors in individuals' decisions to offend. People who have nothing to lose are much more likely to take risks and violate others' rights than those who have relatively more invested in social institutions.[52]

One distinguishing feature of Toby's theory is his emphasis on peer influences in terms of both motivating and inhibiting antisocial behavior depending on whether most peers have low or high stakes in conformity. Toby's stake in conformity has been used effectively in subsequent control theories of crime.

---

[48]Albert Reiss, "Delinquency as the Failure of Personal and Social Controls," *American Sociological Review 16* (1951): 196–207.

[49]For a comprehensive review of studies of low self-control, see Travis Pratt and Frank Cullen, "The Empirical Status of Gottfredson and Hirschi's General Theory of Crime: A Meta-Analysis," *Criminology 38* (2000): 931–964.

[50]See Vold et al., *Theoretical Criminology*, 202–203.

[51]Jackson Toby, "Social Disorganization and Stake in Conformity: Complementary Factors in the Predatory Behavior of Hoodlums," *Journal of Criminal Law, Criminology, and Police Science 48* (1957): 12–17.

[52]Travis Hirschi, *Causes of Delinquency* (Berkeley: University of California Press, 1969); Robert Sampson and John Laub, *Crime in the Making: Pathways and Turning Points in Life* (Cambridge: Harvard University Press, 1993).

Nye's Control Theory

A year after Toby introduced the stake in conformity, F. Ivan Nye proposed a relatively comprehensive control theory that placed a strong focus on the family.[53] Following the assumptions of early control theorists, Nye claimed that there was no significant positive force that caused delinquency because such antisocial tendencies are universal and would be found in virtually everyone if not for certain controls usually found in the home.

Nye's theory consisted of three primary components of control. The first component was internal control, which is formed through social interaction. This socialization, he claimed, assists in the development of a conscience. Nye further claimed that if individuals are not given adequate resources and care, they will follow their natural tendencies toward doing what is necessary to protect their interests.

Nye's second component of control was direct control, which consists of a wide range of constraints on individual propensities to commit deviant acts. Direct control includes numerous types of sanctions, such as jail and ridicule, and the restriction of one's chances to commit criminal activity. Nye's third component of control was indirect control, which occurs when individuals are strongly attached to their early caregivers. For most children, it is through an intense and strong relationship with their parents or guardians that they establish an attachment to conventional society. However, Nye suggested that when the needs of an individual are not met by their caregivers, inappropriate behavior can result.

As shown in Figure 8.3, Nye predicted a U-shaped curve of parental controls in predicting delinquency. Specifically, he argued that either no controls (i.e., complete freedom) or too much control (i.e., no freedom at all) would predict the most chronic delinquency. He believed that a healthy balance of freedom and parental control was the best strategy for inhibiting criminal activity. Some recent research supports Nye's prediction.[54] Contemporary control theories, such as Tittle's **control-balance theory,** draw heavily on Nye's idea of having a healthy balance of controls and freedom.[55]

Reckless's Containment Theory

Another control theory, known as **containment theory,** has been proposed by Walter Reckless.[56] This theory emphasizes both inner containment and outer containment, which can be viewed as internal and external controls. Reckless broke from traditional assumptions of social control theories by identifying predictive factors that push or pull individuals toward antisocial behavior. However, the focus of his theory remained on the controlling elements, which can be seen in the emphasis placed on containment in the theory's name.

Reckless claimed that individuals can be pushed into delinquency by their social environment, such as by a lack of opportunities for education or employment. Furthermore, he pointed out some individual factors, such as brain disorders or risk-taking personalities, could push some people to commit criminal behavior. Reckless also noted that some individuals could be pulled into criminal activity by hanging out with delinquent peers, watching too much violence on television, and so on. All told, Reckless went beyond the typical control theory assumption of inborn tendencies. In addition to these natural dispositions toward deviant behavior, containment theory proposes that extra pushes and pulls can motivate people to commit crime.

---

[53]F. Ivan Nye, *Family Relationships and Delinquent Behavior* (New York: Wiley, 1958).

[54]Ruth Seydlitz, "Complexity in the Relationships among Direct and Indirect Parental Controls and Delinquency," *Youth and Society 24* (1993): 243–275.

[55]Charles Tittle, *Control Balance: Toward a General Theory of Deviance* (Boulder: Westview, 1995).

[56]Walter Reckless, *The Crime Problem,* 4th ed. (New York: Appleton-Century-Crofts, 1967).

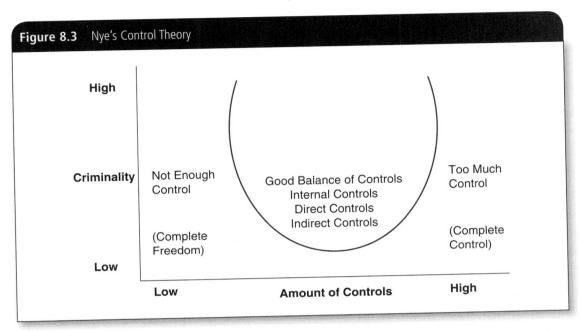

Figure 8.3   Nye's Control Theory

Reckless further claimed that the pushes and pulls toward criminal behavior could be enough to force individuals into criminal activity unless they are sufficiently contained or controlled. Reckless claimed that such containment should be both internal and external. By *internal containment,* he meant building a person's sense of self, which helps the person resist the temptations of criminal activity. According to Reckless, other forms of internal containment include the ability to internalize societal norms. With respect to *external containment,* Reckless claimed that social organizations, such as school, church, and other institutions, are essential in building bonds that inhibit individuals from being pushed or pulled into criminal activity.

Reckless offered a visual image of containment theory, which we present in Figure 8.4. The outer circle (Circle 1) in the figure represents the social realm of pressures and pulls (e.g., peer pressure, etc.), whereas the innermost circle (Circle 4) symbolizes a person's individual-level pushes to commit crime, such as predispositions or personality traits that are linked to crime. In between these two circles are the two layers of controls, external containment (Circle 2) and internal containment (Circle 3). The structure of Figure 8.4 and the examples included in each circle are those specifically noted by Reckless.[57]

While some studies have shown general support for containment theory, others offer more support for some components, such as internalization of rules, than for other factors, such as self-perception, in accounting for variations in delinquency.[58] External factors may be more important than internal ones. Furthermore, some studies have noted weaker support for Reckless's theory among minorities and females, who may be more influenced by their peers or other influences. Thus, the model appears to be most valid for White males, at least according to empirical studies.[59]

---

[57]Ibid., 479.

[58]Richard Lawrence, "School Performance, Containment Theory, and Delinquent Behavior," *Youth and Society 7* (1985): 69–95; Richard A. Dodder and Janet R. Long, "Containment Theory Reevaluated: An Empirical Explication," *Criminal Justice Review 5* (1980): 74–84.

[59]William E. Thompson and Richard A. Dodder, "Containment Theory and Juvenile Delinquency: A Reevaluation Through Factor Analysis," *Adolescence 21* (1986): 365–376.

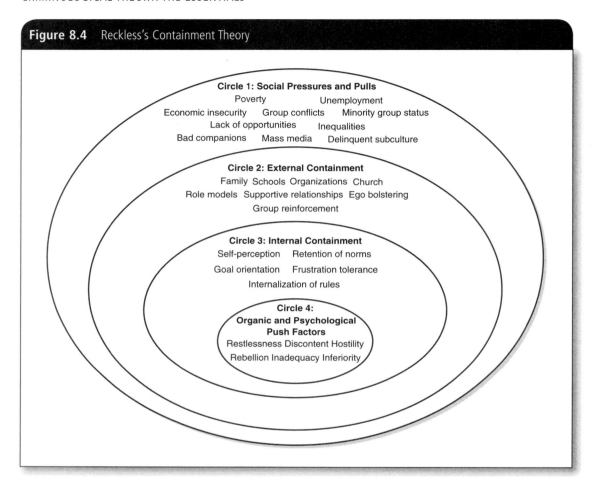

Figure 8.4   Reckless's Containment Theory

One of the problems with containment theory is that it does not go far enough toward specifying the factors that are important in predicting criminality, especially regarding specific groups of individuals. For example, an infinite number of concepts could potentially be categorized either as a push or pull toward criminality or as an inner or outer containment of criminality. Thus, the theory could be considered too broad or vague and not specific enough to be of practical value. To Reckless's credit, however, containment theory has increased the exposure of control theories of criminal behavior. And although support for containment theory has been mixed, there is no doubt that it has influenced other, more recent control theories.[60]

Modern Social Control Theories

As the previous sections attest, control theory has been around in various forms for some time. Modern social control theories build on these earlier versions and add levels of depth and sophistication. Two modern social control theories are Matza's drift theory and Hirschi's **social bonding theory.**

---

[60]Ronald L. Akers, *Criminological Theories: Introduction, Evaluation, and Application,* 3rd ed. (Los Angeles: Roxbury, 2000), 103–104.

Matza's Drift Theory

The theory of drift, or drift theory, presented by David Matza in 1964, claims that individuals offend at certain times in their lives when social controls—such as parental supervision, employment, and family ties—are weakened.[61] In developing his theory, Matza criticized earlier theories and their tendencies to predict too much crime. For example, the Chicago School would incorrectly predict that all individuals in bad neighborhoods will commit crime. Likewise, strain theory predicts that all poor individuals will commit crime. Obviously, this is not true. Thus, Matza claimed that there is a degree of determinism (i.e., Positive School) in human behavior but also a significant amount of free will (i.e., Classical School). He called this perspective **soft determinism,** which is the gray area between free will and determinism. This is illustrated in Figure 8.5.

Returning to the basics of Matza's theory, he claimed that individuals offend at the time in life when social controls are most weakened. As is well known, social controls are most weakened for most individuals during the teenage years. At this time, parents and other caretakers stop having a constant supervisory role, and at the same time, teenagers generally do not have too many responsibilities—such as careers or children—that would inhibit them from experimenting with deviance. This is very consistent with the well-known age–crime relationship; most individuals who are arrested experience this in their teenage years.[62] Once sufficient ties are developed, people tend to mature out of criminal lifestyles.

Matza further claimed that when supervision is absent and ties are minimal, the majority of individuals are the most free to do what they want. Where, then, does the term *drift* come from? During the times when people have few ties and obligations, they will drift in and out of delinquency, Matza proposed. He

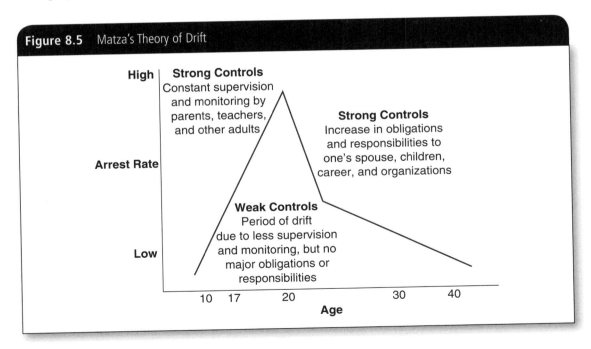

Figure 8.5   Matza's Theory of Drift

---

[61]Matza, *Delinquency and Drift.*

[62]*OJJDP Statistical Briefing Book.* Unpublished arrest data for 1980–1997 from the Federal Bureau of Investigation for 1998, 1999, and 2000 from *Crime in the United States* reports (Washington, DC: Government Printing Office, 1999, 2000, and 2001, respectively).

pointed out that previous theories are unsuccessful in explaining this age–crime relationship: "Most theories of delinquency take no account of maturational reform; those that do often do so at the expense of violating their own assumptions regarding the constrained delinquent."[63]

Matza insisted that drifting is not the same as a commitment to a life of crime. Instead, it is experimenting with questionable behavior and then rationalizing it. The way youths rationalize behavior that they know to be wrong is through learning the techniques of neutralization discussed earlier.

Drift theory goes on to say that individuals do not reject the conventional normative structure. On the contrary, much offending is based on neutralizing or adhering to **subterranean values,** which young people have been socialized to use as a means of circumventing conventional values. This is basically the same as asserting one's independence, which tends to occur with a vengeance during the teenage years.

Subterranean values are quite prevalent and underlie many aspects of our culture, which is why Matza's drift theory is also classified as a learning theory, as discussed earlier in this chapter. For example, while it is conventional to believe that violence is wrong, boxing matches and injury-prone sports are some of the most popular spectator activities. Such phenomena create an atmosphere that readily allows neutralization or rationalization of criminal activity.

We will see other forms of subterranean values when we discuss risk taking and low self-control later in this chapter. In many contexts, such as business, risk taking and aggressiveness are seen as desirable characteristics, so many individuals are influenced by such subterranean values. This, according to Matza, adds to individuals' likelihood of drifting into crime and delinquency.

Matza's theory of drift seems sensible on its face, but empirical research examining the theory has shown mixed results.[64] One of the primary criticisms of Matza's theory, which even he acknowledged, is that it does not explain the most chronic offenders, the people who are responsible for the vast majority of serious, violent crimes. Chronic offenders often offend long before and well past their teenage years, which clearly limits the predictive value of Matza's theory.

Despite its shortcomings, Matza's drift theory appears to explain why many people offend exclusively during their teenage and young adult years but then grow out of it. Also, the theory is highly consistent with several of the ideas presented by control theorists, including the assumption that (1) selfish tendencies are universal, (2) these tendencies are inhibited by socialization and social controls, and (3) the selfish tendencies appear at times when controls are weakest. The theory goes beyond previous control theories by adding the concepts of soft determinism, neutralization, and subterranean values, as well as the idea that, in many contexts, selfish and aggressive behaviors are not wrong but actually desirable.

Hirschi's Social Bonding Theory

Perhaps the most influential social control theory was presented by Travis Hirschi in 1969.[65] Hirschi's model of social bonding theory takes an assumption from Durkheim that "we are all animals, and thus naturally capable of committing criminal acts."[66] However, as Hirschi acknowledged, most humans can be adequately socialized to become tightly bonded to conventional entities, such as families, schools, communities, and

---

[63]"Age-arrest rate curve" is loosely based on data provided by Federal Bureau of Investigation, *Crime in the United States* report (Washington, DC: Government Printing Office, 1997).

[64]See Vold et al., *Theoretical Criminology,* 205–207.

[65]Travis Hirschi, *Causes of Delinquency* (see chap. 6, n. 30).

[66]Ibid., 31, in which Hirschi cites Durkheim.

the like. Hirschi said that the stronger a person is bonded to conventional society, the less prone to engaging in crime he or she will be. More specifically, the stronger the social bond, the less likely it is that an individual will commit criminal offenses.

As shown in Figure 8.6, Hirschi's social bond is made up of four elements: (1) attachment, (2) commitment, (3) involvement, and (4) moral belief. The stronger or more developed a person is in each of the four elements, the less likely he or she will be to commit crime. Let us now consider each element in detail.

The most important factor in the social bond is *attachments*, which consist of affectionate bonds between an individual and his or her significant others. Attachment is vitally important for the internalization of conventional values. Hirschi said, "The essence of internalization of norms, conscience, or superego thus lies in the attachment of the individual to others."[67] Hirschi made it clear, as did Freud, that strong, early attachments are the most important factor in developing a social bond. The other constructs in the social bond—commitment, involvement, and belief—are contingent on adequate attachment to others, he argued. That is, without healthy attachments, especially early in life, the probability of acting inappropriately increases.

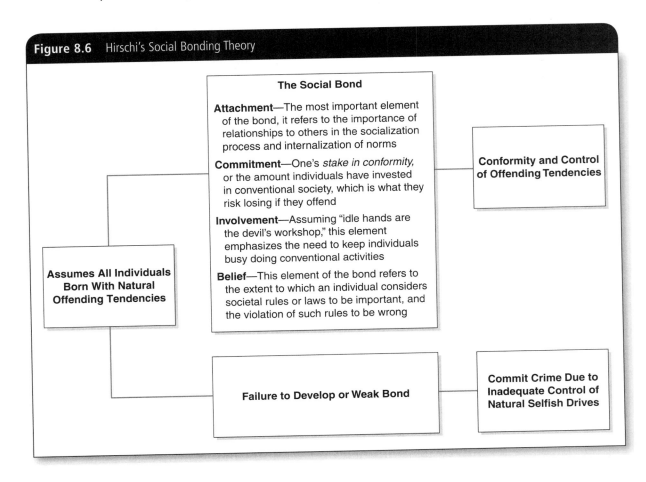

Figure 8.6 Hirschi's Social Bonding Theory

The Social Bond

Attachment—The most important element of the bond, it refers to the importance of relationships to others in the socialization process and internalization of norms

Commitment—One's *stake in conformity,* or the amount individuals have invested in conventional society, which is what they risk losing if they offend

Involvement—Assuming "idle hands are the devil's workshop," this element emphasizes the need to keep individuals busy doing conventional activities

Belief—This element of the bond refers to the extent to which an individual considers societal rules or laws to be important, and the violation of such rules to be wrong

Assumes All Individuals Born With Natural Offending Tendencies

Conformity and Control of Offending Tendencies

Failure to Develop or Weak Bond

Commit Crime Due to Inadequate Control of Natural Selfish Drives

---

[67]Ibid., 18.

Commitment, the second element of Hirschi's social bond, is the investment a person has in conventional society. This has been explained as one's stake in conformity, or what is at risk of being lost if one gets caught committing crime. If people feel they have much to lose by committing crime, they will probably not do so. In contrast, if someone has nothing to lose, what is to prevent that person from doing something he or she may be punished for? The answer is, of course, not much. And this, some theorists claim, is why it is difficult to control so-called chronic offenders. Trying to instill a commitment to conventional society in such individuals is extremely difficult.

Another element of the social bond is *involvement*, which is the time spent in conventional activities. The assumption is that time spent in constructive activities will reduce time devoted to illegal behaviors. This element of the bond goes back to the old adage that "idle hands are the devil's workshop."[68] Hirschi claimed that participating in conventional activities can inhibit delinquent and criminal activity.

The last element of the social bond is *beliefs*, which have generally been interpreted as moral beliefs concerning the laws and rules of society. This is one of the most examined and consistently supported aspects of the social bond. Basically, individuals who feel that a course of action is against their moral beliefs are much less likely to pursue it than individuals who don't see a breach of morality in such behavior. For example, we all probably know some people who see drunk driving as a very serious offense because of the injury and death it can cause. However, we also probably know individuals who don't see a problem with such behavior. The same can be said about speeding in a car, shoplifting from a store, or using marijuana; people differ in their beliefs about most forms of criminal activity.

Hirschi's theory has been tested by numerous researchers and has, for the most part, been supported.[69] However, one criticism is that the components of the social bond may predict criminality only if they are defined in a certain way. For example, with respect to the involvement element of the bond, studies have shown that not all conventional activities are equal when it comes to preventing delinquency. Only academic or religious activities seem to have consistent effects on inhibiting delinquency. In contrast, many studies show that teenagers who date or play sports actually have an increased risk of committing crime.[70]

Another major criticism of Hirschi's theory is that the effect of attachments on crime depends on to whom one is attached. Studies have clearly and consistently shown that attachment to delinquent peers is a strong predictor of criminal activity.

Finally, some evidence indicates that social bonding theory may better explain why individuals start offending than why they continue or escalate in their offending. One reason for this is that Hirschi's theory does not elaborate on what occurs after an individual commits criminal activity. This is likely the primary reason why some of the more complex, integrated theories of crime often attribute the initiation of delinquency to a breakdown in the social bond. However, they typically see other theories (such as differential reinforcement) as being better predictors of what happens after the initial stages of the criminal career.[71]

---

[68]Ibid., 22.

[69]For a review, see Akers, *Criminological Theories*, 105–110.

[70]Ibid.

[71]Delbert Elliott, David Huizinga, and Suzanne Ageton, *Explaining Delinquency and Drug Use* (Beverly Hill: Sage, 1985).

Despite the criticism it has received, Hirschi's social bonding theory is still one of the most accepted theories of criminal behavior.[72] It is a relatively convincing explanation for criminality because of the consistent support that it has found among samples of people taken from all over the world.[73]

Integrated Social Control Theories

Although we will review integrated theories in detail in Chapter 11, it is worthwhile to briefly discuss the two integrated models that most incorporate the control perspective into their frameworks. These two integrated models are control-balance theory and **power-control theory.** Both have received considerable attention in the criminological literature. Other integrated theories that incorporate control theory to a lesser extent include Braithwaite's shaming theory and Sampson and Laub's life-course theory. These will be covered in more detail in Chapters 11 and 12.

Tittle's Control-Balance Theory

Presented by Charles Tittle in 1995, control-balance theory proposes that (1) the amount of control to which one is subjected and (2) the amount of control one can exercise determine the probability of deviance occurring. The balance between these two types of control, he argued, can even predict the type of behavior that is likely to be committed.[74]

Tittle argued that a person is least likely to offend when he or she has a balance of controlling and being controlled. Furthermore, the likelihood of offending increases when these become unbalanced. If individuals are more controlled (Tittle calls this *control deficit*), then the theory predicts they will commit predatory or defiant acts. In contrast, if an individual possesses an excessive level of control (Tittle calls this *control surplus*), then he or she will be more likely to commit acts of exploitation or decadence. Note that excessive control is not the same as excessive self-control. Tittle argues that people who are controlling, that is, who have excessive control over others, will be predisposed toward inappropriate activities.

Initial empirical tests of control-balance theory have reported mixed results, with both surpluses and deficits predicting the same types of deviance.[75] In addition, researchers have uncovered differing effects of the control-balance ratio on two types of deviance that are contingent on gender. This finding is consistent with the gender-specific support found for Reckless's containment theory described earlier in this chapter.[76]

---

[72]Walsh and Ellis, "Political Ideology."

[73]Dennis Wong, "Pathways to Delinquency in Hong Kong and Guangzhou, South China," *International Journal of Adolescence and Youth 10* (2001): 91–115; Alexander Vazsonyi and Martin Killias, "Immigration and Crime Among Youth in Switzerland," *Criminal Justice and Behavior 28* (2001): 329–366; Manuel Eisner, "Crime, Problem Drinking, and Drug Use: Patterns of Problem Behavior in Cross-National Perspective," *Annals of the American Academy of Political and Social Science 580* (2002): 201–225.

[74]Charles Tittle, *Control Balance: Toward a General Theory of Deviance* (Boulder: Westview, 1995).

[75]Alex Piquero and Matthew Hickman, "An Empirical Test of Tittle's Control Balance Theory," *Criminology 37* (1999): 319–342; Matthew Hickman and Alex Piquero, "Exploring the Relationships Between Gender, Control Balance, and Deviance," *Deviant Behavior 22* (2001): 323–351.

[76]Hickman et al., "Exploring the Relationships," 323–351.

Hagan's Power-Control Theory

Power-control theory is an integrated theory that was proposed by John Hagan and his colleagues.[77] The primary focus of this theory is on the level of control and patriarchal attitudes, as well as structure in the household, which are influenced by parental positions in the workforce. Power-control theory assumes that, in households where the mothers and fathers have relatively similar levels of power at work (i.e., balanced households), mothers will be less likely to exert control on their daughters. These balanced households will be less likely to experience gender differences in the criminal offending of the children. However, households in which mothers and fathers have dissimilar levels of power in the workplace (i.e., unbalanced households) are more likely to suppress criminal activity in daughters. In addition, assertiveness and risky activity among the males in the house will be encouraged. This assertiveness and risky activity may be precursors to crime.

Most empirical tests of power-control have provided moderate support for the theory, while more recent studies have further specified the validity of the theory in different contexts.[78] For example, one recent study reported that the influence of mothers, not fathers, on sons had the greatest impact on reducing the delinquency of young males.[79] Another researcher has found that differences in perceived threats of embarrassment and formal sanctions varied between more patriarchal and less patriarchal households.[80] Finally, studies have also started measuring the effect of patriarchal attitudes on crime and delinquency.[81] Power-control theory is a good example of a social control theory in that it is consistent with the idea that individuals must be socialized and that the gender differences in such socialization make a difference in how people will act throughout life.

A General Theory of Crime: Low Self-Control

In 1990, Travis Hirschi, along with his colleague Michael Gottfredson, proposed a general theory of low self-control, which is often referred to as *the general theory of crime*.[82] This theory has led to a significant amount of debate and research in the field since its appearance, more than any other contemporary theory of crime. Like previous control theories of crime, this theory assumes that individuals are born predisposed toward selfish, self-centered activities and that only effective child-rearing and socialization can create self-control.

---

[77] John Hagan, *Structural Criminology* (Newark: Rutgers University Press, 1989); John Hagan, A. Gillis, and J. Simpson, "The Class Structure of Gender and Delinquency: Toward a Power-Control Theory of Common Delinquent Behavior," *American Journal of Sociology 90* (1985): 1151–1178; John Hagan, A. Gillis, and J. Simpson, "Clarifying and Extending Power-Control Theory," *American Journal of Sociology 95* (1990): 1024–1037; John Hagan, J. Simpson, and A. Gillis, "Class in the Household: A Power-Control Theory of Gender and Delinquency," *American Journal of Sociology 92* (1987): 788–816.

[78] Hagen et al., "Class in the Household," 788–816; B. McCarthy and John Hagan, "Gender, Delinquency, and the Great Depression: A Test of Power-Control Theory," *Canadian Review of Sociology and Anthropology 24* (1987): 153–177; Merry Morash and Meda Chesney-Lind, "A Reformulation and Partial Test of the Power-Control Theory of Delinquency," *Justice Quarterly 8* (1991): 347–377; Simon Singer and Murray Levine, "Power-Control Theory, Gender, and Delinquency: A Partial Replication with Additional Evidence on the Effects of Peers," *Criminology 26* (1988): 627–647.

[79] B. McCarthy, John Hagan, and T. Woodward, "In the Company of Women: Structure and Agency in a Revised Power-Control Theory of Gender and Delinquency," *Criminology 37* (1999): 761–788.

[80] Brenda Sims Blackwell, "Perceived Sanction Threats, Gender, and Crime: A Test and Elaboration of Power-Control Theory," *Criminology 38* (2000): 439–488.

[81] Blackwell, "Perceived Sanction Threats," 439–488; Kristin Bates and Chris Bader, "Family Structure, Power-Control Theory, and Deviance: Extending Power-Control Theory to Include Alternate Family Forms," *Western Criminology Review 4* (2003).

[82] Michael Gottfredson and Travis Hirschi, *A General Theory of Crime* (Palo Alto: Stanford University Press, 1990).

Without such adequate socialization (i.e., social controls) and reduction of criminal opportunities, individuals will follow their natural tendencies to become selfish predators. Furthermore, the general theory of crime assumes that self-control must be established by age 10. If it has not formed by that time, then, according to the theory, individuals will forever exhibit low self-control.

Although Gottfredson and Hirschi still attribute the formation of controls as coming from the socialization processes, the distinguishing characteristic of this theory is its emphasis on the individual's ability to control himself or herself. That is, the general theory of crime assumes that people can take a degree of control over their own decisions and, within certain limitations, control themselves.

The general theory of crime is accepted as one of the most valid theories of crime.[83] This is probably because it identifies only one primary factor that causes criminality—low self-control. But, low self-control theory may actually implicate a series of personality traits and behavior, including risk taking, impulsiveness, self-centeredness, short-term orientation, and quick temper. For example, recent research has supported the idea that inadequate child-rearing practices tend to result in lower levels of self-control among children and that these low levels produce various risky behaviors, including criminal activity.[84] Such propensities toward low self-control can manifest in varying forms across an individual's life. For example, teenagers with low self-control will likely hit or steal from peers, and as they grow older, will be more likely to gamble or cheat on taxes.

Psychological Aspects of Low Self-Control

Criminologists have recently claimed that low self-control may be due to the emotional disposition of individuals. For example, one study showed that the effects of low self-control on intentions to commit drunk driving and shoplifting were tied to individuals' perceptions of pleasure and shame. More specifically, the findings of this study showed that individuals who had low self-control had significantly lower levels of anticipated shame but significantly higher levels of perceived pleasure in committing both drunk driving and shoplifting.[85] These results suggest that individuals who lack self-control will be oriented toward gaining pleasure and taking advantage of resources and toward avoiding negative emotional feelings (e.g., shame) that are primarily induced through socialization.

Physiological Aspects of Low Self-Control

Low self-control can also be tied to physiological factors. Interestingly, research has shown that chronic offenders show greater arousal toward danger and risk taking than the possibility of punishment.[86] This arousal has been measured by monitoring brain activity in response to certain stimuli. The research suggests that individuals are encouraged to commit risky behavior due to physiological mechanisms that reward their risk-taking activities by releasing pleasure chemicals in their brains.[87]

---

[83]For an excellent review of studies regarding low self-control theory, see Travis Pratt and Frank Cullen, "The Empirical Status of Gottfredson and Hirschi's General Theory of Crime: A Meta-Analysis," *Criminology 38* (2000): 931–964; for critiques of this theory, see Ronald Akers, "Self-Control as a General Theory of Crime," *Journal of Quantitative Criminology 7* (1991): 201–211; for a study that demonstrates the high popularity of the theory, see Walsh and Ellis, "Political Ideology."

[84]Carter Hay, "Parenting, Self-Control, and Delinquency: A Test of Self-Control Theory," *Criminology 39* (2001): 707–736; Karen Hayslett-McCall and Thomas Bernard, "Attachment, Masculinity, and Self-Control: A Theory of Male Crime Rates," *Theoretical Criminology 6* (2002): 5–33.

[85]Alex Piquero and Stephen Tibbetts, "Specifying the Direct and Indirect Effects of Low Self-Control and Situational Factors in Offenders' Decision Making: Toward a More Complete Model of Rational Offending," *Justice Quarterly 13* (1996): 481–510.

[86]Adrian Raine, *The Psychopathology of Crime* (San Diego: Academic Press, 1993).

[87]Anthony Walsh, *Biosocial Criminology: Introduction and Integration* (Cincinnati: Anderson, 2002).

In a similar vein, recent studies show that chronic gamblers tend to get a physiological high (such as a sudden, intense release of brain chemicals similar to that following a small dose of cocaine) from the activity of betting, particularly when they are gambling with their own money and risking a personal loss.[88] Undoubtedly, a minority of individuals thrive off of risk-taking behaviors significantly more than others. This suggests that physiological as well as psychological differences may explain why certain individuals favor risky behaviors.

Researchers have also found that criminal offenders generally perceive a significantly lower level of internal sanctions (e.g., shame, guilt, embarrassment) than do nonoffenders.[89] So, in summary, a select group of individuals appear to derive physiological and psychological pleasure from engaging in risky behaviors while simultaneously being less likely to be inhibited by internal emotional sanctions. Such a combination, Gottfredson and Hirschi claimed, is very dangerous and helps explain why some impulsive individuals often end up in prison.

Finally, the psychological and physiological aspects of low self-control may help explain the gender differences observed between males and females. Specifically, studies show that females are significantly more likely to experience internal emotional sanctioning for offenses they have committed than are males.[90] In other words, there appears to be something innately different about males and females that helps explain the differing levels of self-control each possesses.

Summary of Control Theories

Control perspectives are among the oldest and most respected explanations of criminal activity. The fundamental assumption that humans have an inborn, selfish disposition that must be controlled through socialization distinguishes control theories from other theories of crime. The control perspective's longevity as one of the most popular criminological theories demonstrates its legitimacy as an explanation of behavior. This is likely due to the dedication and efforts of criminologists who are constantly developing new and improved versions of control theory, many of which we have discussed here.

Policy Implications

Numerous policy implications can be taken from the various types of social learning and control theories presented here. We will concentrate on those that are likely to be most effective and pragmatic in helping to reduce criminal behavior.

A number of policy implications can be drawn from the various learning models. Perhaps their most important suggestion is to supply many opportunities for positive reinforcements, or rewards, for good behavior. Such reinforcements have been found to be far more effective than punishments, especially among criminal offenders.[91] Furthermore, studies show that the most effective rehabilitation programs for offenders

---

[88]Christopher D. Fiorillo, Phillippe N. Tobler, and Wolfram Schultz, "Discrete Coding of Reward Probability and Uncertainty by Dopamine Neurons," *Science 299* (2003): 1898–1902.

[89]Harold Grasmick and Robert Bursik, "Conscience, Significant Others, and Rational Choice: Extending the Deterrence Model," *Law and Society Review 24* (1990): 837–861; Stephen Tibbetts, "Shame and Rational Choice in Offending Decisions," *Criminal Justice and Behavior 24* (1997): 234–255.

[90]Stephen Tibbetts and Denise Herz, "Gender Differences in Factors of Social Control and Rational Choice," *Deviant Behavior 17* (1996): 183–208.

[91]Van Voorhis et al., *Correctional Counseling.*

should be based on a cognitive-behavioral approach, which teaches individuals to think before they act.[92] Furthermore, evaluation studies have shown that simply grouping offenders together for counseling or peer-therapy sessions is not an effective strategy; rather, it appears that such programs often show no effect, or actually increase offending among participants, perhaps because they tend to learn more antisocial attitudes from such sessions.[93] Ultimately, the offender programs that emphasize positive reinforcements and are based on a cognitive-behavioral approach are those that show the greatest success.

Regarding the policy implications of control theories, we will focus on the early social bonding that must take place and the need for more parental supervision to help an individual develop or learn self-control and create healthy, strong bonds to conventional society. Most control theories assume that individuals are predisposed to criminal behavior, so the primary focus of programs should be to reduce this propensity toward such behavior. According to most control perspectives, the most important factor in preventing or controlling this predisposition involves early parenting and building attachments or ties to prosocial aspects of the individual's environment.

Thus, perhaps the most important policy recommendation is to increase the ties between early caregivers or parents and their children. A variety of programs try to increase the relationship and bonding that takes place between infants and young children and their parents, as well as to monitor the supervision that takes place in this dynamic. Such programs have consistently been shown to be effective in preventing and reducing criminality in high-risk children, especially when such programs involve home visitations by health care (e.g., nurses) and social care experts (e.g., social workers).[94] By visiting the homes of high-risk children, workers can provide more direct, personal attention in helping aid and counsel parents about how best to nurture, monitor, and discipline their young children.[95] These types of programs may lead to more control over behavior while building stronger bonds to society and developing self-control among these high-risk individuals.

✂ Conclusion

In this chapter, we have discussed a wide range of theories that may appear to be quite different. However, all of the criminological theories here share an emphasis on social processes as the primary reason for why individuals commit crime. This is true regarding the learning theories, which propose that people are taught to commit crime, as well as the control theories, which claim people offend naturally and rather must be taught not to commit crime. Despite their seemingly opposite assumptions of human behavior, the fact is that learning and control theories both identify socialization, or the lack thereof, as the key cause of criminal behavior.

We also examined some of the key policy recommendations that have been suggested by both of these theoretical perspectives. Specifically, we noted that programs that simply group offenders together only seem to reinforce their tendency to offend, whereas programs that take a cognitive-behavioral approach and

---

[92]Ibid.

[93]Ibid.; also see reviews of such programs in Richard J. Lundman, *Prevention and Control of Juvenile Delinquency* (New York: Oxford University Press, 1993); Akers and Sellers, *Criminological Theories,* 101–108.

[94]Joycen Carmouche and Joretta Jones, "Her Children, Their Future," *Federal Prisons Journal 1* (1989): 23, 26–27.

[95]For a review, see Stephen G. Tibbetts, "Perinatal and Developmental Determinants of Early Onset of Offending: A Two-Peak Model Approach," in *The Development of Persistent Criminality,* ed. Johanna Savage (New York: Oxford University Press, 2009), 179–201.

use many reward opportunities appear to have some effect in reducing recidivism. Also, we concluded that programs that involve home visitations by experts (e.g., nurses, counselors) tend to aid in developing more effective parenting and building of social bonds among young individuals, which helps build strong attachments to society and develop self-control.

Chapter Summary

- First, we discussed what distinguishes learning theories of crime from other perspectives.
- We then discussed Sutherland's differential association theory and how this framework was improved by Akers's differential reinforcement theory.
- We examined in depth the psychological learning model of classical conditioning, as well as its limitations.
- We then explored two other learning models that were the basis of differential reinforcement theory, namely operant conditioning and learning according to modeling or imitation.
- We reviewed the theory of neutralization, including the five original techniques of neutralization presented by Sykes and Matza.
- We also reviewed several early forms of social control theory, such as Hobbes, Freud, and Durkheim.
- Then, we examined the early social control theories of crime, presented by Reiss, Toby, and Nye, along Reckless's containment theory.
- We then examined more modern social control theories, such as Hirschi's social bonding theory.
- Integrated social control theories were briefly examined, including Tittle's control-balance theory and Hagan's power-control theory.
- Finally, we reviewed low self-control theory, from both psychological and physiological perspectives.

KEY TERMS

Classical conditioning	Id	Positive punishment
Containment theory	Imitation–modeling	Positive reinforcement
Control-balance theory	Learning theories	Power-control theory
Control theories	Low self-control theory	Social bonding theory
Differential association theory	Modeling and imitation	Soft determinism
Differential identification theory	Negative punishment	Stake in conformity
Differential reinforcement theory	Negative reinforcement	Subterranean values
Drift theory	Neutralization theory	Superego
Ego	Operant conditioning	*Tabula rasa*

DISCUSSION QUESTIONS

1. What distinguishes learning theories from other criminological theories?

2. What distinguishes differential association from differential reinforcement theory?

3. What did differential identification add to learning theories?

4. Which technique of neutralization do you use or relate to the most? Why?

5. Which technique of neutralization do you find least valid? Why?

6. Which element of Hirschi's social bond do you find you have highest levels of?

7. Which element of Hirschi's social bond do you find you have lowest levels of?

8. Can you identify someone you know who fits the profile of a person with low self-control?

9. Which aspects of the low self-control personality do you think you fit?

10. Regarding Matza's theory of drift, do you think this relates to when you or your friends committed crime in life? Studies show that most people commit crimes when they are in their teens or 20s or, at least, know people who do (e.g., drinking under age 21, speeding).

WEB RESOURCES

Differential Association Theory

http://www.criminology.fsu.edu/crimtheory/sutherland.html

Differential Reinforcement Theory

http://www.criminology.fsu.edu/crimtheory/akers.htm

Social and Self-Control Theory

http://www.criminology.fsu.edu/crimtheory/hirschi.htm

Techniques of Neutralization

http://www.criminology.fsu.edu/crimtheory/matza.htm

9

Social Reaction, Critical, and Feminist Models of Crime

T his chapter will discuss the evolution of social reaction and **labeling theory,** reviewing contributions made by early theorists, as well as the modern developments in this area. We will then discuss social conflict and reaction models of criminal behavior with an emphasis on the foundational assumptions and principles of Marx, as well as the more criminological applications of Marxist and **conflict theory** by Bonger, Turk, Vold, and others. We'll also review modern applications of various forms of feminism theory.

During the 1960s and early 1970s, social reaction and labeling theories, as well as various critical, conflict, and feminist theories, became popular. At the time, society was looking for theories that placed the blame for criminal offending on government authorities, either the police or societal institutions like economic or class structures. Here, we explore these various theories with a special emphasis on how they radically altered the way that crime and law were viewed, as well as how these perspectives highly represented the overall climate in the United States at that time. Specifically, many groups of people—particularly the lower class, minorities, and women—were fighting for their rights during this period, and this manifested itself in criminological theory and research.

⬜ Labeling and Social Reaction Theory

Social reaction theory, otherwise referred to as *labeling theory,* is primarily concerned with how individuals' personal identities are highly influenced by the way that society or authority tends to categorize them as offenders. With such categorization or labeling, an offender becomes a self-fulfilling prophecy, in this perspective, and results in individuals confirming their status as criminals or delinquents by increasing the frequency or seriousness of their illegal activity. Furthermore, this perspective assumes that there is a tendency to put negative labels on lower-class individuals or minorities as offenders significantly more often than on middle- or upper-class White people.[1]

This perspective assumes that people who are labeled offenders have virtually no choice but to conform to the role that they have been "assigned" by society. Thus, social reaction theory claims that recidivism can be reduced by limiting stigmatization by authorities (e.g., law enforcement) and society. This is referred to as the *hands-off policy,* and it became very popular in the 1960s and early 1970s.[2] Policies that became popular during this period were **diversion, decriminalization,** and **deinstitutionalization** (known as the Ds). (See Chapter 12 for a discussion of these policies.) All of these attempted to get youthful or first-time offenders out of the formal justice system as soon as possible to avoid stigmatizing or labeling them as offenders. Today, these very policies have led critics to dismiss labeling theory by claiming that it promotes lenient and ineffective policy in sentencing practices.[3]

Labeling theory was based on seminal work by George Mead and Charles Cooley, which emphasized the importance of the extreme ways that individuals react to and are influenced by the social reaction to their roles and behavior. George Herbert Mead, who was a member of the Chicago School (see Chapter 7), said that a person's sense of self is constantly constructed and reconstructed through the various social interactions a person has on a daily basis.[4] Every person is constantly aware of how he or she is judged by others through social interactions.

Readers can probably relate to this in the sense that they have experienced how differently they are treated in stores or restaurants if they are dressed nicely as opposed to being less well dressed; as you have observed, there is a significant difference in the way one is treated. Also, when growing up, you probably heard your parents or guardians warn that you should not hang out with Johnny or Sally because they were "bad" kids. Or perhaps you were a Johnny or Sally at some point. Either way, you can see how certain individuals can be labeled by authorities or society and then ostracized by mainstream groups. This can lead to isolation and typically results in a person having only other "bad" kids or adults to hang out with. This results in a type of feedback system, in which the person begins associating with others who will only increase their propensity toward illegal activity.[5]

Many strain theorists claim that certain demographic factors, such as social class or the neighborhood where a certain offense took place, may make it more likely that the offender will be caught and labeled by authorities. This claim is quite likely to be true, especially given recent policing strategies that target areas or neighborhoods that have high rates of crime. This is the side of social reaction and labeling theory that deals with the disproportionate rate at which members of the lower class and minorities are labeled as offenders.

Some of the earliest labeling theorists laid the groundwork for this perspective long before it became popular in the 1960s. For example, in the 1930s, Frank Tannenbaum noted the **dramatization of evil** that occurred when youth were arrested and charged with their first offense.[6] Later, other theorists such as Edwin Lemert contributed a highly important causal sequence to how labeling affects criminality among

---

[1]Vold et al., *Theoretical Criminology* (see chap. 2, n. 40).

[2]Edwin Schur, *Radical Nonintervention: Rethinking the Delinquency Problem* (Englewood Cliffs: Prentice Hall, 1973).

[3]Akers and Sellers, *Criminological Theories* (see chap. 8, n. 17).

[4]George H. Mead, *Mind, Self, and Society* (Chicago: University of Chicago, 1934).

[5]Howard S. Becker, *Outsiders: Studies in the Sociology of Deviance* (New York: Free Press, 1963).

[6]Frank Tannenbaum, *Crime and the Community* (Boston: Ginn, 1938).

those who are labeled. Lemert said that individuals, typically youths, commit **primary deviance,** which is not serious (i.e., it is nonviolent) and not frequent, but they happen to be caught by police and are subsequently labeled.[7] The stigma of the label makes them think of themselves as offenders and forces them to associate only with other offenders. This results in what Lemert referred to as **secondary deviance,** in which offending is more serious (often violent) and far more frequent. Thus, the causal model that Lemert describes is illustrated as

Primary Deviance → Caught and Labeled → Secondary Deviance

According to Lemert's model, if the label or stigma was not placed on a young or first-time offender, then the more serious and more frequent offending of secondary deviance would not take place. Therefore, Lemert's model is highly consistent with the labeling approach's hands-off policies, such as diversion, decriminalization, and deinstitutionalization. If you ignore such behavior, Lemert reasons, it will tend to go away. However, since the mid-1970s, the get-tough approach has become highly dominant, so such policies are not emphasized by society or policy makers today.

Research on labeling theory suffered a significant blow in the 1970s and 1980s when empirical findings showed consistently that formal arrests and sanctions did not tend to result in ways that supported tradi-

tional labeling theory.[8] In fact, most people who are arrested once are never arrested again, which tends to support deterrence theory and does not support labeling theory. In addition, some experts have concluded that "the preponderance of research finds no or very weak evidence of [formal] labeling effects. . . . The soundest conclusion is that official sanctions by themselves have neither strong deterrent nor a substantial labeling effect."[9] Furthermore, some theorists have questioned the basic assumptions of labeling theory, pointing out that the label does not cause the initial (or primary) offending and that labeling theorists largely ignore the issue of what is causing individuals to engage in illegal activity in the first place. Also, labeling theorists do not recognize the fact that offenders who are caught tend to be the ones who are committing more crimes than those who are not caught; in fact, there tends to be a strong relationship between being caught and committing multiple offenses.[10]

However, more contemporary research and theorizing have emphasized more informal forms of labeling, such as labeling by the community, parents, or friends. Studies have shown more support for the influence of this informal labeling on individuals' behavior.[11] After all, it only makes sense that informal labeling by

▲ **Image 9.1** Karl Marx

---

[7]Edwin Lemert, *Human Deviance, Social Problems, and Social Control,* 2nd ed. (Englewood Cliffs: Prentice Hall, 1972).

[8]Marvin Wolfgang, Robert Figlio, and Thorsten Sellin, *Delinquency in a Birth Cohort* (Chicago: University of Chicago Press, 1972).

[9]Akers and Sellers, *Criminological Theories,* 142.

[10]Charles Wellford, "Labeling Theory and Criminology," *Social Problems* 22 (1975): 313–332.

[11]Lening Zhang, "Informal Reactions and Delinquency," *Criminal Justice and Behavior* 24 (1997): 129–150.

people with whom you interact on a daily basis (i.e., parents, friends, neighbors, employers, etc.) will have more impact in terms of how you feel about yourself than labeling by police or other authorities, which tends to be temporary or situational.

Marxist Theories of Crime

Based on the writings of Karl Marx, Marxist theories of crime focus on the fact that people from the lower classes (i.e., poor) are arrested and charged with crimes at a disproportionate rate. As in conflict criminology, Marxist theories emphasize the effects of a capitalistic society on how justice is administered, describing how society is divided by money and power.[12] Marx said the law is the tool by which the **bourgeoisie** (the ruling class in a country without a ruling aristocracy; e.g., industrialists and financiers in Western industrialized countries) controls the lower classes (the **proletariat** and the lowest group, the *lumpenproletariat*) and keeps them in a disadvantaged position. In other words, the law is used as a mechanism by which the middle or upper class maintains their dominance over the lower classes. More specifically, Marx claimed that law is used as a tool to protect the economic interests and holdings of the bourgeoisie, as well as to prevent the lower classes from gaining access to financial resources.[13] Thus, Marxist theories propose that economic power can be translated into legal or political power and substantially accounts for the general disempowerment of the majority.

Willem Bonger

One early key theorist who applied Marxist theory to crime was Willem Bonger, who emphasized the relationship between economy and crime but did not believe simply being poor would cause criminal activity. Rather, he thought crime resulted because capitalism caused a difference in the way individuals felt about society and their places in it. In the early 1900s, Bonger said that the contemporary economic structure, particularly capitalism, was the cause of crime in the sense that it promoted a system based on selfishness and greed.[14] Such selfishness manifests itself in competition among individuals, which is obvious in interactions and dealings for goods and resources. This competition and selfishness lead to more isolation, individualism, and egoistic tendencies, which promote a strong focus on self-interests at the expense of communitarianism and societal well-being. Bonger believed that this strong focus on the individual leads to criminal behavior.[15] He also stressed the association between social conditions (largely the result of economic systems) and criminal offending; because of cultural differences, crime can be a normal, adaptive response to social and economic problems, he argued. The poor often develop a strong feeling of injustice, which also contributes to their entering into illegal activity.[16]

Richard Quinney

Although Bonger's theory did not become popular in the early 1900s, when his book was originally published, his ideas received a lot of attention when a neo-Marxist period began in the early 1970s. This

---

[12]Karl Marx, *Selected Works of Karl Marx and Frederick Engels* (Moscow: Foreign Languages Publishing House, 1962).

[13]Vold et al., *Theoretical Criminology*.

[14]Willem Bonger, *Criminality and Economic Conditions* (Bloomington: Indiana University Press, 1969 [originally published in 1905]).

[15]Vold et al., *Theoretical Criminology*; Akers and Sellers, *Criminological Theories*.

[16]Bonger, *Criminality and Economic Conditions*.

renewed interest in Marxist theory was coupled with harsh criticisms leveled at the existing theoretical frameworks, which is why these neo-Marxist theories are often referred to as *critical theories*. This time, Marxist theories of crime became quite popular largely because the social climate desired such perspectives. Whereas notable European theorists in this vein include Ian Taylor, Paul Walton, and Jock Young,[17] one of the key figures in this neo-Marxist perspective in the United States was Richard Quinney.[18]

Like Bonger, Quinney claimed that crime was caused by the capitalistic economic structure and the emphasis on materials that this system produced. One way that Quinney's theory goes beyond Bonger is that Quinney further proposed that even the crimes committed by the upper classes were caused by capitalism. To clarify, Quinney claimed that such acts were crimes of "domination and repression" committed by the elite to keep the lower classes down or to protect their property, wealth, and power.[19] A good example is white-collar crimes, which almost always involve raising profits or income, whether for individual or company advantage; such crimes often result in losses to the relatively lower-income clients or customers.

Evidence Regarding Marxist Theories of Crime

Many critics noted that these seminal Marxist theories of crime were too simplistic, as well as somewhat naïve in the sense that they seemed to claim that the capitalistic economic system was the only reason for crime and that socialism or communism was the only sure way to reduce crime in the United States.[20] Now, even most Marxist theorists reject this proposition. Thus, more modern frameworks have been presented that place more emphasis on factors that stem from capitalism. For example, Colvin and Pauly presented a theoretical model that claims delinquency and crime are the result of problematic parenting, which results from the degrading and manipulative treatment that parents of lower-class children get in the workplace.[21] However, the empirical tests of this more modern Marxist theory have demonstrated rather weak effects regarding the importance of capitalism (or parenting practices resulting from employment positions or social class).[22] Thus, there does not seem to be much empirical support for Marxist or neo-Marxist theories of crime, which is perhaps why this theoretical framework is not one of the primary models currently accepted by most criminologists.[23]

Conflict Theories of Crime

Conflict theories of crime assume that all societies are in a process of constant change and that this dynamic process inevitably creates conflicts among various groups.[24] Much of the conflict is due to the

---

[17]Ian Taylor, Paul Walton, and Jock Young, *The New Criminology: For a Social Theory of Deviance* (New York: Harper & Row, 1973).

[18]Richard Quinney, *Critique of Legal Order: Crime Control in Capitalist Society* (Boston: Little, Brown and Company, 1974).

[19]Quinney, *Critique of Legal Order;* also see Akers and Sellers, *Criminological Theories*, 224–226 for a discussion.

[20]For a discussion, see Akers and Sellers, *Criminological Theories*, 226–231.

[21]Mark Colvin and John Pauly, "A Critique of Criminology: Toward an Integrated Structural Marxist Theory of Delinquency Production," *American Journal of Sociology 89* (1983): 513–551.

[22]See Sally Simpson and Lori Elis, "Is Gender Subordinate to Class? An Empirical Assessment of Colvin and Pauly's Structural Marxist Theory of Delinquency," *Journal of Criminal Law and Criminology 85* (1994): 453–480; see further discussion in Akers and Sellers, *Criminological Theories*.

[23]Lee Ellis and Anthony Walsh, "Criminologists' Opinions About the Causes and Theories of Crime and Delinquency," *The Criminologist 24* (1999): 1–4.

[24]William Chambliss and Robert Seidman, *Law, Order, and Power* (Reading: Addison-Wesley, 1971).

competition to have each group's interests promoted, protected, and often put into law. If all groups were equally powerful and had the same amount of resources, such battles would involve much negotiation and compromise; however, groups tend to differ significantly in the amount of power or resources. Thus, laws can be created and enforced such that powerful groups can exert dominance over the weaker groups. So, like Marxist theories, law is seen as a tool by which some groups gain and maintain dominance over less powerful groups. Furthermore, this state of inequality and resulting oppression creates a sense of injustice and unfairness among members of the less powerful groups, and such feelings are a primary cause of crime.[25]

There are several types of conflict theory, and fittingly for this framework (as well as inherently supportive of the model), theorists of varying types often give scathing reviews of the other types of conflict theory. Marxist theories are one example. Critics have noted that many communistic countries (e.g., Cuba, Russia, etc.) have high rates of crime, whereas some countries that have capitalistic economic structures have very low crime rates, such as Sweden.

Another type of conflict theory is referred to as *pluralistic;* it argues that, instead of one or a few groups holding power over all the other groups, a multitude of groups must compete on a relatively fair playing field.[26] However, this latter type of conflict theory is not one of the more popular versions among critical theorists because it is sometimes seen as rather naïve and idealistic.[27] Some of the key theorists in the **pluralistic (conflict) perspective** are Thorsten Sellin, George Vold, and Austin Turk.

Thorsten Sellin

Thorsten Sellin applied Marxist and conflict perspectives, as well as numerous other types of models, to studying the state of cultural diversity in industrial societies. Sellin claimed that separate cultures will diverge from a unitary, mainstream set of norms and values held by the dominant group in society.[28] Thus, these minority groups that break off from the mainstream will establish their own norms. Furthermore, when laws are enacted, they will reflect only the values and interests of the dominant group, which causes what Sellin referred to as *border culture conflict.* This conflict of values, which manifests itself when different cultures interact, can cause a backlash by the weaker groups, which tend to react defiantly or defensively. According to Sellin, the more unequal the balance of power, the worse the conflict tends to be.[29]

George Vold

Another key conflict theorist was George Vold, who presented his model in his widely used textbook, *Theoretical Criminology.*[30] Vold claimed that people are naturally social and inevitably form groups out of shared needs, values, and interests. Because various groups compete with each other for power and to promote their values and interests, each group competes for control of political processes, including the power to

---

[25]Vold et al., *Theoretical Criminology.*

[26]Akers and Sellers, *Criminological Theories.*

[27]Quinney, *Critique of Legal Order.*

[28]Thorsten Sellin, *Culture Conflict and Crime* (New York: Social Science Research Council, 1938).

[29]Sellin, *Culture Conflict.*

[30]George Vold, *Theoretical Criminology* (New York: Oxford University Press, 1958).

create and enforce laws that can suppress the other groups. Some critics have argued that Vold put too much emphasis on the battle for creation of laws as opposed to the enforcement of laws.[31]

Austin Turk

Like the other conflict theorists, Austin Turk assumed that the competition for power among various groups in society is the primary cause of crime.[32] Turk emphasized the idea that a certain level of conflict among groups can be very beneficial because it reminds citizens to consider whether the status quo or conventional standards can be improved. This type of idea is very similar to Durkheim's proposition that a certain level of crime is healthy for society because it defines moral boundaries and sometimes leads to progress (see Chapter 6). Another aspect of Turk's theorizing, which separates him from other conflict theorists, is that he saw conflict among the various components of the criminal justice system. For example, the police often are at odds with the courts and district attorney's office. Such tension or conflict among formal agencies that should be on the same side of the playing field leads to even more frustration and inefficiency when it comes to fighting crime and ensuring that justice is served.

Evidence Regarding Conflict Theories of Crime

Empirical tests of conflict theories are rare, likely because of the nature of the framework, which lends itself to a global view of societal structure and a perhaps infinite number of interest groups who are constantly in play for power.[33]

However, one notable study found evidence of a relationship between U.S. states that had large numbers of interest groups and violent crime but not property crime.[34] The authors concluded that these findings demonstrated the need for more discussion about how competitiveness in the United States affects criminal behavior, but no other studies have examined the influence of political interest groups on criminal behavior. It is rather difficult to test conflict theory in other ways. Perhaps conflict theory researchers should build an agenda of more rigorous ways to test the propositions of their theoretical perspective; as it stands, it remains quite vague. The few studies do not seem rigorous enough to persuade other criminologists or readers toward accepting the validity of this model.[35]

Despite the lack of empirical research supporting the conflict (and Marxist or critical) theories of crime, there is little doubt that such perspectives have contributed much to the theorizing and empirical studies of criminologists regarding this framework. In fact, the American Society of Criminology (ASC)—which is probably the largest and best known professional society in the discipline—has a special division made up of experts devoted to this area of study. So it is likely that the theorizing and empirical research will be greatly enhanced in the near future. Furthermore, it is clear that criminologists have acknowledged the need to explore the various issues presented by the conflict and Marxist perspectives in research in criminal justice and offending.

---

[31] Akers and Sellers, *Criminological Theories*.

[32] Austin Turk, *Criminality and Legal Order* (Chicago: Rand McNally, 1969).

[33] See discussion by Akers and Sellers, *Criminological Theories*, 210–212.

[34] Gregory G. Brunk and Laura A. Wilson, "Interest Groups and Criminal Behavior," *Journal of Research in Crime and Delinquency 28* (1991): 157–173.

[35] Ellis and Walsh, "Criminologists' Opinions."

Feminist Theories of Crime

About the same time that Marxist theories of crime were becoming popular in the early 1970s, the feminist perspective began to receive attention; this was a key period in the women's rights movement. The Feminist School of criminology began largely as a reaction to the lack of rational theorizing about why females commit crime and why they tend to be treated far differently by the criminal justice system.[36] Prior to the 1970s, theories of why girls and women engage in illegal activities were primarily based on false stereotypes.

Much of the attention of theorists in this area can be broken into two categories: the *gender ratio* issue and the *generalizability* issue. The gender ratio issue refers to theorizing and research that examines why females so often commit less serious, less violent offenses than males. Some experts feel that this does not matter; however, if we understood why females commit far less violence, then perhaps we could apply such knowledge to reducing male offending.[37] The generalizability argument is consistent with the idea some make about the gender ratio issue; specifically, many of the same critics argue that theorists should simply take the findings found for male offending and generalize them to females. However, given the numerous differences found among males and females in what predicts their offending patterns, simply generalizing across gender is not a wise thing to do.[38]

▲ **Image 9.2**   Female Gang Members

Another important issue in feminist research on crime is that women today have more freedom and rights than those in past generations. Seminal theories of female crime in the 1970s predicted that this would result in far higher offending rates for women.[39] However, this has not been seen in serious, violent crimes, rather in observed increases in property and public order crimes typically committed by girls or women who have not benefited from such freedom and rights, for example, those who do not have much education or strong employment records.

Also, there are numerous forms of feminism and, thus, many types of feminist theories of crime, as pointed out by Daly and Chesney-Lind.[40] One of the earliest was **liberal feminism,** which assumes that differences between males and females in offending were due to the lack of female opportunities in education and employment and that, as more females were given such opportunities, they would come to resemble males in terms of offending.

Another major feminist perspective of crime is **critical** or **radical feminism,** which emphasizes the idea that many societies (such as the United States) are based on a structure of patriarchy, in which males

---

[36]Meda Chesney-Lind and Lisa Pasko, *The Female Offender: Girls, Women, and Crime* (Thousand Oaks: Sage, 2004).

[37]For a discussion, see Stephen Tibbetts and Denise Herz, "Gender Differences in Factors of Social Control and Rational Choice," *Deviant Behavior 17* (1996): 183–208.

[38]Tibbetts and Herz, "Gender Differences."

[39]Freda Adler, *Sisters in Crime: The Rise of the New Female Criminal* (New York: McGraw Hill, 1975).

[40]Kathleen Daly and Meda Chesney-Lind, "Feminism and Criminology," *Justice Quarterly 5* (1988): 536–558.

dominate virtually every aspect of society, such as politics, family structure, and economy. It is hard to contest the primary assumption of this theory. Despite the fact that more women than ever hold professional, white-collar jobs, men still get paid a significant amount more than women for the same positions, on average. Furthermore, the U.S. Senate and House of Representatives—and other high political offices, such as president, vice president, cabinet posts, U.S. Supreme Court justices—are still made up primarily (or exclusively) by men. So the United States, like most other countries in the world, appears to be based in patriarchy.

The extent to which this model explains female criminality, however, remains to be seen. Regarding serious crimes, it is not clear why this perspective would expect higher or lower rates of female criminal behavior. Regarding some delinquent offenses, it may partially explain the greater tendency to arrest females for certain offenses. For example, virtually every self-report study ever conducted shows that males run away far more than females; however, FBI data show that female juveniles are arrested for running away far more often than males. This model may provide the best explanation for this difference. Females are more protected—that is, reported and arrested for running away—because they are considered to be more like property in our patriarchal society. This is just one explanation, but it appears to be somewhat valid.

Similar to critical or radical feminism is **Marxist feminism,** which emphasizes men's ownership and control of the means of economic production, thus focusing solely on the economic structure. Marxist feminists point out that men control economic success in our country, as well as in virtually every country in the world, and that this flows from capitalism. One of the primary assumptions of capitalism is *survival of the fittest* or *the best person for the job,* which would seem to favor women. Studies have found that women in the United States do far better, despite our capitalistic system, than those in most other countries. Furthermore, women in countries based on Marxism have a less favorable lifestyle and are no better off economically than those in the United States. Whether or not one believes in a Marxist economic structure, it does not readily explain female criminality.

Another feminist perspective is that of **socialist feminism,** which moved away from economic structure (e.g., Marxism) as the primary detriment for females and placed a focus on control of reproductive systems. This model believes that women should take control of their own bodies and their reproductive functions in order to control their criminality. It is not entirely clear how females' taking charge of their reproductive destinies can increase or reduce their crime rates. Although no one can deny that data show females who reproduce frequently, especially in inner-city, poor environments, tend to offend more often than other females, it appears that other factors mediate these effects. Women who want good futures tend to take more precautions against becoming pregnant; on the other hand, the very females who most need to take precautions against getting pregnant are the least likely to do so, despite the availability of numerous forms of contraception. This is one of the many paradoxes in our field. It is unclear how much socialist feminism has contributed to an understanding of female criminality.

An additional perspective of feminist criminology is that of **postmodern feminism,** which holds that an understanding of women as a group, even by other women, is impossible because every person's experience is unique. If this is true, we should give up discussing female criminal theory and theories of criminality in general—along with all studies of medicine, astronomy, psychology, and so on—because every person interprets each observation subjectively. According to postmodern feminists, there is no point in measuring anything. Thus, this model is based on anti-science and has contributed nothing to the study of female criminality.

In all these variations of feminist perspectives, little emphasis is placed on parental differences in how children are disciplined and raised. Studies have clearly shown that parents, often without realizing it, tend to globally reward young boys for completing a task (e.g., "You are such a good boy"), whereas they tend to tell a young girl that she did a good job. On the other hand, when young boys do not successfully complete a task, most parents tend to excuse the failure (e.g., "It was a hard thing to do; don't worry"), whereas for young girls, the parents will often globally evaluate them for the task (e.g., "Why couldn't you do it?"). Although numerous psychological studies have found this tendency, it has yet to make it into the mainstream criminological theories of crime.

Evidence Regarding Feminist Theories of Crime

As discussed above, there is no doubt that female offenders were highly neglected by traditional models of criminological theory, and given that they make up at least 50% of the population of the world, it is important they be covered and explained by such theories. Furthermore, we also discussed the fact that if we know why females everywhere commit far less violence than men, it would likely go a long way toward policy implications for reducing the extremely higher rates of violence among males. However, in other ways, the feminist theories of crime have not been supported.

For instance, as noted above, the seminal feminist crime theories specifically proposed that, as women became liberated, their rates of crime would become consistent with the rates of male offending.[41] Not only did this fail to occur, but the evidence actually supports the opposite trend; specifically, the females who were given the most opportunities (e.g., education, employment, status, etc.) were the *least* likely to offend, whereas the women who were not liberated or given such opportunities were the *most* likely to engage in criminal behavior.[42]

On the other hand, one major strength of feminist theories of crime is that they have led to a number of studies showing that the factors causing crime in males are different than those for females. For example, females appear to be far more influenced by internal, emotional factors; they are more inhibited by moral emotions, such as shame, guilt, and embarrassment.[43] Ultimately, there is no doubt that feminist theories of crime have contributed much to the discourse and empirical research regarding why females (as well as males) commit crime. In fact, some highly respected criminology and criminal justice journals have been created to deal with that very subject. So, in that sense, the field has recognized and accepted the need to explore the various issues involved in examining feminist theorizing and research in offending and the justice system.

⊠ Policy Implications

A variety of policy implications have come from the theoretical perspectives in this chapter. Regarding social reaction and labeling theory, several policies have evolved, known as the Ds: diversion, decriminalization, and deinstitutionalization. Diversion, which is now commonly used, involves trying to get cases out of the formal justice system as soon as possible. Courts try to get many juvenile cases and, in most recent times, drug possession cases diverted to a less formal, more administrative process (e.g., drug courts, youth accountability boards or teams, etc.). Such diversion programs appear to have saved many billions of dollars on offenders, who otherwise would have been incarcerated, while providing a way for

---

[41]Adler, *Sisters in Crime.*

[42]For a discussion, see Akers and Sellers, *Criminological Theories;* also see discussion in Vold et al., *Theoretical Criminology.*

[43]For a review, see Tibbetts and Herz, "Gender Differences."

first-time or relatively nonserious offenders not to experience the stigmatizing effects of being incarcerated. Although empirical evaluations of such diversion programs are mixed and suffer from methodological problems (i.e., the individuals who volunteer or qualify for such programs are likely the better cases among the sample population), some studies have shown their potential promise.[44]

Regarding decriminalization, there have been numerous examples of reducing the criminality of certain illegal activities. A good example is the legal approach to marijuana possession in California. Unlike other jurisdictions, California does not incarcerate individuals for possessing less than an ounce of marijuana; rather, they receive a citation, similar to a parking ticket. There are many other forms of decriminalization, distinguished from legalization, which makes an act completely legal and not subject to legal sanction. The purpose of decriminalization is to reduce emphasis and resources placed on offenders who pose less danger to society, but in terms of social reaction and labeling theory, decriminalization also reduces the stigmatization of individuals who are relatively minor offenders, who would likely become more serious offenders if they were incarcerated with more chronic offenders.

Another policy implication of this chapter is deinstitutionalization. In the early 1970s, federal laws were passed that ordered all youth status offenders to be removed from incarceration facilities. This has not been accomplished; some are still being placed in such facilities. However, the number and rate of status offenders being placed in incarceration facilities has declined, avoiding any further stigmatization and integration into further criminality. Overall, this deinstitutionalization has kept relatively minor, often first-time offenders from experiencing the ordeals of incarceration.

Additional policy implications that can be inferred from this chapter involve providing more economic and employment opportunities to those who do not typically have access to such options. From a historical perspective, such as the New Deal during the Great Depression of the 1930s, providing more employment opportunities can greatly enhance the well-being of the population, and in that period, there was a very significant decrease in crime and homicide rates (see Chapter 1). Today, perhaps nothing could be more important in our nation than creating jobs; the ability to do this will largely determine future crime rates.

Finally, there are numerous policy implications regarding feminist theory and perspectives of crime. Of primary importance is to include such perspectives in future research and theoretical developments. Furthermore, it is important to realize that females offend far less than males; if criminologists could figure out why, this could be a landmark finding in reducing crime significantly. Ultimately, the policy implications from this chapter indicate that the more attention and resources given to disenfranchised groups (e.g., youth, minorities, women, etc.), the less likelihood that they will offend, and the better we will understand the reasons why they offend.

⊠ Conclusion

This chapter examined the theories that place responsibility for individuals' criminal behavior on societal factors. Specifically, this chapter discussed theories of social reaction or labeling, critical perspectives, and feminist theories of criminal behavior. All of these theories have a common theme: They emphasize the use of legal or criminal justice systems to target or label certain groups of people (poor, women, etc.) as criminals while protecting the interests of those who have power (i.e., White males). Given the evidence discussed here, readers will have enough evidence to make objective conclusions about whether this

---

[44]For a review, see John Worrall, *Crime Control in America: An Assessment of the Evidence* (Boston: Allyn & Bacon, 2006): 228–229.

perspective is valid, as well as which aspects of these theories are more supported by empirical research and which are more in question.

⬚ Chapter Summary

- First, we explored the basic assumptions of social reaction and labeling theory.
- We then reviewed the primary theoretical concepts and propositions of labeling theory, especially the importance of distinguishing primary deviance from secondary deviance.
- We discussed the current state of labeling theory, which emphasizes informal sources of social reaction, not just the formal sources, as in traditional models.
- Then, we examined Marxist theories of crime, as well as numerous subsequent versions of this theory, such as those developed by Bonger, Chambliss, and Quinney.
- We also examined conflict theories, including theories by Sellin, Vold, and Turk.
- Finally, we examined the basic assumptions of feminist perspectives of crime, as well as some of the more notable implications that can be derived from these perspectives.

KEY TERMS

Bourgeoisie	Dramatization of evil	Primary deviance
Conflict theory	Labeling theory	Proletariat
Critical feminism	Liberal feminism	Radical feminism
Decriminalization	Marxist feminism	Secondary deviance
Deinstitutionalization	Pluralist (conflict) perspective	Socialist feminism
Diversion	Postmodern feminism	

DISCUSSION QUESTIONS

1. What are the major assumptions of labeling and social reaction theory, and how does this differ significantly from other traditional theories of crime?

2. According to Lemert, what is the difference between primary deviance and secondary deviance?

3. How can you relate personally to being labeled, even if not for offending?

4. How do Marx's ideas relate to the study of crime? Provide some examples.

5. Which conflict theory do you buy into the most? The least? Why?

6. What are the key assumptions and features of the various feminist perspectives?

7. Which type of feminist theory do you believe is the most helpful for explaining crime?

WEB RESOURCES

Conflict Theories

http://www.criminology.fsu.edu/crimtheory/conflict.htm

Feminist Theories of Crime

http://www.keltawebconcepts.com.au/efemcrim1.htm

Labeling Theory

http://www.criminology.fsu.edu/crimtheory/week_10.htm

Marxist Theory

www.sociology.org.uk/pcdevmx.doc

http://law.jrank.org/pages/819/Crime-Causation-Sociological-Theories-Critical-theories.html

10

Life-Course Perspectives of Criminality

This chapter will discuss the development of the life-course perspective in the late 1970s and its influence on modern research on criminal trajectories. We will focus on explaining the various concepts in the life-course perspective, such as onset, desistence, and frequency, as well as the arguments against this perspective. Finally, we will review the current state of research regarding this perspective.

This chapter will present one of the most current and progressive approaches toward explaining why individuals engage in criminal activity, namely **developmental theories** of criminal behavior. Developmental theories are explanatory models of criminal behavior that follow individuals throughout their life courses of offending, thus explaining the development of offending over time. Such developmental theories represent a break with past traditions of theoretical frameworks, which typically focused on the contemporaneous effects of constructs and variables on behavior at a given point in time. Virtually no theories attempted to explain the various stages (e.g., onset, desistence) of individuals' criminal careers, and certainly no models differentiated the varying factors that are important at each stage. Developmental theories have been prominent in modern times, and we believe that readers will agree that developmental theories have added a great deal to our understanding and thinking about why people commit criminal behavior.

Developmental Theories

Developmental theories, which are also to some extent integrated, are distinguished by their emphasis on the evolution of individuals' criminality over time. Developmental theories tend to look at the individual as the unit of analysis, and such models focus on the various aspects of the onset, frequency, intensity, duration, desistence, and other aspects of the individual's criminal career. The onset of offending is when the offender first begins offending, and desistence is when an individual stops committing crime. Frequency refers to how often the individual offends, whereas intensity is the degree of seriousness of the offenses he or she commits. Finally, duration is the length of an individual's criminal career.

Experts have long debated and examined these various aspects of the development of criminal behavior. For example, virtually all studies show an escalation from minor status offending (e.g., truancy, underage drinking, smoking tobacco) to petty crimes (e.g., shoplifting, smoking marijuana) to far more serious criminal activity, such as robbery and aggravated assault, and then murder and rape. This development of criminality is shown across every study that has ever been performed and demonstrates that, with very few exceptions, people begin with relatively minor offending and progress toward more serious, violent offenses.

Although this trend is undisputed, other issues are not yet resolved. For example, studies have not yet determined when police contact or an arrest becomes *early onset*. Most empirical studies draw the line at age 14, so that any arrest or contact prior to this time is considered early onset.[1] However, other experts would disagree and say that this line should be drawn earlier (say 12 years old) or even later (such as 16 years old). Still, however it is defined, early onset is one of the most important predictors of any of the measures we have in determining who is most at risk for developing serious, violent offending behavior.

Perhaps the most discussed and researched aspect of developmental theory is that of *offender frequency,* which has been referred to as *lambda*. Estimates of lambda, or average frequency of offending by criminals over a year period, vary greatly.[2] Some estimates of lambda are in the high single digits and some are in the triple digits. Given this large range, it does not do much good in estimating what the frequency of most offenders are. Rather, the frequency depends on many, many variables, such as what type of offenses the individual commits. Perhaps if we were studying only drug users or rapists, it would make sense to determine lambda, but given the general nature of most examinations of crime, such estimates are not useful. Still, the frequency of offending, even within crime type, varies so widely across individuals that we question its use in understanding criminal careers.

Before we discuss the dominant models of developmental theory, it is important to discuss the opposing viewpoint, which is that of complete stability in offending. Such counterpoint views assume that the developmental approach is a waste of time because the same individuals who show antisocial behavior at early ages (before age 10) are those who will exhibit the most criminality in their teenage years, 20s, 30s, 40s, and so on. This framework is most notably represented by the theoretical perspective proposed by Gottfredson and Hirschi in their model of low self-control.

Antidevelopmental Theory: Low Self-Control Theory

In 1990, Travis Hirschi, along with his colleague Michael Gottfredson, proposed a general theory of low self-control as the primary cause of all crime and deviance (see prior discussion in Chapter 8); this is often referred to as *the general theory of crime*.[3] This theory has led to a significant amount of debate and research in the field since its appearance, more than any other contemporary theory of crime.

Like other control theories of crime, this theory assumes that individuals are born predisposed toward selfish, self-centered activities and that only effective child-rearing and socialization can create self-control.

---

[1]For more discussion, see Stephen Tibbetts and Alex Piquero, "The Influence of Gender, Low Birth Weight, and Disadvantaged Environment in Predicting Early Onset of Offending: A Test of Moffitt's Interactional Hypothesis," *Criminology* 37 (1999): 843–878; Chris L. Gibson and Stephen Tibbetts, "A Biosocial Interaction in Predicting Early Onset of Offending," *Psychological Reports* 86 (2000): 509–518.

[2]For a review, see Samuel Walker, *Sense and Nonsense about Crime and Drugs: A Policy Guide,* 5th ed. (Belmont: Wadsworth, 2001).

[3]Michael Gottfredson and Travis Hirschi, *A General Theory of Crime* (Palo Alto: Stanford University Press, 1990).

Without such adequate socialization (i.e., social controls) and reduction of criminal opportunities, individuals will follow their natural tendencies to become selfish predators. The general theory of crime assumes that self-control must be established by age 10. If it has not formed by that time, then according to the theory, individuals will forever exhibit low self-control. This assumption of the formation of low self-control by age 10 or before is the oppositional feature of this theory to the developmental perspective. Once low self-control is set by age 10, there is no way to develop it afterward, the authors assert. In contrast, developmental theory assumes that people can indeed change over time.

Like others, Gottfredson and Hirschi attribute the formation of controls to socialization processes in the first years of life; the distinguishing characteristic of this theory is its emphasis on the individual's ability to control himself or herself. That is, the general theory of crime assumes that people can take a degree of control over their own decisions and, within certain limitations, control themselves. The general theory of crime is accepted as one of the most valid theories of crime.[4] This is probably due to the parsimony, or simplicity, of the theory because it identifies only one primary factor that causes criminality—low self-control. But, low self-control may actually consist of a series of personality traits, including risk taking, impulsiveness, self-centeredness, short-term orientation, and quick temper. Recent research has supported the idea that inadequate child-rearing practices tend to result in lower levels of self-control among children and that these low levels produce various risky behaviors, including criminal activity.[5] It is important to note that this theory has a developmental component in the sense that it proposes that self-control develops during early years from parenting practices; thus, even this most notable antidevelopment theory actually includes a strong developmental aspect.

In contrast to Gottfredson and Hirschi's model, one of the most dominant and researched frameworks of the last 20 years, another sound theoretical model shows that individuals can change their life trajectories in terms of crime. Research shows that events or realizations can occur that lead people to alter their frequency or incidence of offending, sometimes to zero. To account for such extreme **transitions,** we must turn to the dominant life-course model of offending, which is that of Sampson and Laub's developmental model of offending.

Sampson and Laub's Developmental Model

Perhaps the best known and researched developmental theoretical model to date is that of Robert Sampson and John Laub.[6] Sampson and Laub have proposed a developmental framework that is largely based on a reanalysis of original data collected by Sheldon and Eleanor Glueck in the 1940s. As a prototype developmental model, individual stability and change are the primary foci of their theoretical perspective.

Most significantly, Sampson and Laub emphasized the importance of certain events and life changes, which can alter an individual's decisions to commit (or not commit) criminal activity. Although based on a

---

[4]For an excellent review of studies regarding low self-control theory, see Travis Pratt and Frank Cullen, "The Empirical Status of Gottfredson and Hirschi's General Theory of Crime: A Meta-Analysis," *Criminology 38* (2000): 931–964; for critiques of this theory, see Ronald Akers, "Self-Control as a General Theory of Crime," *Journal of Quantitative Criminology 7* (1991): 201–211; for a study that demonstrates the high popularity of the theory, see Walsh and Ellis, "Political Ideology," 1, 14 (see chap 1, n. 13).

[5]Carter Hay, "Parenting, Self-Control, and Delinquency: A Test of Self-Control Theory," *Criminology 39* (2001): 707–736; K. Hayslett-McCall and T. Bernard, "Attachment, Masculinity, and Self-Control: A Theory of Male Crime Rates," *Theoretical Criminology 6* (2002): 5–33.

[6]Robert Sampson and John Laub, "Crime and Deviance Over the Life Course: The Salience of Adult Social Bonds," *American Sociological Review 55* (1990): 609–627; Robert Sampson and John Laub, "Turning Points in the Life Course: Why Change Matters to the Study of Crime," *Criminology 31* (1993): 301–326; Robert Sampson and John Laub, *Crime in the Making: Pathways and Turning Points Through Life* (Cambridge: Harvard University Press, 1993).

Figure 10.1 Gottfredson and Hirschi's Theory of Low Self-Control

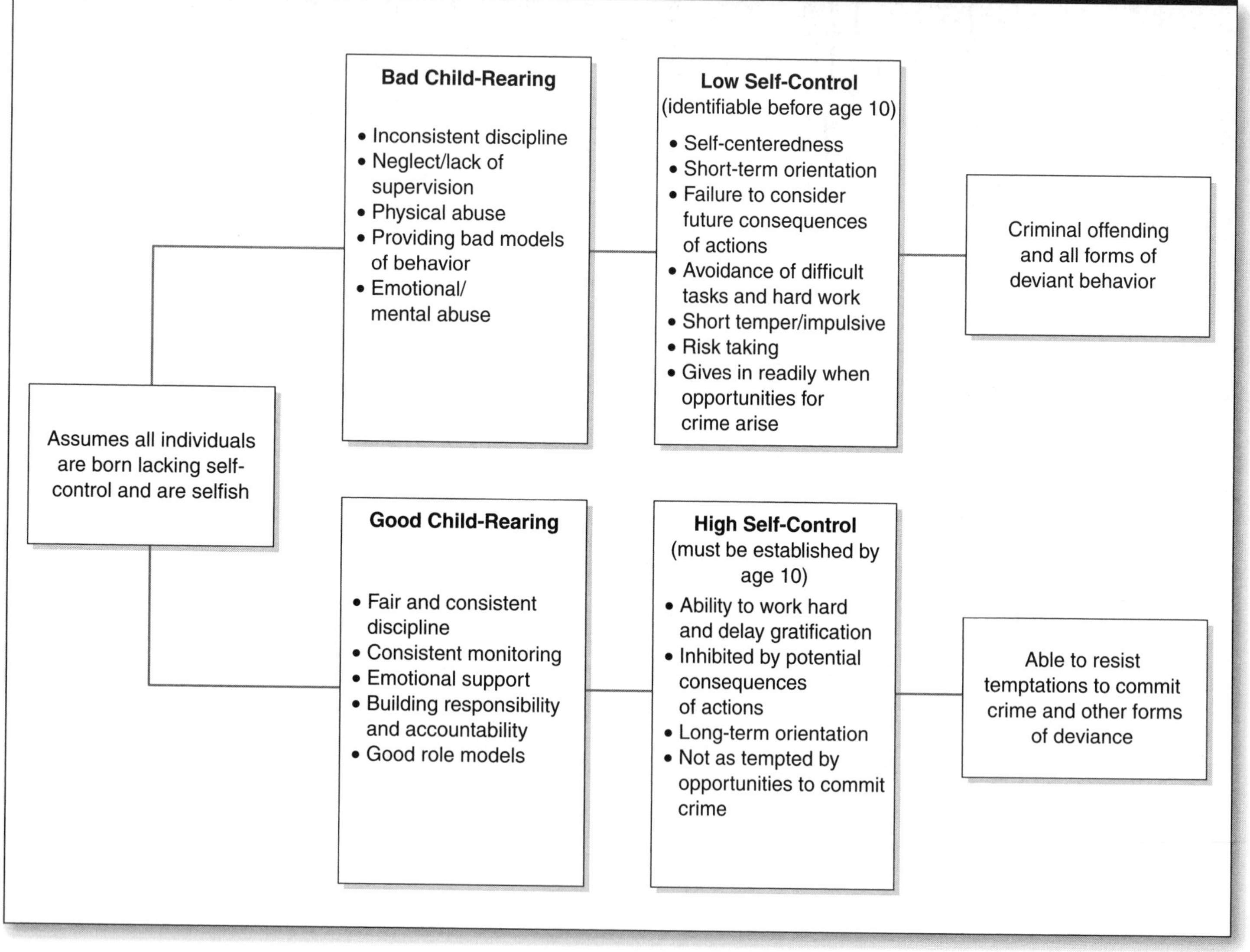

Assumes all individuals are born lacking self-control and are selfish

Bad Child-Rearing

- Inconsistent discipline
- Neglect/lack of supervision
- Physical abuse
- Providing bad models of behavior
- Emotional/mental abuse

Low Self-Control
(identifiable before age 10)

- Self-centeredness
- Short-term orientation
- Failure to consider future consequences of actions
- Avoidance of difficult tasks and hard work
- Short temper/impulsive
- Risk taking
- Gives in readily when opportunities for crime arise

Criminal offending and all forms of deviant behavior

Good Child-Rearing

- Fair and consistent discipline
- Consistent monitoring
- Emotional support
- Building responsibility and accountability
- Good role models

High Self-Control
(must be established by age 10)

- Ability to work hard and delay gratification
- Inhibited by potential consequences of actions
- Long-term orientation
- Not as tempted by opportunities to commit crime

Able to resist temptations to commit crime and other forms of deviance

social control framework, this model contains elements of other theoretical perspectives. First, Sampson and Laub's model assumes, like other developmental perspectives, that early antisocial tendencies among individuals, regardless of social variables, are often linked to later adult criminal offending. Furthermore, some social structure factors (e.g., family structure, poverty, etc.) also tend to lead to problems in social and educational development, which then leads to crime. Another key factor in this development of criminality is the influence of delinquent peers or siblings, which further increases an individual's likelihood for delinquency.

However, Sampson and Laub also strongly emphasize the importance of transitions, or events that are important in altering trajectories toward or against crime, such as marriage, employment, or military service, drastically changing a person's criminal career. Sampson and Laub show sound evidence that many individuals who were once on a path toward a consistent form of behavior, in this case, serious, violent crime, suddenly (or gradually) halted due to such a transition or series of transitions. In some ways, this model is a more specified form of David Matza's theory of drift, which we discussed in Chapter 8, in which individuals tend to grow out of crime and deviance due to the social controls imposed by marriage, employment, and so on. Still, Sampson and Laub's framework contributed much to the knowledge of criminal offending by providing a more specified and grounded framework that identified the ability of individuals to change their criminal trajectories via life-altering transitions, such as the effect that marriage can have on a man or woman, which is quite profound. In fact, recent research has consistently shown that marriage and full-time employment significantly reduced the recidivism of California parolees, and other recent studies have shown similar results from employment in later years.[7]

▲ **Image 10.1**   Life-course theories attempt to uncover points of desistance, critical events that may change a person's life path from deviant behavior back to law-abiding status. Getting married is considered one of those events. Once one is married, free time to hang out with one's friends tends to disappear.

Moffitt's Developmental Taxonomy

Another primary developmental model that has had a profound effect on the current state of criminological thought and theorizing is Terrie **Moffitt's developmental theory (taxonomy)**, proposed in 1993.[8] Moffitt's framework distinguishes two types of offenders: **adolescence-limited** and **life-course persistent offenders**. Adolescence-limited offenders make up most of the general public and include all persons who committed offenses when they were teenagers or young adults. Their offending was largely caused by association with peers and a desire to engage in activities that were exhibited by the adults that they are trying to be. Such activities are a type of rite of passage and quite normal among all people who have normal social interactions with their peers in teenage or young adult years. It should be noted that a very small percentage

---

[7]Alex Piquero, Robert Brame, Paul Mazzerole, and Rudy Haapanen, "Crime in Emerging Adulthood," *Criminology 40* (2002): 137–170; Chris Uggen, "Work as a Turning Point in the Life Course of Criminals: A Duration Model of Age, Employment, and Recidivism," *American Sociological Review 65* (2000): 529–546.

[8]Terrie Moffitt, "Adolescence Limited and Life Course Persistent Antisocial Behavioral: A Developmental Taxonomy," *Psychological Review 100* (1993): 674–701.

(about 1 to 3%) of the population are nonoffenders who quite frankly do not have normal relations with their peers and therefore do not offend at all, even in adolescence.

On the other hand, there exists a small group of offenders, referred to in this model as *life-course persistent offenders*. This small group, estimated to be 4 to 8% of offenders—albeit the most violent and chronic—commit the vast majority of the serious, violent offenses in any society, such as murder, rape, and armed robbery. In contrast to the adolescence-limited offenders, the disposition of life-course persistent offenders toward offending is caused by an entirely different model: an interaction between neurological problems and the disadvantaged or criminological environments in which they are raised.

For example, if an individual had only neurological problems or only a poor, disadvantaged environment, then that individual would be unlikely to develop a life-course persistent **trajectory** toward crime. However, if a person has both neurological problems and a disadvantaged environment, then that individual would have a very high likelihood of becoming a chronic, serious, violent offender. This proposition, which has been supported by empirical studies,[9] suggests that it is important to pay attention to what happens early in life. Because illegal behaviors are normal among teenagers or young adults, more insight can be gained by looking at the years prior to age 12 to determine who is most likely to become chronic, violent offenders. Life-course persistent offenders begin offending very early in life and continue to commit crime far into adulthood, even middle age, whereas adolescence-limited offenders tend to engage in criminal activity only during teenage and young adult years. Moffitt's model suggests that more than one type of development explains criminality. Furthermore, this framework shows that certain types of offenders commit crime due to entirely different causes and factors.

✉ Policy Implications

There are many, perhaps an infinite number of, policy implications that can be derived from developmental theories of criminality. Thus, we will focus on the most important, which is that of prenatal and perinatal stages of life because the most significant and effective interventions can occur during this time. If policy makers hope to reduce early risk factors for criminality, they must insist on universal health care for pregnant women, as well as their newborn infants through the first few years of life. The United States is one of the few developed nations that does not guarantee this type of maternal and infant medical care and supervision. Doing so would go a long way to avoiding the costly damages (in many ways) of criminal behavior among youths at risk.[10]

Furthermore, there should be legally mandated interventions for pregnant women who are addicted to drugs or alcohol. Although this is a highly controversial topic, it appears to be a no-brainer that women who suffer from such addictions may become highly toxic to the child(ren) they carry and should receive closer supervision and more health care. There may be no policy implementation that would have as much influence on reducing future criminality in children as making sure their mothers do not take toxic substances while they are pregnant.[11]

Other policy implications include assigning special caseworkers for high-risk pregnancies, such as those involving low birth weight or low Apgar scores. Another advised intervention would be to have a

---

[9]See Tibbetts and Piquero, "Influence of Gender."

[10]Wright et al., *Criminals in the Making*, 256–257 (see chap. 5, n. 32).

[11]Ibid., 256.

centralized medical system that provides a flag for high-risk infants who have numerous birth or delivery complications, so that the doctors who are seeing them for the first time are aware of their vulnerabilities.[12] Finally, universal preschool should be funded and provided to all young children; studies have shown this leads to better performance once they enter school, both academically and socially.[13]

Ultimately, as the many developmental theories have shown, there are many concepts and stages of life that can have a profound effect on the criminological trajectories that lives can take. However, virtually all of these models propose that the earlier stages of life are likely the most important in determining whether an individual will engage in criminal activity throughout life or not. Therefore, policymakers should focus their efforts on providing care and interventions in this time period.

Conclusion

This chapter presented a brief discussion of the importance of developmental or life-course theories of criminal behavior. This perspective is relatively new, becoming popular in the late 1970s, compared to other traditional theories explored in this text. Ultimately, this is one of the most cutting-edge areas of theoretical development, and life-course theories are likely to be the most important frameworks in the future of the field of criminological theory.

Then, we examined the policy implications of this developmental approach, which emphasized the need to provide universal care for pregnant mothers, as well as their newborn children. Other policy implications included legally mandated interventions for mothers who are addicted to toxic substances (e.g., alcohol, drugs) and assignment of caseworkers for high-risk infants and children, such as those with birth or delivery complications. Such interventions would go a long way toward saving society the many problems (e.g., financial, victimization, etc.) that will persist without such interventions. Ultimately, a focus on the earliest stages of intervention will pay off the most and will provide the biggest bang for the buck.

Chapter Summary

- Developmental or life-course theory focuses on the individual and following such individuals throughout life to examine their offending careers. In-depth consideration of changes during the life course are of highest concern, especially regarding general conclusions that can be made about the factors that tend to increase or decrease the risk that they will continue offending.
- Life-course perspectives emphasize such concepts as onset, frequency of offending, duration of offending, seriousness of offending, desistence of offending, and other factors that play key roles in when individuals offended and why they did so—or didn't do so—at certain times of their lives.
- There are many critics of the developmental or life-course perspective, particularly those who buy into the low self-control model, which is antidevelopmental in the sense that it assumes that propensities for crime do not change over time but rather remain unchanged across life.
- One of the developmental models that has received the most attention is that by Sampson and Laub, which emphasizes transitions in life (e.g., marriage, military service, employment, etc.) that alter trajectories either toward or away from crime.

---

[12]Ibid.

[13]Ibid.

- Moffitt's developmental theory of chronic offenders (which she labeled life-course persistent offenders) versus more normal offenders (which she labeled adolescence-limited offenders) is the developmental model that has received the most attention over the last decade, and much of this research is supportive of the interactive effects of biology and environment combining to create chronic, habitual offenders.

KEY TERMS

Adolescence-limited offenders	Life-course persistent offenders	Trajectory
Developmental theories	Moffitt's developmental theory (taxonomy)	Transitions

DISCUSSION QUESTIONS

1. What characteristic distinguishes developmental theories from traditional theoretical frameworks?

2. What aspects of a criminal career do experts consider important in such a model? Describe all of the aspects they look at in a person's criminal career.

3. Discuss the primary criticisms regarding the developmental perspective, particularly that presented by Gottfredson and Hirschi. Which theoretical paradigm do you consider the most valid? Why?

4. What transitions or trajectories have you seen in your life or your friends' lives that support Sampson and Laub's developmental model? What events encouraged offending or inhibited it?

5. Given Moffitt's dichotomy of life-course persistent offenders and adolescence-limited, which of these should be given more attention by research? Why do you feel this way?

WEB RESOURCES

Developmental Theories of Crime

http://en.wikipedia.org/wiki/life_course_theory/

Sampson and Laub's Model

http://harvardmagazine.com/2004/03/twigs-bent-trees-go-stra.html

Terrie Moffitt's Theory

www.wpic.pitt.edu/research/famhist/PDF_Articles/APA/BF16.pdf

11

Integrated Theoretical Models and New Perspectives of Crime

This chapter will introduce **integrated theories,** those in which two or more traditional theories are merged into one cohesive model. We will then discuss the pros and cons of integrating theories and explain the various ways theories can be integrated. A review of integrated theories will demonstrate the many ways different theories have been merged to form more empirically valid explanations for criminal behavior.

This chapter examines relatively recent changes in criminological development, with virtually all of these advances taking place in the last couple of decades. Specifically, we discuss the types of models that modern explanatory formulations seem to have taken, namely integrated theories. Of course, other unique theoretical frameworks have been presented in the last 20 years, but most of the dominant models presented during this time fall into the category of integrated theories.

Integrated theories attempt to put together two or more traditionally separate models of offending to form one unified explanatory theory. Given the empirical validity of most of the theories discussed in previous chapters, as well as the failure of these previously examined theories to explain all variation in offending behavior, it makes sense that theorists would try to combine the best of several theories, while disregarding or deemphasizing the concepts and propositions that don't seem to be as scientifically valid. Furthermore, some forms of theoretical integration deal with only concepts or propositions, while others vary by level of analysis (micro vs. macro or both). Although such integrated formulations sound attractive and appear to be sure-fire ways to develop sounder explanatory models of behavior, they have a number of weaknesses and criticisms. Here, we explore these issues, as well as the best known and most accepted integrated theories that are currently being examined and tested in the extant criminological literature.

⬚ Integrated Theories

About 30 years ago at a conference at the State University of New York at Albany, leading scholars in criminological theory development and research came together to discuss the most important issues in the growing area of theoretical integration.[1] Some integrated theories go well beyond formulating relationships between two or more traditionally separate explanatory models; they actually fuse the theories into one, all-encompassing framework. The following sections will examine why integrated theories became popular over the last several decades while discussing different types of integrated theories, the strengths and weaknesses of theoretical integration, and several of the better known and respected integrated theories.

The Need for Integrated Theories in Criminology

The emphasis on theoretical integration is a relatively recent development that has evolved due to the need to improve the empirical validity of traditional theories, which suffer from lack of input from various disciplines.[2] This was the result of the history of criminological theory, which we discussed in prior chapters of this book. Specifically, most 19th-century theories of criminal behavior are best described as based on single-factor (e.g., IQ) or limited-factor (e.g., stigmata) reductionism. Later, in the early 1900s, a second stage of theoretical development involved the examination of various social, biological, and psychological factors, which became known as the multiple-factor approach and is most commonly linked with the work of Sheldon and Eleanor Glueck.

Finally, in the latter half of the 20th century, a third stage of theoretical development and research in criminology became dominant, which represented a backlash against the multiple-factor approach. This stage has been called *systemic reductionism,* which refers to the pervasive attempts to explain criminal behavior in terms of a particular system of knowledge.[3] For example, a biologist who examines only individuals' genotypes will likely not be able to explain much criminal behavior because he is missing a lot of information about the environment in which the people live, such as level of poverty or unemployment. For the last 50 years, the criminological discipline in the United States has been dominated by sociologists, which is largely due to the efforts and influence of Edwin Sutherland (see Chapter 8). Thus, as one expert has observed:

> It is not surprising to find that most current explanations of criminal behavior are sociologically based and are attempts to explain variations in the rates of criminal behavior as opposed to individual instances of that behavior. . . . [E]ven when the effort is to explain individual behavior, the attempt is to use exposure to or belief in cultural or social factors to explain individual instances or patterns of criminal behavior. . . . We find ourselves at the stage of development in criminology where a variety of sociological systemic reductionistic explanations dominate, all of which have proven to be relatively inadequate (to the standard of total explanation, prediction, or control) in explaining the individual occurrence or the distribution of crime through time or space.[4]

---

[1]For a compilation of the papers presented at this conference, see Steven F. Messner, Marvin D. Krohn, and Allen E. Liska, eds., *Theoretical Integration in the Study of Deviance and Crime: Problems and Prospects* (Albany: State University of New York Press, 1989).

[2]This discussion is largely drawn from Charles F. Wellford, "Towards an Integrated Theory of Criminal Behavior," in Messner et al., *Theoretical Integration,* 119–127.

[3]Ibid., 120.

[4]Ibid., 120–121.

Other criminologists have also noted this sociological dominance, with some going as far to claim that

Sutherland and the sociologists were intent on turning the study of crime into an exclusively sociological enterprise and that they overreacted to the efforts of potential intruders to capture some of what they regarded as their intellectual turf.[5]

This dominance of sociology over the discipline is manifested in many obvious ways. For example, virtually all professors of criminology have Doctorates of Philosophy (Ph.D.) in sociology or criminal justice, and virtually no professors have a degree in biology, neuropsychology, or other fields that obviously have important influence on human behavior. Furthermore, virtually no undergraduate (or even graduate) programs in criminal justice and criminology currently require students to take a course that covers principles in biology, psychology, anthropology, or economics; rather, virtually all the training is sociology based. In fact, most criminal justice programs in the United States do not even offer a course in biopsychological approaches toward understanding criminal behavior, despite the obvious need for and relevance of such perspectives.

Many modern criminologists now acknowledge the limitations of this state of systemic reductionism, which limits theories to only one system of knowledge, in this case sociology.[6] What resulted is a period of relative stagnation in theoretical development, with experts regarding the 1970s as "one of the least creative periods in criminological history."[7] In addition, the mainstream theories that were introduced for most of the 1900s (such as differential association, strain, social bonding, and labeling, which we reviewed in previous chapters) received limited empirical support, which should not be too much of a surprise given that they were based on principles of only one discipline, namely sociology. Thus, it has been proposed that integrated theories evolved as a response to such limitations and the need to revitalize progress in the area of criminological theory building.[8] After reviewing some of the many integrated theories that have been proposed in the last two decades, we think readers will agree that the approach of combining explanatory models has indeed helped stimulate much growth in the area of criminological theory development. But, before examining these theories, it is important to discuss the varying forms they take.

Different Forms of Integrated Theories

There are several different types of integrated theories, which are typically categorized by the way that their creators propose how the theories fuse together in a useful way. The three most common forms of propositional theoretical integration, meaning that they synthesize theories based on their postulates, are known as end-to-end, side-by-side, and up-and-down integration.[9] First, we will explore each of these types before discussing a few more variations of combining theoretical concepts and propositions.

---

[5]Quote from Brown et al., *Criminology: Explaining Crime,* 251 (see chap. 6, n. 21), regarding claims made by Robert Sampson and John Laub, "Crime and Deviance Over the Life Course: The Salience of Adult Social Bonds," *American Sociological Review 55* (1990): 609–627. Also, see discussion in Richard A. Wright, "Edwin H. Sutherland," *Encyclopedia of Criminology,* Vol. 1, eds. Richard A. Wright and J. Mitchell Miller (New York: Routledge, 2005).

[6]Wellford, "Towards an Integrated Theory"; Sampson and Laub, "Crime and Deviance."

[7]Wellford, "Towards an Integrated Theory," 120, citing Frank P. Williams III, "The Demise of Criminological Imagination: A Critique of Recent Criminology," *Justice Quarterly 1* (1984): 91–106.

[8]Wellford, "Towards an Integrated Theory."

[9]This typology is largely based on Travis Hirschi, "Separate and Unequal Is Better," *Journal of Research in Crime and Delinquency 16* (1979): 34–37.

End-to-End Integration

The first type, **end-to-end theoretical integration,** typically is used when theorists expect that one theory will come before or after another in terms of temporal ordering of causal factors. This type of integration is more developmental in the sense that it tends to propose a certain ordering of the component theories that are being merged. For example, an integrated theory may claim that most paths toward delinquency and crime have their early roots in the breakdown of social attachments and controls (i.e., social bonding theory), but later, the influence of negative peers (i.e., differential association) becomes more emphasized. Thus, such a model would look like this:

Weak Social Bond → Negative Peer Associations → Crime

Such an integrated model is referred to as *end-to-end* (or *sequential*) integration, which is appropriately named because it conveys the linkage of the theories based on the temporal ordering of two or more theories in their causal timing.[10] Specifically, the breakdown of social bonds comes first, followed by the negative peer relations. Another way of saying this is that the breakdown of social bonds is expected to have a more remote or indirect influence on crime, which is mediated by differential peer influences; on the other hand, according to this model, peer influences are expected to have a more immediate or direct effect on crime, with no mediating influences of other variables. This model is hypothetical and presented to illustrate the end-to-end form of integration, but we will see that some established frameworks have incorporated similar propositions regarding social bonding and differential association and reinforcement theories.

Many of the traditionally separate theories that we have examined in previous introductions tend to differ in their focus on either remote or immediate causal factors.[11] For instance, one of the assumptions of differential association theory is that psychological learning of crime is based on day-to-day (or even moment-to-moment) learning, so the emphasis is on more immediate causes of crime, namely interactions with peers and significant others. On the other hand, other theories tend to focus more on remote causes of crime, such as social disorganization and strain theory, which place the emphasis on social structure factors, such as relative deprivation or industrialization, that most experts would agree are typically not directly implicated in situational decisions to engage in an actual criminal act but are extremely important nonetheless.

This seems to be conducive to using end-to-end integration, and often, the theories seem to complement each other quite well, as in our hypothetical example in which social control theory proposes the remote cause (i.e., weakened bonds) and differential association theory contributes the more direct, proximate cause (i.e., negative peer influence). On the other hand, two or more theories that focus only on more immediate causes of crime would be harder to combine because they both claim they are working at the same time, in a sense competing against the other for being the primary direct cause of criminal activity; thus, they would be unlikely to fuse together as nicely and would not complement one another. Also, some theorists have argued that end-to-end integration is simply a form of **theoretical elaboration,** which we will discuss later. Another major limitation of end-to-end integration is the issue of whether the basic assumptions of the included theories are consistent with one another. We will investigate this in the following section, which discusses criticism of integration. But first, we must examine the other forms of theoretical integration.

---

[10]Allen E. Liska, Marvin D. Krohn, and Steven F. Messner, "Strategies and Requisites for Theoretical Integration in the Study of Crime and Deviance," in Messner et al., *Theoretical Integration*, 1–19.

[11]Much of our discussion of these forms of integration is taken from Liska et al., "Strategies and Requisites."

Side-by-Side Integration

Another type of integrated theory is called side-by-side (or horizontal) integration. In the most common form of side-by-side integration, cases are classified by a certain criterion (e.g., impulsive vs. planned), and two or more theories are considered parallel explanations depending on what type of case is being considered. So, when the assumptions or target offenses of two or more theories are different, a side-by-side integration is often the most natural way to integrate them. For example, low self-control theory may be used to explain impulsive criminal activity, whereas rational choice theory may be used to explain criminal behavior that involves planning, such as white-collar crime. Traditionally, many theorists would likely argue that low self-control and rational choice theory are quite different, almost inherently opposing perspectives of crime. However, contemporary studies and theorizing have shown that the two models complement and fill gaps in the other.[12] Specifically, the rational choice and deterrence framework has always had a rather hard time explaining and predicting why individuals often do very stupid things for which there is little payoff and a high likelihood of getting in serious trouble, so the low self-control perspective helps fill in this gap by explaining that some individuals are far more impulsive and are more concerned with the immediate payoff (albeit often small) rather than any long-term consequences.

An illustration of this sort of side-by-side integration of these two theories might look something like this:

For most typical individuals:

High Self-Control → Consideration of Potential Negative Consequences
→ Deterred From Committing Crime

For more impulsive individuals or activities:

Low Self-Control → Desire for Immediate Payoff → Failure to Consider Consequences
→ Decision to Commit Criminal Act

This side-by-side integration shows how two different theories can each be accurate, depending on what type of individual or criminal activity is being considered. As some scholars have concluded, rational choice theory is likely not a good explanation for homicides between intimates (which tend to be spontaneous acts), but it may be very applicable to corporate crime.[13]

---

[12]Daniel Nagin and Raymond Paternoster, "Enduring Individual Differences and Rational Choice Theories of Crime," *Law and Society Review 27* (1993): 467–496; Alex Piquero and Stephen Tibbetts, "Specifying the Direct and Indirect Effects of Low Self-Control and Situational Factors in Offenders' Decision Making: Toward a More Complete Model of Rational Offending," *Justice Quarterly 13* (1996): 481–510; Tibbetts and Herz, "Gender Differences in Factors," 183–208 (chap. 8, n. 90); Stephen Tibbetts and David Myers, "Low Self-Control, Rational Choice, and Student Test Cheating," *American Journal of Criminal Justice 23* (1999): 179–200; Bradley Wright, Avshalom Caspi, and Terrie Moffitt, "Does the Perceived Risk of Punishment Deter Criminally Prone Individuals? Rational Choice, Self-Control, and Crime," *Journal of Research in Crime and Delinquency 41* (2004): 180–213; for a review, see Stephen Tibbetts, "Individual Propensities and Rational Decision-Making: Recent Findings and Promising Approaches," in *Rational Choice and Criminal Behavior: Recent Research and Future Challenges*, eds. Alex Piquero and Stephen Tibbetts (New York: Routledge, 2002), 3–24.

[13]Liska et al., "Strategies and Requisites"; for a review, see Brown et al., *Criminology: Explaining Crime*; for a recent study of rational choice being applied to corporate crime, see Nicole Leeper Piquero, M. Lyn Exum, and Sally S. Simpson, "Integrating the Desire-for-Control and Rational Choice in a Corporate Crime Context," *Justice Quarterly 22* (2005): 252–281.

Up-and-Down Integration

Another way of combining two or more theories is referred to as **up-and-down** (or *deductive*) **integration** and is generally considered the classic form of theoretical integration because it has been done relatively often in the history of criminological theory development, even before it was considered integration. This often involves increasing the level of abstraction of a single theory so that postulates seem to follow from a conceptually broader theory. Up-and-down integration can take two prevalent forms: theoretical reduction and theoretical synthesis.

Theoretical reduction is typically done when it becomes evident "that theory A contains more abstract or general assumptions than theory B and, therefore, that key parts of theory B can be accommodated within the structure of theory A."[14] Regarding theoretical reduction, we have discussed this form of integration previously in this book without actually calling it by this name. For example, in Chapter 8, we discussed how differential reinforcement theory subsumed Sutherland's differential association theory. By equating concepts contained in both theories, the authors of differential reinforcement argued somewhat effectively that the learning that takes place through interactions with primary groups is one type of conditioning.[15] As you will recall, the main point is that the concepts and propositions of differential reinforcement theory are more general than, but entirely consistent with, those of differential association theory, such that the latter is typically a specific form of the former model. In other words, differential reinforcement is a much more broad, general theory, which accounts for not only differential association (i.e., classical conditioning) but also many other theoretical concepts and propositions (operant conditioning, modeling, or imitation).

This same type of theoretical reduction was also discussed in Chapter 6 when we noted that general strain theory subsumed Merton's traditional strain theory. Specifically, traditional strain theory focused on only one type of strain: failure to achieve positively valued goals. In comparison, while general strain theory also places an emphasis on failure to achieve positively valued goals, it is more general and broad because it also focuses on two other forms of strain: loss of positively valued stimuli and exposure to noxious stimuli. Therefore, it seems to make sense that general strain theory would subsume traditional strain theory because the concepts and principles of traditional strain appear to simply represent a specific type of general strain and can be fully accounted for by the more general version of the theory.[16]

Despite the obvious efficiency in theoretical reduction, many theorists have criticized such subsuming of theories by another. Specifically, many experts in the social sciences view such deduction as a form of theoretical imperialism because the theory being deduced essentially loses its unique identity.[17] Therefore, the very phrase *theoretical reduction* generally has a negative connotation among scholars. In fact, one of the most accepted reductions in 20th-century criminological literature, the differential association-differential reinforcement synthesis we referred to above, has been condemned as a "revisionist takeover ... a travesty of Sutherland's position."[18] So, even the most widely known and cited forms of theoretical reduction have been harshly criticized, which gives readers some idea of how frowned upon by social scientists is such subsuming of theories by another.

---

[14]Liska et al., "Strategies and Requisites," 10.

[15]Liska et al., "Strategies and Requisites," 13; Robert L. Burgess and Ronald Akers, "Differential Association-Reinforcement Theory of Criminal Behavior," *Social Problems 14* (1966): 128–146; Akers, *Deviant Behavior* (see chap. 8, n. 17).

[16]Robert Agnew, "Foundation for a General Strain Theory of Crime and Delinquency," *Criminology 30* (1992): 47–87; Robert Agnew and Helen R. White, "An Empirical Test of General Strain Theory," *Criminology 30* (1992): 475–499.

[17]Liska et al., "Strategies and Requisites."

[18]Quote from Ian Taylor, Paul Walton, and Jock Young, *The New Criminology* (New York,: Harper & Row, 1973), 131–132; see discussion in Liska et al., "Strategies and Requisites," 13.

The other form of up-and-down integration is referred to as *theoretical synthesis,* which is "done by abstracting more general assumptions from theories A and B, allowing parts of both theories to be incorporated in a new theory C."[19] This form of up-and-down integration is more uncommon in social science than theoretical reduction, perhaps because it is more difficult to achieve successfully as it, by definition, requires a new theory that essentially necessitates new concepts or hypotheses that are not already found in the original component models. Furthermore, if the constituent theories are competing explanations, then it is quite likely that new terminology will have to be created or incorporated to resolve these differences.[20] However, if theoretical synthesis can be done correctly, it is perhaps the type of integration that provides the most advancement of theory development because, in addition to bringing together previously independent models, it also results in a new theory with predictions and propositions that go beyond the original frameworks.

One of the best known and most accepted (despite its critics) examples of theoretical synthesis is **Elliott's integrated model,** which we review in detail later in this chapter.[21] For our purposes here, it is important only to understand why Elliott's model is considered theoretical synthesis. This is because Elliott's model integrates the concepts and propositions of various theories (e.g., social control and differential association) and contributes additional propositions that did not exist in the component theories.

Levels of Analysis of Integrated Theories

Beyond the variation of types of propositional theoretical integration discussed above, such synthesis of explanatory models also differs in terms of the level of analysis that is being considered. Specifically, the integrated models can include component theories of micro-micro, macro-macro, or even micro-macro (called *cross-level*) combinations. For example, Elliott's integrated theory is a micro-micro level theory, which means that all of the component theories that make up the synthesized model refer to the individual as the unit of analysis. Although these models can provide sound understanding for why certain individuals behave in certain ways, they typically do not account for differences in criminality across groups (e.g., gender, socioeconomic groups, etc.).

On the other hand, some integrated models include theories from only macrolevels. A good example of this type of integration is seen in Bursik's synthesis of conflict theory and the social disorganization framework.[22] Both of these component theories focus only on the macro (group) level of analysis, thus neglecting the individual level, such as explaining why some individuals do or do not commit crime even in the same structural environment.

The most complicated integrated theories, at least in terms of levels of analysis, are those that include both micro and macro models. Such models are likely the most difficult to synthesize successfully because it involves bringing together rather unnatural relationships between individual-based propositions and group-level postulates. However, when done effectively, such a model can be rather profound in terms of

---

[19]Liska et al., "Strategies and Requisites," 10.

[20]David Wagner and Joseph Berger, "Do Sociological Theories Grow?" *American Journal of Sociology* 90 (1985): 697–728.

[21]Delbert Elliott, Suzanne S. Ageton, and Rachelle J. Canter, "An Integrated Theoretical Perspective on Delinquent Behavior," *Journal of Research in Crime and Delinquency* 16 (1979): 3–27; also see discussion in Delbert Elliott, "The Assumption That Theories Can Be Combined With Increased Explanatory Power: Theoretical Integrations," in *Theoretical Methods in Criminology,* ed. Robert F. Meier (Beverly Hills: Sage, 1985), 123–149; discussion of Elliott's model as an example of theoretical synthesis can be found in Charles R. Tittle, "Prospects for Synthetic Theory: A Consideration of Macrolevel Criminological Activity," in Messner et al., *Theoretical Integration,* 161–178.

[22]Robert J. Bursik, "Political Decision Making and Ecological Models of Delinquency, Conflict and Consensus," in Messner et al., *Theoretical Integration,* 105–118.

explanation of crime. After all, one of the primary objectives of a good theory is to explain behavior across as many circumstances as possible. Thus, a theory that can effectively explain why crime occurs in certain individuals as well as certain groups would be better than a theory that explains crime across only individuals or groups.

An example of this is **Braithwaite's reintegrative shaming theory,** which begins with an individual (or micro) level theory of social control and bonding—which he refers to as **interdependency**—and then relates levels of this concept to a group or community (or macro) level theory of bonding—which he refers to as **communitarianism.**[23] This theory will be discussed in far more detail later, but it is important here to acknowledge the advantages of explaining both the micro (individual) and macro (group) level of analysis in explanations of criminal behavior, such as in the theory of reintegrative shaming.

Ultimately, the levels of analysis of any component theories are important considerations in the creation of integrated models of crime. It is particularly important to ensure that the merging of certain theories from within or across particular levels makes rational sense and, most important, advances our knowledge and understanding of the causes of crime.

Additional Considerations Regarding Types of Integration

Beyond the basic forms of propositional integration models and levels of analysis, there are two additional types of integration—**conceptual integration** and theoretical elaboration—that are quite common, perhaps even more common than the traditional forms discussed above. Conceptual integration involves the synthesis of models in which "the theorist equates concepts in different theories, arguing that while the words and terms are different, the theoretical meanings and operations of measurement are similar."[24] Essentially, the goal in such a formulation is to take the primary constructs of two or more theories and merge them into a more general framework that aids understanding in explaining behavior by unifying terms that fundamentally represent similar phenomena or issues. Such formulations are considered less intrusive on the component theories than the propositional integrations we discussed previously.

One of the first and most cited examples of conceptual integration was provided by Pearson and Weiner in 1985.[25] As shown in Table 11.1, Pearson and Weiner attempted to map the various concepts of numerous criminological theories. This is done by creating a category that numerous concepts from various theories appear to fit into through their inclusion in explanatory models of criminal behavior and across level of explanation. Although based on a social learning and differential reinforcement perspective, Pearson and Weiner's model attempts to include concepts of virtually all existing theories of crime and delinquency models of behavior. A particular strength is that this model includes feedback or behavioral consequence elements, as well as classifying each model by unit of measurement and analysis.

For example, as illustrated in Table 11.1, the conceptual model shows that differential association theory has concepts that apply to all internal aspects and one of the two concepts under consequences or feedback at the microlevel but none of the aspects under the external microlevel or any of the four macrolevel concepts. On the other hand, Marxist-critical theory is shown to apply to three of the internal micro

---

[23]John Braithwaite, *Crime, Shame and Reintegration* (Cambridge: Cambridge University, 1989).

[24]Liska et al., "Strategies and Requisites," 15; definition adapted from Frank S. Pearson and Neil Alan Weiner, "Toward an Integration of Criminological Theories," *Journal of Criminal Law and Criminology* 76 (1985): 116–150.

[25]Frank S. Pearson and Neil Alan Weiner, "Toward an Integration"; see review in Thomas J. Bernard and Jeffrey B. Snipes, "Theoretical Integration in Criminology," in *Crime and Justice Review*, Vol. 20, ed. Michael Tonry (Chicago: University of Chicago Press, 1996), 301–348.

Table 11.1 Mapping of the Selected Criminological Theories Into the Integrative Structure

	INTEGRATIVE CONSTRUCTS											
	MICROLEVEL								MACROLEVEL			
	ANTECEDENT						CONSEQUENCES OR FEEDBACK		SOCIAL STRUCTURAL PRODUCTION AND DISTRIBUTION OF:			
	INTERNAL				EXTERNAL							
Selected Criminological Theories	Utility Demand (Deprivation)	Behavioral Skill	Rules of Expedience	Rules of Morality	Signs of Favorable Opportunities (Descriminative Stimuli)	Behavioral Resources	Utility Reception	Information Acquisition	Utilities	Opportunities	Rules of Expedience and Morality	Belief About Sanctioning Practices
Differential association	✓	✓	✓	✓			✓					
Negative labeling	✓	✓		✓			✓					
Social control:												
1. Attachment	✓		✓									
2. Commitment	✓		✓		✓			✓				
3. Involvement					✓			✓				
4. Belief				✓								
Deterrence	✓		✓		✓		✓	✓				✓
Economic	✓		✓		✓		✓	✓	✓	✓	✓	
Routine activities	✓	✓	✓		✓	✓			✓	✓		
Neutralization		✓		✓	✓				✓	✓	✓	
Relative deprivation	✓		✓	✓			✓	✓	✓	✓	✓	
Strain	✓		✓	✓		✓	✓	✓	✓	✓	✓	
Normative (culture) conflict				✓			✓	✓	✓	✓	✓	
Generalized strain and normative (culture) conflict	✓		✓		✓		✓	✓	✓	✓	✓	
Marxist-critical/ group conflict	✓		✓	✓			✓		✓	✓		

and one of the consequences or feedback level concepts and to most of the macrolevel concepts as well. Although some of the theories that we discuss in this book are included, largely because Pearson and Weiner's conceptual integration was done in the mid-1980s, this model includes most of the dominant theoretical frameworks that were prevalent in the criminological literature at the time of their formulation. Thus, this conceptual model remains as a sort of prototype for future attempts to conceptually integrate established theories explaining criminal conduct.

One notable strength of their conceptual integration is the inclusion of most mainstream theories—differential association, labeling, social control, deterrence, economic, routine activities, neutralization, strain, normative (culture) conflict, generalized strain, and Marxist-critical group conflict—at the time in which they created their integrated framework. Another strength of their integrated framework is the fact that they account for concepts at different levels of analysis, namely micro (individual) and macro (group) levels, as shown in Table 11.1. So, the framework clearly does a good job at creating links between the most prominent theories in terms of the concepts that they propose as the primary causes of crime.

However, Pearson and Weiner's conceptually integrated framework does have several notable weaknesses. One of the most prominent criticisms of the conceptual model is that it is based on a single theory, specifically social learning theory (tied to Akers's differential reinforcement and social learning theory discussed in Chapter 8), and the authors never really provide a strong argument for why they chose this particular base framework for their model. Another major weakness of this conceptual framework is that it completely neglects many biological and biosocial factors—hormones, neurotransmitters, toxins—that have been consistently shown to have profound effects on human behavior. Still, despite the criticisms of this integrated framework, respected theorists have noted that Pearson and Weiner's "integration of these concepts into a consistent, coherent framework is impressive," but the model has never received much attention in the criminological literature.[26]

However, despite the obvious tendency to simplify formulations through conceptual integration, many critics have claimed that such attempts are simply a means toward propositional integration. Thus, despite its categorization as conceptual, this type of integration is not necessarily seen as such, and it may be seen by many as a form of deductive integration. However, many experts have noted that conceptual integration is actually not a form of side-by-side integration nor is it end-to-end integration. Rather, it is what it says it is: conceptual integration. It is nothing more and nothing less.[27] Therefore, it appears that conceptual integration is a distinct derivative form of integration and that it can and should occur independently if it helps to advance understanding of why people commit crime, particularly for integrated models of offending. As the most knowledgeable experts recently concluded, "establishing some conceptual equivalence is necessary for deductive integration."[28]

An additional variation of integration is theoretical elaboration, which is a strategy that takes an existing theory that is arguably underdeveloped and then further develops it by adding concepts and propositions to include other components from other theories. Many critics of traditional theoretical integration have argued that theoretical elaboration is a more attractive strategy because existing theories of offending are not developed enough to fully integrate them.[29] An example of theoretical integration can be seen in the

---

[26]Akers and Sellers, *Criminological Theories*, 288 (see chap. 8, n. 17).

[27]Liska et al., "Strategies and Requisites," 15.

[28]Ibid., 16.

[29]Liska et al., "Strategies and Requisites," 16; Travis Hirschi, "Exploring Alternatives to Integrated Theory," in Messner et al., *Theoretical Integration,* 37–50; Terence P. Thornberry, "Reflections on the Advantages and Disadvantages of Theoretical Integration," in Messner et al., *Theoretical Integration,* 51–60; Robert F. Meier, "Deviance and Differentiation," in Messner et al., *Theoretical Integration,* 199–212.

expansion of rational choice theory, which had traditionally focused on ratios of perceived costs and benefits, to include deontological constructs, such as moral beliefs. Specifically, studies have consistently shown that ethical constraints condition the influence of expected consequences on criminal behavior, ranging from violence to academic dishonesty to white-collar crime.[30] Although they are not without critics, most rational choice scholars appear to be in agreement that such elaboration advanced the theoretical framework and made it a more accurate explanation of human behavior. In sum, theoretical integration, like conceptual integration, is an option for merging and improving theoretical models without completely synthesizing two or more entire paradigms.

Criticisms and Weaknesses of Integrated Theories

A number of criticisms have been leveled at theoretical integration. Perhaps one of the most obvious and prevalent, not to mention extremely valid, criticisms is the argument that caution should be taken in attempting to integrate theories that have apparent contradictions or inconsistencies in their postulates.[31] As we have seen, different theories are based on varying, often opposite, perspectives of human nature. While most versions of strain theory (e.g., Merton's theory) assume that human beings are born with a natural tendency toward being good, most variations of control theory (e.g., Hirschi's social bonding and self-control theories) assume that humans are innately selfish and hedonistic. At the same time, most versions of learning theory (e.g., Sutherland's theory of differential association and Akers's differential reinforcement theory) assume that humans are born with a blank slate, in other words, people are born with neither good nor bad tendencies but rather learn their morality moment-to-moment from social interaction.

Obviously, attempts to integrate these theories face the problematic hurdles of dealing with such obvious contradictions, and many formulations that do merge some of these perspectives simply do not deal with this issue. The failure to acknowledge, let alone explain, such inconsistencies is likely to result in regression of theoretical development instead of leading to progress in understanding, which is the primary goal of integrating theories in the first place.

Other experts have argued that any attempts to integrate theories must unite three different levels of analysis, which include individual (micro), group (macro), and microsituational, which merges the micro-level with spontaneous context.[32] While we agree that this would be the ideal for theory formulation, we know of no theory that does so. Therefore, this proposition is more of an ideal that has not yet been attempted. So, we are inclined to go with the best that has been offered by expert theorists. Still, in the future, it is hoped that such an integrated model will be offered that addresses all of these aspects in an explanatory model.

Another argument against theoretical integration is the stance that explanatory perspectives of crime are meant to stand alone. This is the position taken by Travis Hirschi, who is one of the most cited and

---

[30]Amitai Etzioni, *The Moral Dimension: Toward a New Economics* (New York: Free Press, 1988); Ronet Bachman, Raymond Paternoster, and Sally Ward, "The Rationality of Sexual Offending: Testing a Deterrence/Rational Choice Conception of Sexual Assault," *Law and Society Review 26* (1992): 401–419; Raymond Paternoster and Sally Simpson, "Sanction Threats and Appeals to Morality: Testing a Rational Choice Model of Corporate Crime," *Law and Society Review 30* (1996): 378–399; Stephen G. Tibbetts, "College Student Perceptions of Utility and Intentions of Test Cheating," PhD dissertation, University of Maryland—College Park (Ann Arbor: UMI, 1997); for a review and discussion, see Sally Simpson, Nicole Leeper Piquero, and Raymond Paternoster, "Rationality and Corporate Offending Decisions," in Piquero and Tibbetts, *Rational Choice*, 25–40.

[31]Much of this discussion is taken from Brown et al., *Criminology: Explaining Crime,* and Messner et al., *Theoretical Integration.*

[32]James Short, "The Level of Explanation Problem in Criminology," in *Theoretical Methods in Criminology,* ed. Robert Meier (Beverly Hills: Sage, 1985), 42–71; see also James Short, "On the Etiology of Delinquent Behavior," *Journal of Research in Crime and Delinquency 16* (1979): 28–33.

respected scholars in criminology over the last 40 years.[33] As others have noted, Hirschi informed his position on this debate unequivocally when he stated that "separate and unequal is better" than integrating traditionally independent theoretical models.[34]

This type of perspective is also called *oppositional tradition* or *theoretical competition,* because the separate theories are essentially pitted against one another in a form of battle or opposition. Although scientists are trained always to be skeptical of their own beliefs and open to other possibilities, especially the desire to refine theoretical models that are shown to be invalid, it is surprising that such a position exists. Despite the rather unscientific nature of this stance, such a position of oppositional tradition has many supporters. Specifically, one of the most respected and cited criminologists, Travis Hirschi claimed:

> The first purpose of oppositional theory construction is to make the world safe for a theory contrary to currently accepted views. Unless this task is accomplished, there will be little hope for the survival of the theory and less hope for its development. Therefore, oppositional theorists should not make life easy for those interested in preserving the status quo. They should instead remain at all times blind to the weaknesses of their own position and stubborn in its defense. Finally, they should never smile.[35]

Unfortunately, this position against theoretical integration is presented in a very unscientific tone. After all, scientists should always be critical of their own views and theories. By stating that theorists should be "blind" to opposing viewpoints and "stubborn" in their own perspective's defense, Hirschi advocates a position that is absolutely against science. Still, this statement, albeit flawed, shows the extreme position against theoretical integration and is favorable toward having each independent theory standing opposed to others. It is our position that this stand does not have much defense, which is shown by Hirschi's lack of rational argument. Furthermore, it is generally acceptable to smile when presenting any theoretical or empirical conclusions, even when they involve opposition to or acceptance of integrated theoretical models of criminal behavior.

Perhaps one of the most important criticisms against Hirschi's and others' criticisms of integrated models is that most traditional models alone only explain a limited amount of variation in criminal activity. Elliott and colleagues have claimed the following:

> Stated simply, the level of explained variance attributable to separate theories is embarrassingly low, and, if sociological explanations for crime and delinquency are to have any significant impact upon future planning and policy, they must be able to demonstrate greater predictive power.[36]

While some put this estimate at 10 to 20% of the variance in illegal activities, this is simply an average across different theories and various forms of deviant behavior.[37] However, this range is an overestimate

---

[33]Bernard and Snipes, "Theoretical Integration in Criminology"; Hirschi, "Exploring Alternatives."

[34]Hirschi, "Separate but Unequal."

[35]Hirschi, "Exploring Alternatives," 45.

[36]Delbert Elliott, David Huizinga, and Suzanne Ageton, *Explaining Delinquency and Drug Use* (Beverly Hills: Sage, 1985), 125, as quoted in Bernard and Snipes, "Theoretical Integration in Criminology," 306.

[37]This estimate can be found in Bernard and Snipes, "Theoretical Integration in Criminology," 306, but is based on the estimates of others, as discussed in this work.

regarding many studies on the accuracy of separate theories, which tend to show weak support (well under 10% explained variation in offending), particularly for some tests of social bonding and strain models, as well as others.[38]

On the other hand, this estimated range of explained variance underestimates the empirical validity of some theoretical frameworks that consistently show high levels of explained variation in certain criminal behaviors. For example, a large number of studies examining Akers's differential reinforcement and social learning theory (discussed in Chapter 8), which examined not only a wide range of samples (in terms of age, nationality, and other demographic characteristics) but also a large range of deviant activities (e.g., cigarette smoking to drug usage to violent sexual crimes), consistently account for more than 20% of variation in such behaviors.[39] Specifically, most of these studies estimate that Akers's social learning model explains up to 68% or more of the variation in certain deviant behaviors, with the lowest estimate around 30%.

Obviously, not all independent theories of crime lack empirical validity, so this does not support critics' claims that traditionally separate theories of crime do not do a good job in explaining criminal behavior. However, it is also true that many of the theories that do the best job in empirical tests for validity are those that are somewhat integrated in the sense that they often have been formed by merging traditional theories with other constructs and propositions, much like Akers's differential reinforcement theory, which added more modern psychological concepts and principles (e.g., operant conditioning and modeling) to Sutherland's traditional theory of differential association (see Chapter 8). So, in a sense, an argument can be made that theoretical integration (or at least theoretical elaboration) had already occurred, which made this theory far more empirically valid than the earlier model.

Another example of the high level of empirical validity of existing models of offending can be found in some models of rational choice, which have been revised through theoretical elaboration and have explained more than 60% of the explained variation in deviant behavior.[40] However, much of the explanatory power of such frameworks relies on incorporating the constructs and principles of other theoretical models, which is what science is based on; specifically, they revise and improve theory based on what is evident from empirical testing. After all, even some of the harshest critics of theoretical integration admit that traditional theories do not own variables or constructs.[41] For example, Hirschi claimed that

---

[38]See review in Akers and Sellers, *Criminological Theories,* particularly 97.

[39]Ronald Akers, Marvin Krohn, Lonn Lanza-Kaduce, and Maria Radosevich, "Social Learning and Deviant Behavior: A Specific Test of a General Theory," *American Sociological Review 44* (1979): 635–655; Marvin Krohn and Lonn Lanza-Kaduce, "Community Context and Theories of Deviant Behavior: An Examination of Social Learning and Social Bonding Theories," *Sociological Quarterly 25* (1984): 353–371; Lonn Lanza-Kaduce, Ronald Akers, Marvin Krohn, and Marcia Radosevich, "Cessation of Alcohol and Drug Use Among Adolescents: A Social Learning Model," *Deviant Behavior 5* (1984): 79–96; Ronald Akers and John Cochran, "Adolescent Marijuana Use: A Test of Three Theories of Deviant Behavior," *Deviant Behavior 6* (1985): 323–346; Ronald Akers and Gang Lee, "Age, Social Learning, and Social Bonding in Adolescent Substance Use," *Deviant Behavior 19* (1999): 1–25; Marvin Krohn, William Skinner, James Massey, and Ronald Akers, "Social Learning Theory and Adolescent Cigarette Smoking: A Longitudinal Study," *Social Problems 32* (1985): 455–473; Ronald Akers and Gang Lee, "A Longitudinal Test of Social Learning Theory: Adolescent Smoking," *Journal of Drug Issues 26* (1996): 317–343; Ronald Akers, Anthony La Greca, John Cochran, and Christine Sellers, "Social Learning Theory and Alcohol Behavior Among the Elderly," *Sociological Quarterly 30* (1989): 625–638; Scot Boeringer, Constance Shehan, and Ronald Akers, "Social Contexts and Social Learning in Sexual Coercion and Aggression: Assessing the Contribution of Fraternity Membership," *Family Relations 40* (1991): 558–564; Sunghyun Hwang and Ronald Akers, "Adolescent Substance Use in South Korea: A Cross-Cultural Test of Three Theories," in *Social Learning Theory and the Explanation of Crime: A Guide for the New Century,* eds. Ronald Akers and Gary Jensen (New Brunswick: Transaction, 2003), 39–64.

[40]For example, see Tibbetts, "College Student Perceptions," which showed that an elaborated rational choice model explained more than 60% of variation in test cheating among college students.

[41]See discussion in Bernard and Snipes, "Theoretical Integration in Criminology," 306–307.

integrationists somehow conclude that variables appear . . . with opposition theory labels attached to them. This allows them to list variables by the theory that owns them. Social disorganization theory . . . might own economic status, cultural heterogeneity, and mobility. . . . Each of the many variables is measured and . . . the theories are ranked in terms of the success of their variables in explaining variation in delinquency . . . such that integration is in effect *required* by the evidence and surprisingly easily accomplished.[42]

This is the way that science and theoretical development and revision are supposed to work, so in our opinion, this is exactly as it should be. All scientists and theoreticians should constantly be seeking to improve their explanatory models and be open to ways to do so, as opposed to being staunch supporters of one position and blind to existing evidence.

Despite the criticisms against theoretical integration, a strong argument has been made that theoretical competition and oppositional tradition is generally pointless.[43] A big reason for this belief of proponents of theoretical integration is that various theories tend to explain different types of crime and varying portions of the causal processes for behavior. For example, some theories focus more on property crimes, while others place emphasis on violent crimes, and some theories emphasize the antecedent or root causes (e.g., genetics, poverty) of crime, while others focus on more immediate causes (e.g., current social context at the scene). Given that there are multiple factors that contribute to crime and that different factors are more important for various types of crime, it only makes sense that a synthesis of traditionally separate theories must come together to explain the wide range of criminal activity that occurs in the real world.

Ultimately, there are both pros and cons regarding whether to integrate theories. It is our belief that theoretical integration is generally a good thing, as long as there is caution and attention given to merging models that have opposing assumptions, such as the natural state of human beings (e.g., good vs. bad vs. blank slate). But, only after considerable empirical research will the true validity of integrated models be tested, and many have already been put to the test. We will now examine a handful of integrated theories that have been proposed over the last couple of decades, as well as the studies that have examined their empirical validity. Not surprisingly, some integrated and elaborated theories appear to be more valid than others, with most adding considerably to our understanding of human behavior and contributing to explaining the reasons why certain individuals or groups commit criminal behavior more than others.

Examples of Integrated Criminological Theory

We have already discussed the advantages and disadvantages of theoretical integration, as well as the ways in which traditionally separate explanatory models are combined to form new, synthesized frameworks. We will now review a number of the most prominent examples of theoretical integration that have been proposed in the last 30 years, which is largely the time period when most attempts at integration have been presented. We hope that readers will critique each theory based on the criteria that we have already discussed, particularly noting the empirical validity of each model based on scientific observation and the logical consistency of its propositions.

---

[42]Hirschi, "Exploring Alternatives," 41, as cited (revised) in Bernard and Snipes, "Theoretical Integration in Criminology," 307.

[43]Quote from Bernard and Snipes, "Theoretical Integration in Criminology," 306, based on the rationale provided by Elliott et al., *Explaining Delinquency.*

Elliott's Integrated Model

Perhaps the first and certainly the most prominent integrated model is that proposed by Delbert Elliott and his colleagues in 1979, which has become known as Elliott's integrated model.[44] In fact, this model "opened the current round of debate on integration," because it was essentially the first major perspective proposed that clearly attempted to merge various traditionally separate theories of crime.[45] Elliott's integrated framework attempts to merge strain, social disorganization, control, and social learning and differential association-reinforcement perspectives for the purpose of explaining delinquency, particularly in terms of drug use, as well as other forms of deviant behavior (see Figure 11.1).

As can be seen in Figure 11.1, the concepts and propositions of strain and social disorganization, as well as inadequate socialization, are considered antecedent (or root) causes of delinquency. In other words, failing to achieve one's goals (i.e., strain theory) or coming from disadvantaged neighborhoods (i.e., social disorganization) are key causes for predisposing people to criminal behavior. Furthermore, the fact that many low-income households tend to lack adequate socialization, such as when a single parent has to work two or three jobs to make ends meet, is also a major root cause of delinquency.

Because this model clearly shows some models and constructs coming first (e.g., strain, social disorganization) and others coming later (e.g., weak bonding and then affiliations with delinquents) that lead to criminality, this is a good example of an end-to-end form of theoretical integration. This is an end-to-end form of integration because some models or concepts, such as strain, occur first, which then lead chronologically to other models and concepts, such as weak conventional bonding or strong delinquent bonding, which then lead to crime.

The notable ways in which this perspective becomes a true integrated model is seen in the mediating or intervening variables. Although some antecedent variables (such as strain) can lead directly toward delinquent activity, most of the criminal activity is theoretically predicted through a process that would include a breakdown of conventional bonding (i.e., social control and bonding theory), which occurs in many individuals who experience strain or social disorganization in their neighborhoods along with inadequate socialization. Furthermore, individuals who have such a breakdown in conventional bonding tend to be more

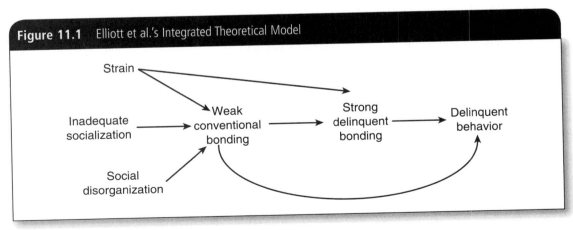

Figure 11.1 Elliott et al.'s Integrated Theoretical Model

[44]Delbert Elliott, S. Ageton, and R. Cantor, "An Integrated Theoretical Perspective on Delinquent Behavior," *Journal of Research in Crime and Delinquency 16* (1979): 3–27; for further elaboration and refinement of this theory, see Elliott et al., *Explaining Delinquency.*

[45]Bernard and Snipes, "Theoretical Integration in Criminology," 310.

highly influenced by the associations that they make in the streets among their peers (i.e., differential association-reinforcement and social learning theory). According to Elliott's integrated theory, this factor—strong delinquent bonding—most directly results in delinquent behavior among most juvenile offenders.

One of the notable features of this theoretical model is that it allows for various types of individuals to become criminal. In other words, unlike traditionally separate frameworks that assume offenders expect not to achieve their goals (e.g., strain theory) or come from bad neighborhoods (e.g., Chicago or social disorganization theory), Elliott's integrated theory allows for a variety of possibilities when it comes to causal paths that explain how crime and delinquency develop in certain people and groups. This is what makes integrated models so much more powerful; namely, they bring several valid explanations together and allow for various possibilities to occur, all of which explain some criminality. Whereas traditional theories largely provide for only one causal process, Elliott and his colleagues showed right from the first major integrated theory that different types of trajectories, or paths to crime, are possible.

One of the major criticisms of merging such theories, particularly strain (which assumes individuals are born relatively good) and control or social bonding (which clearly assumes that people are born relatively bad [i.e., selfish, greedy, aggressive, etc.]), is that they tend to have extremely different, even opposite, assumptions of human behavior. As a recent review of theoretical integration noted, Elliott and his colleagues attempted to circumvent this obvious contrast in basic assumptions by claiming that the model allows for variation in individual motivations for why people engage in delinquency and crime.[46] For example, they claim that failure to achieve one's goals (i.e., strain theory) is not always the motivation for crime; rather, crime can result from inadequate socialization or coming from a disadvantaged neighborhood.

Furthermore, Elliott's integrated model also allows for different forms of control or social bonding, with not all of the delinquents being required to have weak social bonds with conventional society. For instance, as can be seen in Figure 11.1, a person who has experienced strain (or failure to achieve one's goals) can move directly to the social learning and differential association-reinforcement variables of strong delinquent bonding, so that the weak conventional bonding construct is not a required causal process in this theoretical model. So, while some critics claim that Elliott and his colleagues have combined theories that simply cannot be synthesized due to contrasting assumptions, we believe that this integrated theory did, in fact, find a logical and consistent way of showing how both models can be merged in a way that makes a lot of sense in the real world. On the other hand, the model does indeed claim that strain directly causes weak conventional bonding, and the theoretical framework implies that the probability of delinquency is highest when an individual experiences both strain and weak conventional bonding, so to some extent, the critics make an important point regarding the logical consistency of the full model, which Elliott and his colleagues have not adequately addressed.[47]

Despite the presence of elements of four traditionally separate theories (strain, social disorganization, social control or bonding, and social learning or differential association-reinforcement), Elliott and his colleagues identify this integrated theory with social control, as opposed to any of the other perspectives, which they argue is a more general theory and can explain crime and delinquency across different levels of explanation. They also note that the social control perspective is more sociological in the sense that it places more importance on the role of institutional structures in controlling criminal behavior.[48] Perhaps another reason why they identified their model with social control or bonding theory was because both intervening

---

[46]Ibid.

[47]Ibid.

[48]Ibid., 311–312.

constructs represent types of bonds formed or not formed with others: weak conventional bonding and strong delinquent bonding. However, it is important to keep in mind that these constructs actually represent two traditionally separate theories, namely social bonding theory and differential association-reinforcement or social learning, respectively.

The authors of this textbook believe that when an integrated theory (which claims to be such an integrated model) chooses a primary or dominant theory as its basis, it does not help in selling it to the scientific community, let alone others. A similar problem was seen in Pearson and Weiner's conceptual integration in their identifying a single traditionally separate theory—social learning or differential association-reinforcement—as the foundation or basis for their framework. Obviously, the first reaction by many theorists, even those that are not inherently against theoretical integration, would be somewhat cautious or even resistant. After all, why would a theory that claims to be integrated outright claim a single explanatory framework as its basis?

Rather, such models may be seen more so as examples of theoretical elaboration, which tends to start with the assumptions and concepts of one theory and draw from other models to improve the base model. Still, despite the criticisms toward identifying a single model as a basis for developing an integrated framework for explaining criminality, it is apparent that latitude has been shown for this flaw. It is important to note that Elliott's model is considered the first true attempt at theoretical integration and is still widely respected as the prototype and example of what an integrated model could and should be.

Regarding the empirical evidence for and against Elliott's integrated model, much of the evidence for it has been provided by Elliott and his colleagues through their testing of the theory. Specifically, much of the testing they have done has been via the National Youth Survey (NYS), a national survey collected and synthesized by criminologists at the University of Colorado, Boulder, which is where Elliott and most of the colleagues with whom he works are professors. This longitudinal measure of delinquency has been administered and analyzed for several decades, and it represents perhaps the most systematic collection of information from youths regarding key developmental variables and delinquency rates that has ever existed.

Most of the evidence from the NYS shows general strong support for Elliott's model.[49] However, some evidence showed few direct effects of strain and social control concepts, which is surprising given that the basis for the model was the social control and bonding elements of the theory.[50] In fact, the original hypothesis in the model—that strain and social control or bonding theories would have a direct effect on delinquency—was not observed.[51] Rather, only bonding to delinquent peers had a significant and strong effect on future criminality, which supports the social learning and differential association-reinforcement theory. This strongly supports the social learning variables in the model and diminishes the claim by Elliott and his colleagues about the fundamental theoretical perspective of social control and bonding.

Furthermore, a critical review of Elliott's framework, as presented in his 1985 book (with Huizinga and Ageton), notes that a major problem with the integrated framework is that

the most puzzling feature of the theory is the inclusion of social disorganization in it as a causal factor, in the absence of any attempt to measure or test the importance of this factor. Presumably, the authors wished to claim that their theory was "more sociological than psychological."[52]

---

[49]See review in Akers and Sellers, *Criminological Theories*, 273–276.

[50]Elliott et al., *Explaining Delinquency*.

[51]Akers and Sellers, *Criminological Theories*, 275.

[52]David Farrington, book review of Elliott et al., *Explaining Delinquency*, in *British Journal of Addiction 81* (1986): 433; embedded quote taken from 67 in Elliott et al.'s book.

This point is particularly important, given that virtually none of the tests of Elliott's integrated model have included social disorganization factors in their studies, even after this critical review was published more than 20 years ago. So, Elliott and his colleagues should either drop this portion from consideration in the model or provide an adequate test of it in relation to the rest of the framework. Although we would opt for the latter, one of these two alternatives must be chosen.

However, from the presentation that Elliott gave as his presidential address to the American Society of Criminology in 1993, it appears that the former alternative was chosen. Specifically, in his model of the onset of serious violent offending, all of the antecedent variables could be explained by the other traditionally separate theoretical models in his framework. He included two bonding constructs (family and school), parental sanctions, stressful life events (i.e., strain), and early exposure to crime and victimization. Perhaps the last factor could be construed as related to social disorganization theory, but it is probably regarded more as a social learning and differential association-reinforcement variable, as is suggested from the original model.

In addition, this critique noted that the existing evidence shows that most delinquent activity is committed among groups of juveniles, and thus, youths who tend to commit illegal acts naturally associate with delinquents. Yet, the affiliation with delinquent peers is not considered the basic foundation of the theory—as tests of this model demonstrated—but rather an intervening variable in the model. The author of this critique claims that Elliott's model does not emphasize this point strongly enough and thereby does not provide convincing evidence that delinquent peers cause or facilitate offending."[53]

In light of these findings, even Elliott and his colleagues acknowledged that their integrated model best fit the data as a social learning and differential association-reinforcement framework of delinquency.[54] However, they chose to retain social control and bonding theory as the primary foundation of their integrated model, stating that "it is not clear that a social learning model would have predicted a conditional relationship between conventional bonding . . . and deviant bonding."[55] Elliott and colleagues went on to say that they did not attribute most of the explained variation in the model on social learning and differential reinforcement as being the most important construct because they claimed that such variables did not play a strong enough part in the indirect effects that were seen in the estimated models.

This is actually a logical position, given the fact that their integrated framework would, in fact, predict that social learning and differential association-reinforcement variables were the primary direct effects on delinquency, and that social control, strain, and social disorganization variables were considered primarily indirect all along, according to their model. However, some critics, especially the proponents of social learning and differential reinforcement theory, have claimed that

in our opinion, even with the addition of the interactive effects of conventional bonding, the final model reported by Elliott et al. is more of a variation on social learning theory (with bonding modifications) than it is a variation on social bonding theory (with learning modifications).[56]

After all, the measures of delinquent peer bonds used by Elliott and his colleagues are essentially measures that have been used by theorists and researchers of social learning theorists over the last few decades, regarding differential associations, reinforcements, and modeling.[57]

---

[53]Farrington, book review, 433.

[54]Ibid.

[55]Ibid., based on quote from Elliott et al., *Explaining Delinquency*, 137.

[56]Akers and Sellers, *Criminological Theories*, 275.

[57]Ibid., 276.

Additional analyses using the NYS have tended to agree with the critics regarding the importance of social learning and differential reinforcement variables in predicting crime and delinquency. Specifically, one relatively recent reanalysis of NYS data showed that social learning variables appeared to predict more variation in deviance than did the other models or constructs in the model.[58] Some have noted that the social bonding assumption—that strong attachments to others, regardless of who they are—is not supported by empirical research (see Chapter 8). This is true, and all future integrated models must address this issue if they include social control or bonding propositions or constructs in their models.

Despite these criticisms and empirical observations, it appears that Elliott's integrated model has contributed to our understanding of the development of delinquent behavior. In the least, it inspired other theoretical frameworks, some of which we review here. However, it should also be obvious that there are some valid criticisms of this model, such as claiming that it is based on social control and bonding theory while depending heavily on the strong delinquent bonding that takes place in most cases of criminality, clearly implicating social learning and differential reinforcement theory. Again, we want to stress that a true integrated model should not place any emphasis on a particular theory; otherwise, the critics will have a sound argument when findings show that one theory is more influential than another.

Thornberry's Interactional Theory

After Elliott's integrated model, presented in 1979, the next major integrated framework was that of Terrence Thornberry in 1987.[59] This model incorporated empirical evidence drawn since Elliott's presentation to create a unique and insightful model of criminality, which addressed an extremely important aspect that had never been addressed previously in criminological theory. Specifically, Thornberry's interactional theory was the first to emphasize reciprocal, or feedback effects, in the causal modeling of the theoretical framework.

As a basis for his model, Thornberry combined social control and social learning models. According to Thornberry, both of these theories try to explain criminality in a straightforward, causal process and are largely targeted toward a certain age population.[60] Thornberry uniquely claims that the processes of both social control and social learning theory affect each other in a type of feedback process.

Thornberry's integrated model incorporates five primary theoretical constructs, which are synthesized in a comprehensive framework to explain criminal behavior. These five concepts include the following: commitment to school; attachment to parents; belief in conventional values (these first three are taken from social control and bonding theory); adoption of delinquent values; and association with delinquent peers (these last two are drawn from social learning and differential association-reinforcement theory). These five constructs, which most criminologists would agree are important in the development of criminality, are obviously important in a rational model of crime, so at first, it does not appear that Thornberry has added much to the understanding of criminal behavior. Furthermore, Thornberry's model clearly points out that

---

[58]Robert Agnew, "Why Do They Do It? An Examination of the Intervening Mechanisms Between 'Social Control' Variables and Delinquency," *Journal of Research in Crime and Delinquency 30* (1993): 245–266.

[59]Terrence Thornberry, "Toward an Interactional Theory of Delinquency," *Criminology 25* (1987): 863–887; see also Thornberry, "Reflections"; Terrence Thornberry, Alan Lizotte, Marvin Krohn, Margaret Farnworth, and Sung Jang, "Testing Interactional Theory: An Examination of Reciprocal Causal Relationships Among Family, School and Delinquency," *Journal of Criminal Law and Criminology 82* (1991): 3–35; Terrance Thornberry, Alan Lizotte, Marvin Krohn, Margaret Farnworth, and Sung Joon Jang, "Delinquent Peers, Beliefs, and Delinquent Behavior: A Longitudinal Test of Interactional Theory," *Criminology 32* (1994): 47–83.

[60]Much of this discussion is taken from Bernard and Snipes, "Theoretical Integration in Criminology," 314–316; and also Akers and Sellers, *Criminological Theories,* 278.

different variables will have greater effects at certain times; for example, he claims that association with delinquent peers will have more effect in mid-teenage years than at other ages.

What Thornberry adds beyond other theories is the idea of reciprocity or feedback loops, which no previous theory had mentioned, much less emphasized. In fact, much of the previous criminological literature had spent much time debating whether individuals become delinquent and then start hanging out with similar peers or whether individuals start hanging out with delinquent peers and then begin engaging in criminal activity. This has been the traditional chicken or egg question in criminology for most of the 20th century; namely, which came first, delinquency or bad friends. It has often been referred to as the *self-selection* versus *social learning* debate; in other words, do certain individuals select themselves to hang out with delinquents by their previous behavior, or do they learn criminality from delinquents with whom they start associating. One of the major contributions of **Thornberry's interactional model** is that he directly answered this question.

Specifically, Thornberry noted that most, if not all, contributors to delinquency (and criminal behavior itself) are related reciprocally. Thus, Thornberry postulated that engaging in crime leads to hanging out with other delinquents and that hanging out with delinquents leads to committing crimes. It is quite common for individuals to commit crime and then start hanging out with other peers who are doing the same, and it is also quite common for people to start hanging out with delinquent peers and then start committing offenses. Furthermore, it is perhaps the most likely scenario for a person to be offending and to be dealing with both the influences of past experiences and peer effects as well.

As mentioned previously, Thornberry considers the social control and bonding constructs, such as attachments to parents and commitment to school, the most essential predictors of delinquency. Like previous theoretical models of social bonding and control, Thornberry's model puts the level of attachments and commitment to conventional society ahead of the degree of moral beliefs that individuals have toward criminal offending. However, lack of such moral beliefs leads to delinquent behavior, which in turn negatively affects the level of commitment or attachments an individual may have built in his or her development. As Thornberry claimed,

> while the weakening of the bond to conventional society may be an initial cause of delinquency, delinquency eventually becomes its own indirect cause precisely because of its ability to weaken further the person's bonds to family, school, and conventional beliefs.[61]

Thus, the implications of this model are that variables relating to social control or bonding and other sources cause delinquency, which then becomes, in itself, a predictor and cause for the breakdown of other important causes of delinquency and crime.

For example, consider a person we shall call Johnny, who has an absent father, and a mother who uses inconsistent discipline and sometimes harsh physical abuse of her son. He sees his mother's state of constant neglect and abuse as proof that belief in conventional values is wrong, and he becomes indifferent toward governmental laws; after all, his main goal is to survive and be successful. Because of his mother's psychological and physical neglect, Johnny pays no attention to school and rather turns to his older peers for guidance and support. These peers guide him toward behavior that gives him both financial reward (selling what they steal) and status in their group (respect from performing well in illegal acts). At some point, Johnny gets caught, and this makes the peers who taught him how to engage in crime very proud,

---

[61]Thornberry, "Toward an Interactional Theory," 876, as quoted by Bernard and Snipes, "Theoretical Integration in Criminology," 315.

while alienating him from the previous bonds he had with his school, where he may be suspended or expelled, and with his mother, who further distances herself from him. This creates a reciprocal effect or feedback loop to the previous factors, which were lack of attachment to his mother and commitment to school. The lowered level of social bonding and control with conventional institutions and factors (mother, school) and increased influence by delinquent peers then leads Johnny to commit more frequent and more serious crimes.

Such a model, although complex and hard to measure, is logically consistent, and the postulates are sound. However, the value of any theory has to be determined by the empirical evidence that is found regarding its validity. Much of the scientific evidence regarding Thornberry's empirical model has been contributed by Thornberry and his colleagues.

Although the full model has yet to be tested, the researchers "have found general support for the reciprocal relationships between both control concepts and learning concepts with delinquent behavior."[62] One test of Thornberry's model used the longitudinal Rochester Youth Development Study to test its postulates.[63] This study found that the estimates of previous unidirectional models (nonreciprocal models) did not adequately explain the variation in the data. Rather, the results supported the interactional model, with delinquent associations leading to increases in delinquency, delinquency leading to reinforcing peer networks, and both directional processes working through the social environment. In fact, this longitudinal study demonstrated that, once the participants had acquired delinquent beliefs from their peers, the effects of these beliefs had further effects on their future behavior and associations, which is exactly what Thornberry's theory predicts.[64]

Perhaps the most recent test of Thornberry's interactional model examined the age-varying effects of the theory.[65] This study incorporated hierarchical linear modeling in investigating a sample of the NYS. The results showed that, while the effects of delinquent peers were relatively close to predictions, peaking in the mid-teenage years, the predictions regarding the effects of family on delinquency were not found to be significant in the periods that were expected, although family was important during adolescence.

Unlike other authors of integrated theories, Thornberry specifically noted that he prefers to see his approach as theoretical elaboration and not full theoretical integration. While we commend Thornberry for addressing this concern, virtually all criminologists still consider his framework a fully integrated model. And, in many ways, Thornberry's interactional model is far more integrated than others discussed here because it gives equal weight to the traditionally separate theoretical frameworks that are combined into his model in the sense that both are considered antecedent and reciprocal in their effects on criminal behavior.

Braithwaite's Theory of Reintegrative Shaming

A unique integrated model that proposed the synthesis of several traditionally separate theories was presented in 1989 in a book titled *Crime, Shame, and Reintegration*.[66] The theory of reintegrative shaming merges constructs and principles from several theories, primarily social control or bonding theory and

---

[62]Bernard and Snipes, "Theoretical Integration in Criminology," 316.

[63]Thornberry et al., "Delinquent Peers."

[64]Ibid.

[65]Sung Joon Jang, "Age-Varying Effects of Family, School, and Peers on Delinquency: A Multilevel Modeling Test of Interactional Theory," *Criminology* 37 (1999): 643–685.

[66]John Braithwaite, *Crime, Shame, and Reintegration* (New York: Cambridge University Press, 1989).

labeling theory, with elements of strain theory, subculture and gang theory, and differential association theory. All of these theories are synthesized in a clear and coherent framework that is presented in both descriptive and graphic form. We will spend extra time discussing Braithwaite's theory because it discusses not only U.S. culture but also cultural and justice tendencies in Japanese culture.

Braithwaite's idea for the theory was obviously inspired by the cultural differences he observed in Eastern (particularly Japanese) culture, in terms of both socialization practices and the justice system, as compared to the Western world. (Note that Braithwaite is from Australia, which uses the same Western practices as England and the United States.) Specifically, he emphasizes the Japanese focus on the aggregate, such as family, school, or business, to which the individual belongs. In contrast, in many Western cultures, epitomized by U.S. culture, the emphasis is clearly placed on the individual.

This contrast has often been referred to as the *we* culture versus the *me* culture (Eastern versus Western emphases, respectively). Although this is seen in virtually all aspects of culture and policy practices, it is quite evident in people's names. In most Eastern cultures, people are known by their family name, which is placed first in the ordering. This shows the importance that is placed on the group to which they belong. In contrast, Western societies list individual names first, implying a focus on the individual him- or herself. These naming practices are a manifestation of a virtually all-encompassing cultural difference regarding group dynamics and social expectations across societies, especially their justice systems. For example, it is quite common in Japan to receive a sentence of apologizing in public, even for the most serious violent crimes.[67] In his book, Braithwaite points out that, after World War II, Japan was the only highly industrialized society that showed a dramatic decrease in its crime rate.

Criminological theory would predict Japan to have an increasing crime rate, given the extremely high density of urban areas due to rapid industrialization, especially on such a small amount of land, Japan being a large island. As Braithwaite describes it, the Japanese suffered from anomie after the war in the sense of a general breakdown of cultural norms, but the nation was able to deal with this anomic state despite the odds. Japan definitively decided not to follow the Western model of justice after the war; rather, they rejected the Western system of stigmatizing convicted felons. Instead, the Japanese implemented a system in which they reintroduce (hence *reintegration*) the offenders via a formal ceremony in which citizens accept the offenders back into conventional society. In contrast, in the United States, we typically give our ex-cons about $200 on average and make them promise always to identify themselves as felons on legal documents.

In other cultural contrasts, Japan is extremely lenient in sentencing offenders to prison. In contrast, by rate of incarceration, the United States is the most punitive developed nation. In Japan, Braithwaite notes that

> prosecution only proceeds in major cases ... where the normal process of apology, compensation, and forgiveness by the victim breaks down. Fewer than 10 percent of those offenders who are convicted receive prison sentences, and for two-thirds of these, prison sentences are suspended. Whereas 45 percent of those convicted of a crime serve a jail sentence in the U.S., in Japan the percentage is under two.[68]

Public apology is the most common punishment among the Japanese, which strongly reflects the nature of honor in Japanese society, as well as pointing out the fundamental differences between how we

---

[67]Ibid., 62.

[68]Ibid., citing findings from J. Haley, "Sheathing the Sword of Justice in Japan: An Essay on Law Without Sanctions," *Journal of Japanese Studies 8* (1982): 265–281.

deal with offenders. Braithwaite claims that this cultural and political difference has a huge impact on why crime rates in both nations have experienced such differential trends.

Most developing Western nations, including the United States, experienced a rising crime rate in the 1950s, 1960s, and 1970s. Braithwaite argues that this was likely due to culture and the differential treatment of its offenders, namely "it might be argued that this [the downward trend in crime rates after World War II in Japan] was a result of the re-establishment of cultural traditions of shaming wrongdoers, including the effective coupling of shame and punishment."[69] In contrast, Braithwaite claims that "one contention . . . is that the uncoupling of shame and punishment manifested in a wide variety of ways in many Western countries is an important factor in explaining the rising crime rates in those countries."[70]

Furthermore, in contrast to American society, the Japanese are typically less confrontational with authority. For example, some scholars have noted that the Japanese accept the authority of law, and police officers are considered similar to "elder brothers" who rely "on positive rather than negative reinforcement," when it comes to crime control.[71] This difference in the way officers and other authority figures are considered is likely due to the way that the Japanese view society in terms of neighborhood, community, school, and work—the informal institutions of control that we have mentioned earlier in this book as having more effect on the crime rate than formal institutions (i.e., police, courts, prisons).

Beyond discussing the cultural differences, Braithwaite's integrated theory addresses some of the most notable scientific observations regarding what types of individuals and groups are likely to commit crime. Specifically, Braithwaite states that most crime is committed by young, poor, single males who live in urban communities that have high levels of residential mobility; they are not likely attached to or do poorly in school, have low educational or occupational aspirations, are not strongly attached to their parents, associate with delinquent peers, and do not have moral beliefs against violating law.[72]

It is obvious from this list that Braithwaite incorporated some of the major theories and corresponding variables into his theoretical perspective. The emphasis on poor people who do not have high educational or occupational aspirations obviously supports strain theory, whereas the inclusion of urban individuals who live in communities with high residential mobility reflects the Chicago School or social disorganization theory. At the same time, Braithwaite clearly highlights the predisposition of people who have limited moral beliefs and weak attachments, which conjures images of social bonding and control theory; individuals having delinquent peers obviously supports differential association and reinforcement theory. Thus, a handful of theories are important in the construction of Braithwaite's integrated theory.

Braithwaite notes that much of the effectiveness of the Japanese system of crime control depends on two constructs: interdependency and communitarianism. Interdependency is the level of bonds that an individual has to conventional society, such as the degree to which they are involved or attached to conventional groups in society, which would include employment, family, church, and organizations. According to Braithwaite, interdependency is the building block of the other major theoretical construct in his integrated model: communitarianism. Communitarianism is the macro (group) level of interdependency, meaning that it is the cumulative degree of the bonds that individuals have with conventional groups and institutions (e.g., family, employment, organizations, church, etc.) in a given society. Obviously, these theoretical

---

[69]Braithwaite, *Crime Shame*, 61, brackets are authors' paraphrasing.

[70]Ibid., 61.

[71]David H. Bayley, *Forces of Order: Police Behavior in Japan and the United States* (Berkeley: University of California Press, 1976).

[72]Braithwaite, *Crime, Shame*, 44–53.

constructs are mostly based on the theory of social bonding and control in the sense that they are based on attachments, commitment, and involvement in conventional society.

Braithwaite's model has a causal ordering that starts with largely demographic variables, such as age, gender, marital status, employment, and aspirations, on the individual level and urbanization and residential mobility on the macro (group) level. All of these factors are predicted to influence the level of interdependency and communitarianism in the model, which as we previously discussed, is largely based on social control and bonding theory. Depending on which type of culture is considered, or what forms of shaming are used in a given jurisdiction, the various types of shaming that are administered are key in this integrated model.

According to Braithwaite, societies that emphasize reintegrative shaming, such as Japan, will reduce the rates of crime in their societies. When an offender in Japan completes his or her sentence or punishment for committing a crime, the government will often sponsor a formal ceremony in which the offender is reintroduced or reintegrated back into conventional society. According to Braithwaite, we do not reintegrate offenders into our society after shaming them but rather stigmatize them, which leads them to associate only with people from criminal subcultures (e.g., drug users, gang members). Braithwaite claims that this leads to the formation of criminal groups and the grasping of illegitimate opportunities offered in the local community. Ultimately, this means that people who are not reintegrated into conventional society will inevitably be stigmatized and labeled as offenders, preventing them from becoming productive members of the community, even if their intentions are to do such.

Most empirical studies of Braithwaite's theory show mixed results, with most in favor of this theory, especially regarding its implementation for policy. Some tests have shown that reintegrative ideology regarding violations of law can have a positive impact on future compliance with the law; others have found that high levels of shaming in parental practices do not increase offending.[73] Studies in other countries, such as China and Iceland, show partial support for Braithwaite's theory.[74] While some studies have found encouraging results,[75] other recent studies have also found weak or no support for the theory.[76]

The most recent reports regarding the effects of shame show that outcomes largely depend on how shame is measured, which Braithwaite's theory largely ignores. There are, for example, episodic or situational shame states as well as long-term shame traits or propensities. Recent reviews of the literature and studies of how different types of shame are measured show that certain forms of shame are positively correlated with offending, whereas other forms of shame tend to inhibit criminal behavior.[77] If individuals persistently feel shame, they are more likely to commit criminal activity, but if persons who are not predisposed to feel shame perceive that they would feel shame for doing a given illegal activity, then they

[73]Toni Makkai and John Braithwaite, "Reintegrative Shaming and Compliance with Regulatory Standards," *Criminology 32* (1994): 361–386; Carter Hay, "An Exploratory Test of Braithwaite's Reintegrative Shaming Theory," *Journal of Research in Crime and Delinquency 38* (2001): 132–153.

[74]Lu Hong, Zhang Lening, and Terance Miethe, "Interdependency, Communitarianism, and Reintegrative Shaming in China," *Social Science Journal 39* (2002): 189–202; Eric Baumer, Richard Wright, Kristrun Kristinsdottir, and Helgi Gunnlaugsson, "Crime, Shame, and Recidivism: The Case of Iceland," *British Journal of Criminology 42* (2002): 40–60.

[75]Nathan Harris, Lode Walgrave, and John Braithwaite, "Emotional Dynamics in Restorative Conferences," *Theoretical Criminology 8* (2004): 191–210; Eliza Ahmed and Valerie Braithwaite, "'What, Me Ashamed'? Shame Management and School Bullying," *Journal of Research in Crime and Delinquency 41* (2004): 269–294.

[76]Bas Van Stokkom, "Moral Emotions in Restorative Justice Conferences: Managing Shame, Designing Empathy," *Theoretical Criminology 6* (2002): 339–361; Charles Tittle, Jason Bratton, and Marc Gertz, "A Test of a Microlevel Application of Shaming Theory," *Social Problems 50* (2003): 592–617; Lening Zhang and Sheldon Zhang, "Reintegrative Shaming and Predatory Delinquency," *Journal of Research in Crime and Delinquency 41* (2004): 433–453.

[77]Stephen Tibbetts, "Shame and Rational Choice in Offending Decisions," *Criminal Justice and Behavior 24* (1996): 234–255.

would be strongly inhibited from engaging in such activity.[78] This is consistent with findings from another recent study that demonstrated that the effect of reintegrative shaming had an interactive effect on delinquency.[79]

Furthermore, Braithwaite's theory does not take into account other important self-conscious emotions, such as guilt, pride, embarrassment, and empathy, that are important when individuals are deciding whether to commit criminal behavior. Although some theorists claim that shame is the key social emotion, studies show that they are clearly wrong. Rather, many emotions, such as guilt and embarrassment, are based on social interaction and self-consciousness; in many ways, they are just as inhibitory or rehabilitating as shame.[80] The literature examining rational choice theory is also an indication of the effects of emotions other than shame in influencing decisions on offending.[81]

Still, Braithwaite's reintegrative theory provides an important step ahead in theoretical development, particularly in terms of combining explanatory models to address crime rates, as well as correctional policies and philosophies, across various cultures. Specifically, Braithwaite's theory makes a strong argument that Eastern (particularly Japanese) policies of reintegration and apology for convicted offenders are beneficial for their culture. Would this system work in Western culture, especially the United States? The answer is definitely unknown, but it is unlikely that such a model would work well in the United States. After all, most of the chronic, serious offenders in the United States would not be highly deterred by having to apologize to the public, and many might consider such an apology or reintegration ceremony as an honor, or a way to show that they weren't really punished. This may not be true, but we would not know this until such policies were implemented fully in our system. However, programs along these lines have been implemented in our country, and some of these programs will be reviewed later in this book.

Tittle's Control-Balance Theory

One of the more recently proposed models of theoretical integration is that of Charles Tittle's control-balance theory. Presented by Tittle in 1995, control-balance integrated theory proposes that (1) the amount of control to which one is subjected and (2) the amount of control that one can exercise determine the probability of deviance occurring. In other words, the balance between these two types of control, he argued, can predict the type of behavior that is likely to be committed.[82]

In this integrated theoretical framework, Tittle claimed that a person is least likely to offend when he or she has a balance of controlling and being controlled. On the other hand, the likelihood of offending will increase when these become unbalanced. If individuals are more controlled by external forces, which Tittle calls *control deficit*, then the theory predicts they will commit predatory or defiant criminal behavior. In contrast, if an individual possesses an excessive level of control by external forces, which Tittle refers to as *control surplus*, then that individual will be more likely to commit acts of exploitation or decadence. It is important to realize that this excessive control is not the same as excessive self-control, which would be

---

[78]Stephen Tibbetts, "Self-Conscious Emotions and Criminal Offending," *Psychological Reports 93* (2003): 201–231.

[79]Hay, "An Exploratory Test."

[80]For an excellent review, see June Price Tangney and K. Fischer, *Self-Conscious Emotions: The Psychology of Shame, Guilt, Embarrassment, and Pride* (New York: Guilford Press, 1995).

[81]See Harold Grasmick, Brenda Sims Blackwell, and Robert Bursik, "Changes Over Time in Gender Differences in Perceived Risk of Sanctions," *Law and Society Review 27* (1993): 679–705; Harold Grasmick, Robert Bursik, and Bruce Arneklev, "Reduction in Drunk Driving as a Response to Increased Threats of Shame, Embarrassment, and Legal Sanctions," *Criminology 31* (1993): 41–67; Tibbetts, "College Student Perceptions."

[82]Tittle, *Control Balance* (see chap. 8, n. 55).

covered by other theories examined in this chapter. Rather, Tittle argues that people who are controlling, that is, who have excessive control over others, will be predisposed toward inappropriate activities.

Early empirical tests of control-balance theory have reported mixed results, with both surpluses and deficits predicting the same types of deviance.[83] Furthermore, researchers have uncovered differing effects of the control-balance ratio on two types of deviance that are contingent on gender. This finding is consistent with the gender-specific support found for Reckless's containment theory and with the gender differences found in other theoretical models.[84] Despite the mixed findings for Tittle's control-balance theory of crime, this model of criminal offending still gets a lot of attention, and most of the empirical evidence has been in favor of this theory.

Hagan's Power-Control Theory

The final integrated theory that we will cover in this chapter deals with the influence of familial control and how that relates to criminality across gender. Power-control theory is another integrated theory that was proposed by John Hagan and his colleagues.[85] The primary focus of this theory is on the level of patriarchal attitudes and structure in the household, which are influenced by parental positions in the workforce.

Power-control theory assumes that, in households where the mother and father have relatively similar levels of power at work (i.e., balanced households), mothers will be less likely to exert control on their daughters. These balanced households will be less likely to experience gender differences in the criminal offending of the children. However, households in which mothers and fathers have dissimilar levels of power in the workplace (i.e., unbalanced households) are more likely to suppress criminal activity in daughters. In such families, it is assumed that daughters will be taught to be less risk taking than the males in the family. In addition, assertiveness and risky activity among the males in the house will be encouraged. This assertiveness and risky activity may be a precursor to crime, which is highly consistent with the empirical evidence regarding trends in crime related to gender. Thus, Hagan's integrated theory seems to have a considerable amount of face validity.

Most empirical tests of power-control theory have provided moderate support, while more recent studies have further specified the validity of the theory in different contexts.[86] For example, one recent study reported that the influence of mothers, not fathers, on sons had the greatest impact on reducing the delinquency of young males.[87] Another researcher has found that differences in perceived threats of embarrassment and formal sanctions varied between more patriarchal and less patriarchal households.[88] Finally, studies have also started measuring the effect of patriarchal attitudes on crime and delinquency.[89] However, most of the empirical studies that have shown support for the theory have been done by Hagan or his

---

[83]Piquero and Hickman, "An Empirical Test," 319–342 (see chap. 8, n. 75); Hickman and Piquero, "Exploring the Relationships," 323–351 (see chap. 8, n. 75).

[84]Hickman and Piquero, "Exploring the Relationships"; for contrast and comparison, see Grasmick et al., "Changes Over Time"; Grasmick et al., "Reduction in Drunk Driving"; Tibbetts, "College Student Perceptions."

[85]Hagan, *Structural Criminology* (see chap. 8, n. 77); Hagan et al., "The Class Structure of Gender," 1151–1178 (see chap. 8, n. 77); Hagan et al., "Clarifying and Extending Power-Control," 1024–1037 (see chap. 8, n. 77); Hagan et al., "Class in the Household," 788–816 (see chap. 8, n. 77).

[86]John Hagan et al., "Class in the Household"; McCarthy and Hagan, "Gender, Delinquency," 153–177 (see chap. 8, n. 78); Morash and Chesney-Lind, "A Reformulation," 347–377 (see chap. 8, n. 78); Singer and Levine, "Power-Control Theory, Gender, and Delinquency," 627–647 (see chap. 8, n. 78).

[87]McCarthy et al., "In the Company of Women," 761–788 (see chap. 8, n. 79).

[88]Blackwell, "Perceived Sanction Threats," 439–488 (see chap. 8, n. 80).

[89]Blackwell, "Perceived Sanction Threats"; Bates et al., "Family Structure," 170–190 (see chap. 8, n. 81).

colleagues. Still, power-control theory is a good example of a social control theory in that it is consistent with the idea that individuals must be socialized and that the gender differences in such socialization make a difference in how people will act throughout life.

➤ Programs which prepare high-risk youth to start school

Policy Implications

Programs which Improve the ability of parents to do their jobs well

Many policy implications can be drawn from the various integrated theoretical models that we have discussed in this chapter. We will focus on the concepts that were most prominent, which are parenting and peer influences. Regarding the former, numerous empirical studies have examined programs for improving the ability of parents and expecting parents to be effective.[90] Such programs typically involve training high school students—or individuals or couples who are already parents—on how to be better parents. Additional programs include Head Start and other preschool programs that attempt to prepare high-risk youth for starting school; these have been found to be effective in reducing disciplinary problems.[91]

Regarding peer influences, numerous programs and evaluations have examined reducing negative peer influences regarding crime.[92] Programs that emphasize prosocial peer groups are often successful, whereas others show little or no success.[93] The conclusion from most of these studies is that the most successful programs of this type are those that focus on learning life skills and prosocial skills and use a curriculum based on cognitive-behavioral approaches.[94] This approach includes reinforcing positive behavior, clarifying rules of behavior in social settings, teaching life and thinking skills, and perhaps most important, thinking about the consequences of a given behavior before acting (hence, the cognitive-behavioral approach). Studies consistently show that programs using a cognitive-behavioral approach (i.e., think before you act) are far more successful than programs that emphasize interactions among peers or use psychoanalysis or other forms of therapy.[95]

Many other policy implications can be derived from integrated theories explaining criminal behavior, but parenting practices and peer influences are the primary constructs in most integrated models. Thus, these are the two areas that should be targeted for policy interventions, but they must be done correctly. For a start, policy makers could review the findings from empirical studies and evaluations and see that the earlier parenting programs start, particularly for high-risk children, the more effective they can be. Regarding the peer influence programs, the emphasis on cognitive-behavior therapy and training in life skills appear to be more effective than other approaches.

Conclusion

In this chapter, we have reviewed what determines the types of integrated theories and what criteria make an integrated theory a good explanation of human behavior. We have also examined some examples of integrated theories that have been proposed in the criminological literature in the last 20 years. All of the

---

[90]See review in Brown et al., *Criminology: Explaining Crime*, 425.

[91]Ibid.

[92]For a review, see Akers and Sellers, *Criminological Theories*, 102–108.

[93]Ibid.

[94]Ibid., 108.

[95]Ibid.

examples represent the most researched and discussed integrated theories, and they demonstrate both the advantages and disadvantages of theoretical integration or elaboration. We hope that readers will be able to determine for themselves which of these integrated theories are the best in explaining criminal activity.

In this chapter, we have examined the various ways in which theoretical integration can be done, including forms of conceptual integration and theoretical elaboration. Furthermore, the criticisms of the different variations of integration and elaboration have been discussed. In addition, numerous examples of theoretical integration have been presented, along with the empirical studies that have been performed to examine their validity.

Finally, we discussed the policy implications that can be recommended from such integrated theories. Specifically, we concluded that the influence of early parenting and peer influences are the two most important constructs across these theoretical models. Furthermore, we concluded that the earlier, the better regarding parenting programs. We also concluded that a cognitive-behavioral approach, which includes life skills, is most effective for peer influence programs.

▧ Chapter Summary

- Theoretical integration is one of the more contemporary developments in criminological theorizing. This approach brings with it many criticisms, yet arguably many advantages.
- Types of theoretical integration include end-to-end, side-by-side, and up-and-down.
- Conceptual integration appears to be a useful, albeit rarely explored, form of theoretical integration.
- Theoretical elaboration is another form of integration, which involves using one theory as the base or primary model and then incorporating concepts and propositions from other theories to make the primary model stronger.
- A number of seminal integrated theories have been examined using empirical evidence and appear to have enhanced our understanding of criminal behavior.
- Some of the theoretical models that have been proposed (such as Elliott et al.'s model) have been supported by empirical research.
- Theoretical integration models have many critics, who claim that the assumptions, concepts, and propositions of the mixed theories are counterintuitive.

KEY TERMS

Braithwaite's reintegrative shaming theory

Communitarianism

Conceptual integration

Elliott's integrated model

End-to-end theoretical integration

Integrated theories

Interdependency

Side-by-side (or horizontal) integration

Theoretical elaboration

Thornberry's interactional model

Up-and-down integration

DISCUSSION QUESTIONS

1. What is the definition of theoretical integration, and why can such theories be beneficial?

2. Describe what end-to-end theoretical integration is, and provide an example of such integration.

3. Describe what side-by-side theoretical integration is, and provide an example of such integration.

4. Describe what up-and-down theoretical integration is, and provide an example of such integration.

5. Discuss the difference between theoretical elaboration, theoretical integration, and conceptual integration.

6. What are the major strengths and weaknesses of theoretical integration?

7. In your opinion, what is the best of the integrated models? Why do you believe this is the best integrated model?

8. What do you believe is the weakest integrated model? Why?

WEB RESOURCES

Integrated Theories of Crime

http://en.wikipedia.org/wiki/integrative_criminology

John Braithwaite/Reintegrative Theory

http://www.realjustice.org/library/braithwaite06.html

Various Integrated Theories of Crime

http://law.jrank.org/pages/821/Crime-Causation-Sociological-Theories-Integrated-theories.html

12

Applying Criminological Theory to Policy

This chapter will discuss how criminological theories that we have discussed throughout the book have been used to make current criminal justice policies more effective in preventing and reducing crime. Given that we have examined many of the policy implications derived from theoretical models at the end of each chapter throughout this book, here we will provide only a brief review regarding each category of theories. Thus, this chapter is meant to summarize the way policy has been altered by criminological research in U.S. history, as well as suggestions for future policy changes based on the current state of research in criminology and criminal justice.

Criminological theories have been used in informing and development of policy ever since the father of deterrence, Cesare Beccaria, developed the first rational theory of crime in the mid-18th century. We will focus on the policies that have received the most attention by researchers and policy makers. Readers should realize that this chapter is not meant to be a conclusive review of all policies that can be applied from a particular type of theory; on the contrary, this is only a short list of the numerous policies that can be derived from a given theoretical framework.

✉ Policy Applications Derived From the Classical School: Deterrence Theory

Although most modern criminologists and theoreticians in the field do not consider deterrence theory to be one of the more robust models for understanding or reducing illegal activity,[1] there is no doubt that policy making in our criminal justice system is dominated by the ideas of the Classical School, especially

---

[1]Ellis Walsh, "Criminologists' Opinions," 1–4 (see chap. 1, n.6).

deterrence theory.[2] This is seen in most policies used throughout the system by its three major components: law enforcement, courts, and corrections.

Law Enforcement

Perhaps the most important assumption in applying deterrence theory to policy making in the criminal justice system is that individuals are rational. Unfortunately, much research suggests that people often engage in behaviors that they know are irrational and that individuals' cognitive differences can limit rational decision making.[3] Criminologists often refer to this limitation as *bounded rationality*.[4] Therefore, it is not surprising that many attempts by police and other criminal justice authorities to deter potential offenders do not seem to have much effect in preventing crime.

Of course, the simple existence of a policing authority tends to deter people to some extent. When policing breaks down in given jurisdictions, riots and looting often follow, and it can become a free-for-all among a large mob, which typically contains otherwise law-abiding people. This is one of the reasons why it is typically illegal for police unions to conduct a strike that would leave no police officers available for responding to calls. Looking at deterrence another way, studies show that the vast majority of drivers speed on highways. For example, one recent study by researchers at Purdue University revealed that 21% of motorists think it's perfectly safe to exceed the speed limit by 5 mph, another 43% saw no risk in going 10 mph over the limit, and another 36% saw no risk in driving 20 mph over the speed limit.[5] This finding is consistent with other studies on drivers' attitudes regarding speeding and limits.

However, when police cruisers are seen, most people will slow down and be more careful, which is in a sense a form of deterrence. For example, most individuals' heart rates will go up a few notches when they see a police cruiser or highway patrol car, even if they are not doing anything wrong. This is a good example of how deterrence can result from simply seeing a police presence. It is likely we can all relate to this, especially if we are speeding or engaging in other illegal behavior (e.g., talking on the cell phone while driving, etc.).

On the other hand, research is less clear about whether increasing the number of police in a given area, often referred to as *saturated patrol,* can significantly reduce crime. Some early studies, especially one key experiment involving increased random patrol in certain beats in Kansas City, found no deterrent effect from saturated patrol.[6] However, a recent review of the extant literature, hereafter referred to as the Maryland Report (because all key authors of the report were professors at University of Maryland), concluded that extra police who focus on specific locations (i.e., hot spots), as opposed to larger beat areas, can prevent crime.[7] This strategy of cracking down on specific locations applies to deterrence theory and also incorporates another type of theory that assumes a rational individual: routine activities theory. That theory assumes that motivated offenders will be less likely to commit an offense if there is a presence of guardianship, so adding

---

[2]Vold et al., *Theoretical Criminology* (see chap. 1, n. 4).

[3]For a review of the extant research on this topic, see Piquero and Tibbetts, *Rational Choice and Criminal Behavior* (see chap. 3, n. 51).

[4]For more recent discussion on the complexity of developing policies based on deterrence and rational choice models, see Travis Pratt, "Rational Choice Theory, Crime Control Policy, and Criminological Relevance," *Criminology and Public Policy 7* (2008): 43–52.

[5]For an example, see a news report regarding a study by researchers at Purdue University in which the vast majority of drivers believed going over the speed limit was acceptable and safe: http://www.themotorreport.com.au/12358/study-shows-motorists-ambivalent-to-speed-limits/

[6]George Kelling, *The Kansas City Preventive Patrol Experiment* (Washington, DC: Police Foundation, 1974).

[7]Sherman et al., *Preventing Crime* (see chap. 2, n. 47).

police to a specific location reduces the likelihood that they will have the opportunity to commit criminal activity at that hot spot.

This same Maryland Report also noted that proactive arrests for drunk driving have consistently been found to reduce such behavior.[8] The effectiveness of reducing drunk driving this way probably involves deterring individuals from continuing such activity; that effect is not surprising given the fact that drunk driving often is committed by middle- and upper-class people who have a lot to lose if they don't alter their behavior. The Maryland Report also concludes that the same type of phenomenon is seen with arresting employed (but not unemployed) individuals for domestic violence; to clarify, the employed people have more to lose, so they tend to be deterred from recidivating for domestic violence.[9]

The Maryland Report also offers conclusions about which policing strategies do not seem to prevent or deter crime: more rapid response time to 911 calls, arrests of juveniles for minor crimes, arrest crackdowns at drug markets, and community policing programs with no clear focus on crime risk factors.[10] Finally, it is interesting to note that the Maryland Report acknowledges that the government invests far more federal funds in policing—$2 billion per year—than in any other crime prevention institution, despite the fact that there are still many questions about how effective this funding strategy is, given the wide variation in how each police agency uses this funding and the departments' lack of understanding about what types of programs may be effective versus those that aren't.[11] Thus, it would be very beneficial if policy makers and policing authorities would examine the findings of recent empirical research to direct their resources to policies that actually work in reducing crime.

Courts, Diversion Programs, and Corrections

Numerous policies implemented by courts and diversion strategies are based, at least to a large extent, on deterrence. One of the most notorious of the diversion programs is *scared straight*, which became popular in the 1970s.[12] These programs involved having prisoners talk to youth about the realities of their lives, which was supposed to scare them into refraining from recidivating and make them decide to go straight. Unfortunately, virtually all of these programs were found to be ineffective, and some evaluations showed that young participants had significantly higher rates of recidivism.[13] Perhaps the intense experience of being scared by actual inmates had the effect of stigmatizing the youths who went through these programs. Regardless of the reasons, empirical research has consistently found that scared straight programs do not deter or prevent future offending, and the Maryland Report concluded this after reviewing the evaluation literature.

In quite a different strategy, some judges have moved toward using shaming strategies to deter offenders from recidivating.[14] For example, many cities have begun posting the names (and sometimes pictures) of men who are arrested for soliciting prostitutes. Also, some judges have started forcing individuals convicted of drunk driving to place glow-in-the-dark bumper stickers on their cars as a condition of their

---

[8]Ibid.

[9]Ibid.

[10]Ibid.

[11]Ibid.

[12]For a review of these programs and evaluations of them, see Richard Lundman, *Prevention and Control of Juvenile Delinquency,* 2nd ed. (Oxford: Oxford University Press, 1993).

[13]Lundman, *Prevention and Control.*

[14]For a variety of examples of judges using such shaming sanctions, see link to story at: http://www.msnbc.msn.com/id/19957110/

[handwritten: Shaming: No empirical support]

▲ **Image 12.1**   Scared straight programs were designed to expose delinquents to heart-to-heart talks with inmates with the aim of literally scaring them into becoming straight or nondelinquent.

probations, which state "BEWARE: Convicted Drunk Driver," or have shoplifters walk in front of stores with signs that say, "I am a thief. I stole from Wal-Mart!" These are just some of the examples of a growing trend that attempts to deter individuals by threatening them with shaming practices if they choose to engage in illegal behavior. Such sentences apply elements of rational choice theory, particularly its ideas about the strong deterrent influence of informal factors, such as community, family, employment, and church. Unfortunately, there have been virtually no empirical evaluations of the effectiveness of such shaming penalties.

Another strategy courts have used for deterring individuals is to develop special sentencing enhancements for habitual offenders, such as California's *three strikes you're out*. Using the common term from baseball, the three-strikes-you're-out policy refers to sentencing schemes that mandate life imprisonment for three-time felons, although each state differs in how this formula works. Washington state was the first to implement a three-strikes law, and today, most states have some version of it.[15] As a specific deterrent, such a strategy clearly works because a third-time felon will be in prison for life; however, the three-strikes law was also intended to be a general deterrent and to prevent offenders from recidivating.

[handwritten: The Law: No clear result]

Empirical evaluations of the deterrent effect of this law provide no clear result. One study of California's three-strikes law suggests that it reduced crime,[16] whereas several other studies showed negligible effects. In some cases, three-strikes laws may have actually increased violence,[17] perhaps due to the desperation and nothing-left-to-lose mentality of offenders who know they face a life sentence if apprehended.[18] Also, the vast majority of offenders prosecuted under the three-strikes penalty are nonviolent (i.e., drug offenders, property offenders, etc.). This is true in California, one of the states that most frequently uses this type of sentencing policy. For example, a man who had no violent crime on his record stole some DVDs from a store and was given a life sentence, which was upheld by the U.S. Supreme Court because of two prior offenses. In some cases, these types of sentencing policies have actually forced the release of violent offenders to make room for habitual nonviolent drug and property offenders. This issue was also mentioned as a serious problem in the Maryland Report.

---

[15]David Shichor and Dale K. Sechrest, eds., *Three Strikes and You're Out: Vengeance as Social Policy* (Thousand Oaks: Sage, 1996).

[16]Joanna Shepherd, "Fear of the First Strike: The Full Deterrent Effect of California's Two- and Three-Strikes Legislation," *Journal of Legal Studies 31* (2002): 159–201.

[17]Lisa Stolzenberg and Stewart J. D'Alessio, "Three-Strikes-You're-Out: The Impact of California's New Mandatory Sentencing Law on Serious Crime Rates," *Crime and Delinquency 43* (1997): 457–469; Mike Males and Dan Macallair, "Striking Out: The Failure of California's Three-Strikes and You're Out Law," *Stanford Law and Policy Review 11* (1999): 65–72.

[18]Tomislav Kovandzic, John J. Sloan III, and Lynne M. Vieraitis, "Unintended Consequences of Politically Popular Sentencing Policy: The Homicide-Promoting Effects of 'Three Strikes' in U.S. Cities (1980–1999)," *Criminology and Public Policy 1* (2002): 399–424.

Thus far, successful deterrent policies in the courts and corrections components of the criminal justice system seem rare. According to the Maryland Report, one of the only court-mandated policies that seems promising in deterring offending behavior is orders of protection for battered women.[19] Still, far more brainstorming and evaluations are needed of deterrence-based policies.

Policy Applications Derived From Biosocial Theories

Although a multitude of policy strategies are related to the biosocial perspective (see Chapter 5), there has been a reluctance to claim that these policies derive from such a theoretical framework due to the stigma of any reference to biology in curbing crime. However, the Maryland Report suggests the importance of identifying and quickly treating any head and bodily trauma early in development among infants and toddlers, concluding that some of the most consistently supported early-intervention programs for such physiological problems are those that involve weekly infant home visitation.[20] Experts have recently suggested many more strategies, such as mandatory health insurance for pregnant mothers and their children, which is likely one of the most efficient ways to reduce crime in the long run.[21] Furthermore, all youths should be screened for levels of toxins (e.g., lead), as well as abnormal levels of neurotransmitters (e.g., serotonin) and hormones (e.g., testosterone).[22] Furthermore, greater efforts should be made to ensure that any type of abuse or other form of head trauma gets special attention.

Ultimately, despite the neglect that biosocial models of crime receive in terms of both recognition and policy implications, there is no doubt that this area is crucial if we hope to advance our understanding and create more efficient policies regarding criminal behavior. It is time that all criminologists recognize the degree to which human behavior results from physiological disorders. We all have brains, each about 3 pounds in weight, that determine the choices we make. Criminologists must acknowledge the influence of biological or physiological factors that influence this vital organ, or the discipline will be behind the curve in terms of understanding why people commit (or do not commit) criminal offenses.

Policy Applications Derived From Social Structure Theories

Since the introduction of Shaw and McKay's theory of social disorganization (see Chapter 7), there have been numerous attempts to prevent crime in lower-class, urban areas, where crime is concentrated. One of the most straightforward strategies involves forming a neighborhood watch in high-risk neighborhoods.[23] After reviewing the extant evaluations of such neighborhood watch programs, the Maryland Report concluded that they do not seem to work in reducing crime in neighborhoods. Perhaps this is because the cohesive communities that form such watches already have low crime rates; the neighborhoods that have the highest crime rates tend to be those that are the least cohesive or are in the worst areas regarding crime. Furthermore, the report explicitly concludes that community mobilization against crime in high-crime, inner-city, poor areas does not work, probably for the same reasons. Specifically, the most crime-ridden

---

[19]Sherman et al., *Preventing Crime.*

[20]Ibid.

[21]Wright et al., *Criminals in the Making* (see chap. 5, n. 32).

[22]Ibid.

[23]Sherman et al., *Preventing Crime.*

areas tend to have a high turnover of poor residents, who have virtually no investment in the neighborhood; they plan on moving out of the area as soon as they can.

Also related to Shaw and McKay's theory of social disorganization, the Maryland Report concluded that some promising strategies (although there was not enough information to make strong conclusions) included gang violence prevention programs that focus on reducing gang cohesion, as well as volunteer mentoring programs, such as the Big Brothers Big Sisters program, which appears to be especially good for reducing substance abuse among youth.[24] Thus, it appears that focusing on high-risk youth may be one way of effectively targeting resources and strategies for reducing crime, as opposed to trying to address the crime issues of an entire neighborhood or area of a city. This is consistent with meta-analyses of evaluation studies, which show that the most effective programs are those that focus on the individuals that have the highest risk for offending or recidivism.[25] It is also consistent with studies showing that diversion and rehabilitation programs that target the offenders who are most at risk tend to have more effect than those targeting medium- or low-risk offenders, probably due to the fact that the lower risk offenders likely won't offend again anyway.[26]

In regard to Merton's strain theory (see Chapter 6), there have been significant attempts to address the more global issues of deprivation that result from poverty in U.S. society. Perhaps the best example is the War on Poverty, a federal program introduced by President Lyndon B. Johnson during the 1960s. Although an enormous amount of money and other resources were spent in trying to aid the poor in order to reduce crime rates in America, crime rates soared during this period. Many experts noted that the federal investment was not efficiently handed out, but the rise in crime during this period still casts doubt on the effectiveness of such a strategy.

Rather, the Maryland Report concludes that vocational training for adult male offenders has consistently lowered offending rates, as have intensive residential training programs for at-risk youth (such as Job Corps).[27] This conclusion appears consistent with our previous finding that a focus on certain high-risk individuals is more effective than programs addressing an entire group, such as all types of offenders. Perhaps this is because not all offenders have the target characteristic, but such programs tend to be effective for those who do. For example, some offenders are thrust into drug programs even though they have never used drugs or alcohol. Also, some offenders are coerced into engaging in vocational programs when they already have skills in a certain trade. Thus, it makes sense to focus resources (e.g., vocational programs, drug programs, etc.) on the people who are most at risk for the given risk factor.

▨ Policy Applications Derived From Social Process Theories

Related to Chapter 8, ever since the introduction of Sutherland's differential association theory in the 1930s, numerous programs to reduce crime and delinquency have been based on the assumption that youths must be provided with prosocial interactions with their parents and peers. Many of these programs have shown poor results, probably due to the fact that mixing offenders with other offenders only leads to more offending, ironically due to associations with other offenders. However, many of these programs have shown success or at least promise.

---

[24]Ibid.

[25]Patricia Van Voorhis, Michael Braswell, and David Lester, *Correctional Counseling and Rehabilitation* (Cincinnati: Anderson, 1999).

[26]Ibid.

[27]Sherman et al., *Preventing Crime*.

According to the Maryland Report, evidence consistently supports programs that involve long-term, frequent home visitation combined with preschool, as well as family therapy by clinical staff for at-risk youth.[28] Furthermore, the Maryland Report stresses the success of school programs that aim at clarifying and communicating norms about various behaviors by establishing school rules and consistently enforcing them through positive reinforcement and school-wide campaigns.[29] Other effective school programs focus on various competency skills, such as developing self-control, stress management, responsible decision making, problem solving, and communication skills.[30] On the other hand, the Maryland Report noted some school programs that do not reduce delinquent behavior, such as counseling students in a peer-group context (see reasoning above), as well as substance use programs that simply rely on information dissemination or fear arousal (such as the traditional DARE program).[31] Ultimately, programs that teach youths to develop cognitive or thinking skills are beneficial, as opposed to those that simply focus on information about drugs or criminal offending. The bottom line for successful programs appears to be that they must instruct youths to deal with everyday influences in a healthy way (according to a cognitive-behavioral model), as opposed to just giving them information about crime and drugs.

Policy Applications Derived From Social Reaction and Conflict Theories

Most of the policy implications that were derived from labeling and other forms of social reaction theories (see Chapter 9) are often referred to as *the Ds* because they include diversion, decriminalization, and deinstitutionalization, which all became very popular in the late 1960s and early 1970s, when the popularity of labeling theory peaked. Consistent with labeling theory, all three emphasize the hands-off doctrine, which seeks to reduce the stigmatization that may occur from being processed or incarcerated by the formal justice system. Diversion includes a variety of programs, such as drug courts or community service programs, which attempt to remove an offender from the traditional prosecutorial and trial process as soon as possible, often with the promise of the charge being expunged or removed from their permanent records. Despite empirical research showing mixed results, some of these diversion programs, such as drug courts, have shown promise.[32]

Decriminalization is the lessening of punishments, which is typically recommended for nonviolent crimes to reduce the stigma and penalty related to an offense. This can be distinguished from legalization, which makes an act completely legal; rather, when a certain behavior is decriminalized, it remains illegal, but the penalties are significantly reduced. For example, the possession of small amounts of marijuana is decriminalized in some states (e.g., California): Instead of being arrested and possibly jailed, a person found holding only a small amount is given a ticket, similar to a speeding ticket, which requires payment of a fine. Similarly, deinstitutionalization refers to the avoidance of incarcerating minor, nonviolent offenders, especially if they are juveniles committing status offenses (e.g., running away, truancy, etc.). Some have argued that this policy would lead to higher rates of recidivism, but this has not become a problem according to statistics from police and FBI reports.

---

[28]Ibid.

[29]Ibid.

[30]Ibid.

[31]Ibid.

[32]Ibid.

The bulk of policies that have been derived from conflict theories have emphasized attempting to make laws, enforcement, and processing through the justice system more equitable in terms of social class. For example, the federal sentencing guidelines were recently reviewed for drug possession because it became clear that they were unfair in how they punished drug possession offenders. Specifically, according to the 1980s guidelines, possession of an ounce of crack cocaine earned a defendant a sentence 100 times longer than an offender who possessed an ounce of powder cocaine. Not surprisingly, crack cocaine is typically used and sold by lower class minorities, whereas cocaine powder is typically used and sold by upper- or middle-class Whites. The disparity in this form of sentencing, which clearly penalizes a certain group of people more than another, has recently become a target for reform by the Obama administration, so there are current efforts to address this issue. This is just one example of how laws and processing are significantly disproportionate in the sense that certain groups (typically lower class minorities) are arrested, punished, and locked up far more than well-to-do offenders. Studies showing higher conviction rates and longer sentences for offenders who must rely on public defenders to represent them consistently support the high degree of disparities in the criminal justice system, which policies derived from conflict theory attempt to address.

Recently, these hands-off policies have been less attractive to authorities and politicians who want to be seen as hard on criminal offenders so they will be elected or retained in office. Still, due to recent economic crises in many state and local communities, it is likely that many jurisdictions will reconsider these approaches and find alternative approaches for dealing with nonviolent offenders via alternative or diversion strategies for these individuals.

⚅ Policy Applications Derived From Integrated and Developmental Theories

Few policy strategies have been derived from integrated or developmental theories, probably because most policy makers are either unaware of such theoretical frameworks or because the theories are so complicated that it is difficult for practitioners to apply them. However, one exception to this rule is Braithwaite's theory of shaming and reintegration (see Chapter 11). Braithwaite has gone to great lengths to apply his theory in Australia, and some evaluations of this approach have shown promise, even if not in the United States. Interventions stress the restorative concepts of his theory, particularly the idea of reintegrating offenders back into society, as opposed to stigmatizing them.[33] This strategy emphasizes holding offenders accountable, yet it tries to bring them back into conventional society as quickly and efficiently as possible. In contrast, the U.S. system typically gives prisoners a nominal amount of funds (in most jurisdictions, about $200), and they must report themselves as convicted felons for the rest of their lives on all job applications. Would Braithwaite's model work here in the United States? We don't know, but it seems that there could be a better way to incorporate offenders back into society in such a way that they are not automatically set up to fail.

Another area of policy implications from developmental theory is that involving early childhood. Recent research has shown that the earlier resources are provided—for maternal care during pregnancy, as well as infant care, especially for high-risk infants—the less likely that such infants will be disposed toward engaging in criminal offending.[34] Thus, far more resources and funding should be given to early development of youths, especially for high-risk mothers who are pregnant.

---

[33]John Braithwaite, "Restorative Justice: Assessing Optimistic and Pessimistic Accounts," in *Crime and Justice: A Review of Research*, Vol. 25, ed. Michael Tonry (Chicago: University of Chicago Press, 1999), 1–27.

[34]Wright et al., *Criminals in the Making*.

⊠ Conclusion

This chapter briefly examined some of the policy implications that can be derived from the theories discussed throughout this text. Other policy strategies have been tried, but we have focused on those that have received the most attention. It is up to each reader to consider what other types of programs can or should be derived from the various theories explored in this book.

⊠ Chapter Summary

- Ever since the beginning of rational criminological theorizing in the 18th century, policy implications have always been derived from the popular theories of a given time period.
- Policies derived from Classical School or deterrence theory have seen mixed results, largely because of the limited rationality of people due to psychological and physiological propensities. However, some of these policy implications appear to have some success, such as proactive policing and targeting high-crime locations, known as hot spots. On the other hand, many of the court-implemented policies based on deterrence principles, such as three strikes you're out, have not shown much promise as they are currently administered.
- Policies derived from biosocial theories emphasize the need for more medical care, especially during the pre- and perinatal stages of life. The earlier biological and social risk factors can be addressed, the better.
- Many policies can be derived from social structure theories, but these have not always been found successful, for example, the War on Poverty in the 1960s and various forms of neighborhood watch programs in the last few decades. However, some programs do appear to work, such as those that provide vocational training so people can pull themselves out of poverty.
- Social process theories also have numerous policy applications, many of them successful, such as visitations by professionals to make sure the family environment is not problematic, as well as programs that teach individuals (typically youths) cognitive-behavioral skills on how to handle peer pressure and other social situations.
- Social reaction theories have inspired policies such as diversion and decriminalization. Conflict theories have inspired policies such as revising sentencing guidelines to make them more equitable according to the seriousness of a given act.
- Developmental and integrative theories have not inspired too many policies yet, due to the complexity of such models. However, it is quite likely that, in the near future, such theoretical frameworks will be able to inform policy makers.

DISCUSSION QUESTIONS

1. What types of policies have been derived from the Classical School or deterrence theory? Which have been found to be successful, and which have not?

2. What types of policies have been derived from biosocial theories?

3. What types of policies have been derived from social structure theories?

4. What types of policies have been derived from social process theories?

5. What types of policies have been derived from social reaction and conflict theories?

6. What types of policies have been derived from integrated and developmental theories?

7. What pattern can be seen from the successful programs across all the theories discussed in this book? What patterns do you see in the failure of programs across all theories in this book?

WEB RESOURCES

The University of Maryland Report of What Works and What Doesn't

http://www.ncjrs.org/works/

Policy Implications Derived From Criminological Theories

http://www.cjpf.org/crime/crime.html

http://www.crimepolicy.org/

Glossary

Actus reus: In legal terms, whether the offender actually engaged in a given criminal act. It can be contrasted with *mens rea,* which is a concept regarding whether the offender had the intent to commit a given act. This concept is important, especially in situations in which juveniles or mentally disabled individuals engage in offending.

Adaptations to strain: As proposed by Merton, the five ways that individuals deal with feelings of strain; see *conformity, innovation, rebellion, retreatism,* and *ritualism.*

Adolescence-limited offenders: A type of offender labeled in Moffitt's developmental theory; such offenders commit crimes only during adolescence and desist from offending once they reach their twenties or adulthood.

Adoption studies: Studies that examine the criminality of adoptees as compared to the criminality of their biological and adoptive parents; such studies consistently show that biological parents have more influence on children's criminal behavior than the adoptive parents who raise them.

Age of Enlightenment: A period of the late 17th century to 18th century in which philosophers and scholars began to emphasize the rights of individuals in society. This movement focused on the rights of individuals to have a voice in their government and to exercise free choice; it also included the idea of the social contract and other important assumptions that influenced our current government and criminal justice system.

Anomie: A concept originally proposed by Durkheim, which meant *normlessness* or the chaos that takes place when a society (e.g., economic structure) changes very rapidly. This concept was later used by Merton in his strain theory of crime, where he redefined it as a disjunction between the emphasis placed on conventional goals and the conventional means used to achieve such goals.

Atavism: See *atavistic.*

Atavistic: The belief that certain characteristics or behaviors of a person are throwbacks to an earlier stage of evolutionary development.

Autonomic nervous system (ANS): The portion of the nervous system that consists of our anxiety levels, such as the fight or flight response, as well as our involuntary motor activities (e.g., heart rate). Studies consistently show that lower levels of ANS functioning are linked to criminality.

Bourgeoisie: A class or status that Karl Marx assigned to the dominant, oppressing owners of production, who are considered the elite class due to ownership of companies, factories, and so on. Marx proposed that this group created and implemented laws that helped retain their dominance over the proletariat, or working class.

Braithwaite's reintegrative shaming theory: This integrated theoretical model merges constructs and principles from several theories, primarily social control or bonding theory and labeling theory, with elements of strain theory, subculture or gang theory, and differential association theory.

Brutalization effect: The predicted tendency of homicides to increase after an execution, particularly after high-profile executions.

Central nervous system (CNS): The portion of the nervous system that largely consists of the brain and spinal column and is responsible for our voluntary motor activities. Studies consistently show that low functioning of the CNS (e.g., slower brain wave patterns) is linked to criminal behavior.

Cerebrotonic: The type of temperament or personality associated with an ectomorphic (thin) body type; these people tend to be introverted and shy.

Certainty of punishment: Certainty of punishment is one of the key elements of deterrence; the assumption is that, when people commit a crime, they will perceive a high likelihood of being caught and punished.

Chicago School of criminology: A theoretical framework of criminal behavior that is often referred to as the Ecological School or the theory of social disorganization; it emphasizes the environmental impact of living in a high-crime neighborhood and asserts that this increases criminal activity. This model applies ecological principles of invasion, domination, and succession in explaining how cities grow and the implications this has on crime rates. Also, this model emphasizes the level of organization (or lack thereof) in explaining crime rates in a given neighborhood.

Classical conditioning: A learning model that assumes that animals, as well as people, learn through associations between stimuli and responses; this model was primarily promoted by Pavlov.

Classical School: The Classical School of criminological theory is a perspective that is considered the first rational model of crime, one that was based on logic rather than supernatural or demonic factors; it assumes that crime occurs after a rational individual mentally weighs the potential good and bad consequences of crime and then makes a decision about whether to engage in a given behavior; this model is directly tied to the formation of deterrence theory and assumes that people have free will to control their behavior.

Clearance rate: The percentage of crimes reported to police that result in an arrest or an identification of a suspect who cannot be apprehended (due to death of suspect, or fleeing, etc.). In other words, the authorities essentially have a very good idea of who committed the crime, so it is considered solved. The clearance rate can be seen as a rough estimate of the rate at which crimes are solved.

Collective conscience: According to Durkheim, the extent of similarities or likeness that people in a society share. The theory assumes that the stronger the collective conscience, the less crime in that community.

College boy: A type of lower class male youth, identified by Cohen, who has experienced the same strains and status frustration as his peers but responds to his disadvantaged situation by dedicating himself to overcoming the odds and competing in the middle-class schools despite the unlikely chances for success.

Communitarianism: A concept in Braithwaite's theory of reintegration, which is a macrolevel measure of the degree to which individuals are connected or interdependent on mainstream society (via organizations or groups).

Concentric circles theory: A model proposed by Chicago School theorists; assumes that all cities grow in a natural way that universally has the same five zones (or circles or areas). For example, all cities have a central Zone I, which contains basic government buildings, as well as a Zone II, which was once residential but is being invaded by factories. The outer three zones involve various forms of residential areas.

Conceptual integration: A type of theoretical integration in which a theoretical perspective consumes or uses concepts from many other theoretical models.

Concordance rates: Rates at which twin pairs either share a trait (e.g., criminality) or the lack of the trait; for example, either both twins are criminal or neither is criminal; discordant would be if one of the pair is criminal and the other is not.

Conflict gangs: A type of gang identified by Cloward and Ohlin that tends to develop in neighborhoods with weak stability and little or no organization; gangs are typically relatively disorganized and lack the skills and knowledge to make a profit through criminal activity. Thus, their primary illegal activity is violence, which is used to gain prominence and respect among themselves and the neighborhood.

Conflict perspective: Theories of criminal behavior that assume that most people disagree on what the law should be and that law is used as a tool by those in power to keep down other groups.

Conflict theory: See *conflict perspective*.

Conformity: In strain theory, an adaptation in which an individual buys into the conventional means of success and also buys into conventional goals.

Consensual perspective: Theories that assume virtually everyone is in agreement on the laws and therefore assumes no conflict in attitudes regarding the laws and rules of society.

Containment theory: A control theory proposed by Reckless in the 1960s, which presented a model that emphasized internal and social pressures to commit crime, which range from personality predispositions to peer influences, as well as internal and external constraints, ranging from personal self-control to parental control, that determine whether an individual will engage in criminal activity. This theory is often criticized as being too vague or general, but it advanced criminological theory by providing a framework in which many internal and external factors were emphasized.

Control-balance theory: An integrated theory originally presented by Tittle, which assumes that the amount of control to which one is subjected, as compared to the amount of control that one can exercise, determines the probability of deviant behavior and the types of deviance that are committed by that individual. In other words, the balance or imbalance between these two types of control can predict the amount and type of behavior likely to be committed.

Control theories: A group of theories of criminal behavior that emphasize the assumption that humans are born selfish and have tendencies to be aggressive and offend and that individuals must be controlled, typically by socialization and discipline or from internalized self-control that has been developed in their upbringing.

Corner boy: A type of lower class male youth, identified by Cohen, who has experienced the same strains and status frustration as others but responds to his disadvantaged situation by accepting his place in society as someone who will somewhat passively make the best of life at the bottom of the social order. As the label describes, they often hang out on corners.

Correlation: A criterion of causality that requires a change in a predictor variable (X) to be consistently associated with some change (either positive or negative) in the explanatory variable (Y). An example would be unemployment (X) being related to criminal activity (Y).

Covariation: See *correlation*.

Craniometry: The field of study that emphasized the belief that the size of the brain or skull reflected superiority or inferiority, with larger brains and skulls being considered superior.

Criminal gangs: A type of gang identified by Cloward and Ohlin that forms in lower class neighborhoods with an organized structure of adult criminal behavior. Such neighborhoods are so organized and stable that their criminal networks are often known and accepted by conventional citizens. In these neighborhoods, adult gangsters mentor neighborhood youths and take them under their wings. Such gangs tend to be highly organized and stable.

Criminology: The scientific study of crime and the reasons why people engage (or don't engage) in criminal behavior, as well as the study of why certain trends occur or groups of people seem to engage in criminal behavior more than others.

Critical feminism: A perspective of feminist theory that emphasizes the idea that many societies (such as the United States) are based on a structure of patriarchy, in which males dominate virtually every aspect of society, such as law, politics, family structure, and the economy.

Cross-sectional studies: A form of research design model in which a collection of data is taken at one point in time (often in survey format).

Cultural and subcultural theory: A perspective of criminal offending that assumes that many offenders believe in a normative system distinctly different than, and often at odds with, the norms accepted by conventional society.

Cytogenetic studies: Studies of crime that focus on the genetic makeups of individuals, with a specific focus on abnormalities in chromosomal makeups. An example is XYY instead of the normal XX (females) and normal XY (males).

Dark figure: The vast majority of major crime incidents that never get reported to police due to the failure of victims to file a police report; covers most criminal offending, with the exception of homicide and motor vehicle theft, which are almost always reported to police.

Decriminalization: A policy related to labeling theory, which proposes less harsh punishments for some minor offenses, such as the possession of small amounts of marijuana; for example, in California, an offender gets a ticket or fine for this offense rather than being officially charged and prosecuted.

Deinstitutionalization: A policy related to labeling theory; proposes that juveniles or those accused of relatively minor offenses should not be locked up in jail or prison.

Delinquent boy: A type of lower class male youth, identified by Cohen, who responds to strains and status frustration by joining with similar others in a group to commit crime.

Determinism: The assumption that human behavior is caused by factors outside of free will and rational decision making (e.g., biology, peer influence, poverty, bad parenting); it is the distinctive, primary assumption of positivism (as opposed to the Classical School of criminological theory, which assumes free will and free choice).

Deterrence theory: The theory of crime associated with the Classical School, which proposes that individuals will make rational decisions regarding their behavior. This theory focuses on three concepts: the individual's perception of (1) certainty of punishment, (2) severity of punishment, and (3) the swiftness of punishment.

Developmental theories: Perspectives of criminal behavior, which are also to some extent integrated but are distinguished by their emphasis on the evolution of individuals' criminality over time. Specifically, developmental theories tend to look at the individual as the unit of analysis, and such models focus on the various aspects of the onset, frequency, intensity, duration, desistence, and other aspects of individuals' criminal careers.

Deviance: Behaviors that are not normal; includes many illegal acts, as well as activities that are not necessarily criminal but are unusual and often violate social norms, such as burping loudly at a formal dinner or wearing inappropriate clothing.

Differential association theory: A theory of criminal behavior that emphasizes the association with significant others (peers, parents, etc.) in learning criminal behavior. This theory was originally presented by Sutherland.

Differential identification theory: A theory of criminal behavior that is very similar to differential association theory; the major difference is that differential identification theory takes into account associations with persons and images that are presented in the media (e.g., movies, TV, sports, etc.). This model was originally proposed by Glaser.

Differential reinforcement theory: A theory of criminal behavior that emphasizes various types of social learning, specifically classical conditioning, operant conditioning, and imitation or modeling. This theory was originally presented by Burgess and Akers and is one of the most supported theories according to empirical studies.

Diversion: A set of policies related to labeling theory that attempt to get an offender out of the formal justice system as quickly as possible; an offender might perform a service or enter a rehabilitation program instead of serving time in jail or prison. Often, if an offender successfully completes the contract, the official charge or conviction for the crime is expunged or eliminated from the official record.

Dizygotic twins: Also referred to as fraternal or nonidentical twins, these are twin pairs that come from two separate eggs (zygotes) and thus share only 50% of the genetic makeup that can vary.

Dopamine: A neurotransmitter that is largely responsible for good feelings in the brain; it is increased by many illicit drugs (e.g., cocaine).

Dramatization of evil: A concept proposed by Tannenbaum in relation to labeling theory, which states that, often when relatively minor laws are broken, the community tends to overreact and make a big deal out of it (dramatizing it). A good example is when a very young offender sprays graffiti on a street sign, and the neighborhood ostracizes that youth.

Drift theory: A theory of criminal behavior in which the lack of social controls in the teenage years allows for individuals to experiment in various criminal offending, often due to peer influence, without the individuals buying into a criminal lifestyle; this theory was introduced by David Matza.

Ecological School: See *Chicago School of criminology*.

Ectoderm: A medical term for the outer layer of tissue in our bodies (e.g., skin, nervous system).

Ectomorphic: The type of body shape associated with an emphasis on the outer layer of tissue (ectoderm) during development; these people are disposed to thinness.

Ego: The only conscious domain of the psyche; according to Freud, it functions to mediate the battle between id and superego.

Elliott's integrated model: Perhaps the first major integrated perspective proposed that clearly attempted to merge various traditionally separate theories of crime. Elliott's integrated framework attempts to merge strain, social disorganization, control, and social learning or differential association-reinforcement perspectives for the purpose of explaining delinquency, particularly in terms of drug use, as well as other forms of deviant behavior.

Empirical validity: Refers to the extent to which a theoretical model is supported by scientific research. In criminology, empirical research has consistently supported a number of theories and consistently refuted others.

Endoderm: A medical term for the inner layer of tissue in our bodies (e.g., digestive organs).

Endomorphic: The type of body shape associated with an emphasis on the inner layer of tissue (endoderm) during development; these people are disposed to obesity.

End-to-end theoretical integration: A type of theoretical integration that conveys the linkage of the theories based on the temporal ordering of two or more theories in their causal timing. This means that one theory (or concepts from one theory) precedes another theory (or concepts from another theory) in terms of causal ordering or timing.

Equivalency hypothesis: A mirror image tendency, which is the observed phenomenon that virtually all studies have shown; the characteristics, such as young, male, urban, poor, or minority, tend to have the highest rates of criminal offending and the highest rates of victimization. This hypothesis has one important exception, namely lower class offenders tend to have higher rates of theft against middle- to upper-class households/individuals.

Eugenics: The study of and policies related to the improvement of the human race via discriminatory control over reproduction.

Experiential effect: The extent to which individuals' previous experiences have effects on their perceptions of how certain or severe criminal punishment will be when they are deciding whether or not to offend again.

Family studies: Studies that examine the clustering of criminality in a given family.

Feeblemindedness: A technical, scientific term in the early 1900s meaning those who had significantly below average levels of intelligence.

Focal concerns: The primary concept of Walter Miller's theory, which asserts that all members of the lower class focus on a number of concepts they deem important: fate, autonomy, trouble, toughness, excitement, and smartness.

Formal controls: Factors that involve the official aspects of criminal justice, such as police, courts, and corrections (e.g., prisons, parole, probation).

Frontal lobes: A region of the brain that is, as its name suggests, located in the frontal portion of the brain; most of the executive functions of the brain, such as problem solving, take place here, so it is perhaps the most vital portion of the brain and what makes us human.

General deterrence: Punishments given to an individual are meant to prevent or deter other potential offenders from engaging in such criminal activity in the future.

General strain theory: Although derived from traditional strain theory, this theoretical framework assumes that people of all social classes and economic positions deal with frustrations in routine daily life, to which virtually everyone can relate; includes more sources of strain than traditional strain theory.

Hot spots: Specific locations, such as businesses, residences, or parks, that experience high concentrations of crime incidents; a key concept in routine activities theory.

Hypotheses: Specific predictions that are based on a scientific theoretical framework and tested via observation.

Id: A subconscious domain of the psyche, according to Freud, with which we are all born; it is responsible for our innate desires and drives (such as libido [sex drive]) and it battles the moral conscience of the superego.

Imitation–modeling: A learning model that emphasizes that humans (and other animals) learn behavior simply by observing others; this model was most notably proposed by Bandura.

Index offenses: Also known as Part I offenses, according to the FBI Uniform Crime Report, these are eight common offenses: murder, forcible rape, aggravated assault, robbery, burglary, motor vehicle theft, larceny, and arson. All reports of these crimes, even when they do not result in an arrest, are recorded to estimate crime in the nation and various states and regions.

Informal controls: Factors like family, church, or friends that do not involve official aspects of criminal justice, such as police, courts, and corrections (e.g., prisons).

Innovation: In strain theory, an adaptation to strain in which an individual buys into the conventional goals of success but does not buy into the conventional means for getting to the goals.

Integrated theories: Theories that combine two or more traditional theories into one combined model.

Interdependency: A concept in Braithwaite's theory of reintegration, which is a microlevel measure of the degree to which the individuals are connected or interdependent on mainstream society (via organizations or groups).

Interracial: When an occurrence (such as a crime event) involves people of different races or ethnicities, such as a White person committing crime against a Black person.

Intraracial: When an occurrence (such as a crime event) involves people of the same race or ethnicity, such as a White person committing crime against another White person.

Labeling theory: A theoretical perspective that assumes that criminal behavior increases because certain individuals are caught and labeled as offenders; their offending increases because they have been stigmatized as offenders. Most versions of this perspective also assume that certain people (e.g., lower class or minorities) are more likely to be caught and punished. Another assumption of most versions is that, if such labeling did not occur, the behavior would stop; this assumption led to numerous policy implications, such as diversion, decriminalization, and deinstitutionalization.

Learning theories: Theoretical models that assume criminal behavior of individuals is due to a process of learning from others the motivations and techniques for engaging in such behavior. Virtually all of the variations of the learning theory, or perspective, propose that the processes involved in a person learning how and why to commit crimes are the same as those involved in learning to engage in conventional activities (e.g., riding a bike, playing basketball).

Legalistic approach: A way of defining behaviors as crime; includes only acts that are specifically against the legal codes of a given jurisdiction. The problem with such a definition is that what is a crime in one jurisdiction is not necessarily a crime in other jurisdictions.

Liberal feminism: One of the areas of feminist theories of crime that emphasizes the assumption that differences between males and females in offending were due to the lack of female opportunities in education, employment, etc., as compared to males.

Life-course persistent offenders: A type of offender, as labeled by Moffitt's developmental theory; such people start offending early and persist in offending through adulthood.

Logical consistency: Refers to the extent to which concepts and propositions of a theoretical model make sense in terms of both face value and regarding the extent to which the model is consistent with what is readily known about crime rates and trends.

Low self-control theory: A theory that proposes that individuals either develop self-control by the time they are about age 10 or do not. Those who do not develop self-control will manifest criminal or deviant behaviors throughout life. This perspective was originally proposed by Gottfredson and Hirschi.

Macrolevel of analysis: Theories that focus on group or aggregated scores and measures as the unit of analysis, as opposed to individual rates.

Mala in se: Acts that are considered inherently evil and that virtually all societies consider to be serious crimes; an example is murder.

Mala prohibita: The many acts that are considered crimes primarily because they have been declared bad by the legal codes in that jurisdiction; in other places and times, they are not illegal; examples are gambling, prostitution, and drug usage.

Marxist feminism: A perspective of crime that emphasizes men's ownership and control of the means of economic production; similar to critical or radical feminism but distinguished by its reliance on the sole concept of economic structure, in accordance with Marx's theory.

Mechanical societies: In Durkheim's theory, these societies were rather primitive with a simple distribution of labor (e.g., hunters and gatherers) and thus a high level of agreement regarding social norms and rules because nearly everyone is engaged in the same roles.

Mens rea: In legal terms, this means *guilty mind* or intent. This concept involves whether offenders actually knew what they were doing and meant to do it.

Mesoderm: A medical term for the middle layer of tissue in our bodies (e.g., muscles, tendons, bone structure).

Mesomorphic: The type of body shape associated with an emphasis on the middle layer of tissue (mesoderm) during development; these people are disposed to athleticism or strength.

Microlevel of analysis: Theories that focus on individual scores and measures as the unit of analysis, as opposed to group or aggregate rates.

Minor physical anomalies (MPAs): Physical features, such as asymmetrical or low-seated ears, which are believed to indicate developmental problems, typically problems in the prenatal stage in the womb.

Modeling and imitation: A major factor in differential reinforcement theory, which proposes that much social learning takes place via imitation or modeling of behavior, for example, when adults or parents say bad words, their children begin using those words; a primary, early proponent of this model of learning was Bandura.

Moffitt's developmental theory (taxonomy): A theoretical perspective proposed by Moffitt, in which criminal behavior is believed to be caused by two different causal paths: (1) adolescence-limited offenders commit their crimes during teenage years due to peer pressure; and (2) life-course persistent offenders commit antisocial behavior throughout life, starting very early and continuing on throughout their lives, because of the interaction between their neuropsychological deficits and criminogenic environments in their upbringing.

Monozygotic twins: Also referred to as identical twins, these are twin pairs that come from a single egg (zygote) and thus share 100% of their genetic makeup.

National Crime Victimization Survey (NCVS): One of the primary measures of crime in the United States; collected by the DOJ and the Census Bureau, based on interviews with victims of crime. This measure started in the early 1970s.

Natural areas: The Chicago School's idea that all cities naturally contain identifiable clusters, such as a Chinatown or Little Italy, and neighborhoods that have low or high crime rates.

Negative punishment: A concept in social learning in which people are given a punishment by removing something that they enjoy or like (e.g., taking away driving privileges for a teenager).

Negative reinforcement: A concept in social learning in which people are given a reward by removing something that they dislike (e.g., not being on curfew or not having to do their chores).

Neoclassical School: Neoclassical School of criminological theory is virtually identical to the Classical School (both assume free will, rationality, social contract, deterrence, etc.), except that it assumes that aggravating and mitigating circumstances should be taken into account for purposes of sentencing and punishing an offender.

Neurotransmitters: Nervous system chemicals in the brain and body that help transmit electric signals from one neuron to another, thus allowing healthy communication in the brain and to the body.

Neutralization theory: A theory of criminal behavior that emphasizes the excuses or neutralization techniques that are used by offenders to alleviate the guilt of (or to excuse) their behavior when they know that their behavior is immoral; this theory was originally presented by Gresham Sykes and David Matza. In their theory, they presented five key techniques, ways that offenders alleviate their guilt or excuse their behavior that they know is wrong; because they presented this idea in the 1960s, other techniques have been added, especially regarding white-collar crime.

Nonindex offenses: Also known as Part II offenses, more than two dozen crimes that are considered relatively less serious than index crimes and must result in an arrest to be recorded by the FBI; therefore, the data on such results are far less reliable because the vast majority of reports of these crimes do not result in an arrest and are not in the annual FBI report.

Operant conditioning: The learning model that takes place in organisms (such as humans), based on the association between an action and feedback that occurs after it has taken place; for example, a rat running a maze can be trained to run the maze faster based on rewards (reinforcement), such as cheese, as well as punishments, such as electric shocks; introduced and promoted by B. F. Skinner.

Organic societies: In the Durkheimian model, those societies that have a high division of labor and thus a low level of agreement about societal norms, largely because everyone has such different roles in society, leading to very different attitudes about the rules and norms of behavior.

Paradigm: A unique perspective of a phenomenon; has an essential set of assumptions that significantly oppose those of other existing paradigms or explanations of the same phenomenon.

Parsimony: Essentially means *simple;* a characteristic of a good theory, meaning that it explains a certain phenomenon, in our case criminal behavior, with the fewest possible propositions or concepts.

Phenotype: An observed manifestation of the interaction of genotypical traits with the environment, such as height (which depends largely on genetic disposition and diet, as exhibited by Asians or Mexicans who are raised in the United States and thus grow taller).

Phrenology: The science of determining human dispositions based on distinctions (e.g., bumps) in the skull, which is believed to conform to the shape of the brain.

Physiognomy: The study of facial and other bodily aspects to identify developmental problems, such as criminality.

Pluralist (conflict) perspective: A theoretical assumption that, instead of one dominant and other inferior groups in society, there are a variety of groups that lobby and compete to influence changes in law; most linked to Vold's theoretical model.

Policy implications: The extent to which a theory can be used to inform authorities about how to address a given phenomenon; in this case, ways to help law enforcement, court, and prison officials reduce crime and recidivism.

Positive punishment: A concept in social learning in which an individual is given a punishment by doing something they dislike (e.g., spanking, time-out, grounding, etc.).

Positive reinforcement: A concept in social learning in which an individual is given a reward by providing something they like (e.g., money, extending curfew, etc.).

Positive School: The Positive School of criminological theory is a perspective that assumes individuals have no free will to control their behavior. Rather, the theories of the Positive School assume that criminal behavior is determined by factors outside of free choices made by the individual, such as peers, bad parenting, poverty, or biology.

Postmodern feminism: A perspective that says women as a group cannot be understood, even by other women, because every person's experience is unique; therefore, there is no need to measure or research such experiences. This perspective has been criticized as being anti-science and, thus, irrelevant; from this perspective, there is no need to investigate any case because the findings cannot be generalized beyond the specific individual.

Power-control theory: An integrated theory of crime that assumes that, in households where the mother and father have relatively similar levels of power at work (i.e., balanced households), mothers will be less likely to exert control on their daughters. These balanced households will be less likely to experience gender differences in the criminal offending of the children. However, households in which mothers and fathers have dissimilar levels of power in the workplace (i.e., unbalanced households) are more likely to suppress criminal activity in daughters, but more criminal activity is likely in the boys of the household.

Primary deviance: A concept in labeling theory originally presented by Lemert; the type of minor, infrequent offending that people commit before they are caught and labeled as offenders. Most normal individuals commit this type of offending due to peer pressure and normal social behavior in their teenage years.

Proletariat: In Marx's conflict theory, the proletariat is the oppressed group of workers who are exploited by the bourgeoisie, an elite class that owns the means of production; according to Marx, the proletariat will never truly profit from their efforts because the upper class owns and controls the means of production.

Radical feminism: See *critical feminism*.

Rational choice theory: A modern, Classical School-based framework for explaining crime that includes the traditional formal deterrence aspects, such as police, courts, and corrections, and adds other informal factors that studies show consistently and strongly influence behavior, specifically informal deterrence factors (such as friends, family, community, etc.) and also the benefits of offending, whether they be monetary, peer status, or physiological (the rush of engaging in deviance).

Reaction formation: A Freudian defense mechanism applied to Cohen's theory of youth offending, which involves adopting attitudes or committing behaviors that are opposite of what is expected, for example, by engaging in malicious behavior as a form of defiance; youths buy into this antinormative belief system so that they will feel less guilt for not living up to the standards they are failing to achieve and so they can achieve status among their delinquent peers.

Rebellion: In strain theory, an adaptation to strain in which an individual buys into the idea of conventional means and goals of success but does not buy into the current conventional means or goals.

Reintegrative shaming theory: See *Braithwaite's reintegrative shaming theory*.

Relative deprivation: The perception that results when relatively poor people live in close proximity to relatively wealthy people. This concept is distinct from poverty in the sense that a poor area could mean that nearly everyone is poor, but relative deprivation inherently suggests that there is a notable amount of wealth and poor in a given area.

Retreatism: In strain theory, an adaptation to strain in which an individual does not buy into the conventional goals of success and also does not buy into the conventional means.

Retreatist gangs: A type of gang identified by Cloward and Ohlin that tends to attract individuals who have failed to succeed in both the conventional world and the criminal or conflict gangs of their neighborhoods. Members of retreatist gangs are no good at making a profit from crime, nor are they good at using violence to achieve status, so the primary form of offending in retreatist gangs is usually drug usage.

Ritualism: In strain theory, an adaptation to strain in which an individual buys into the conventional means of success (e.g., work, school, etc.) but does not buy into the conventional goals.

Routine activities theory (lifestyle theory): An explanation of crime that assumes that most crimes are committed during the normal daily activities of people's lives; it assumes that crime and victimization are highest in places where three factors come together in time and place: motivated offenders, suitable or attractive targets, and absence of a guardian; this perspective assumes a rational offender who picks targets due to opportunity.

Scenario (vignette) research: Studies that involve providing participants with specific hypothetical scenarios and then asking them what they would do in each situation; typically, they are also asked about their perceptions of punishment and other factors related to each particular situation.

Scientific method: The method used in all scientific fields to determine the most objective results and conclusions regarding empirical observations. This method involves testing hypotheses via observation and data collection and then making conclusions based on the findings.

Scope: Refers to the range of criminal behavior that a theory attempts to explain, which in our case, can be seen as the amount of criminal activity a theory can account for, such as only violent crime, or only property crime, or only drug usage; if a theory has a very large scope, it would attempt to explain all types of offending.

Secondary deviance: A concept in labeling theory originally presented by Lemert; the type of more serious, frequent offending that people commit after they get caught and are labeled offenders. Individuals commit this type of offending because they have internalized their status as offenders and often have resorted to hanging out with other offenders.

Selective placement: A criticism of adoption studies, arguing that adoptees tend to be placed in households that resemble that of their biological parents; thus, adoptees from rich biological parents are placed in rich adoptive households.

Self-report data (SRD): One of the primary ways that crime data are collected, typically by asking offenders about their own offending; the most useful for examining key causal factors in explaining crime (e.g., personality, attitudes, etc.).

Serotonin: A neurotransmitter that is key in information processing and most consistently linked to criminal behavior in its deficiency; low levels are linked to depression and other mental illnesses.

Severity of punishment: One of the key elements of deterrence; the assumption is that a given punishment must be serious enough to outweigh any potential benefits gained from a crime (but not too severe so that it causes people to commit far more severe offenses to avoid apprehension); in other words, this theoretical concept advises graded penalties that increase as the offender recidivates or reoffends.

Side-by-side (or horizontal) integration: A type of theoretical integration in which cases are classified by a certain criteria (e.g., impulsive versus planned) and two or more theories are considered parallel explanations based on what type of case is being considered. Thus, there are two different paths in which a case is predicted to go, typically based on an initial variable (such as low or high self-control).

Social bonding theory: A control theory proposed by Hirschi in 1969 that assumes that individuals are predisposed to commit crime and that conventional bonds that are formed prevent or reduce their offending. This bond is made up of four constructs: attachments, commitment, involvement, and moral beliefs regarding committing crime.

Social contract: An Enlightenment ideal or assumption that stipulates there is an unspecified arrangement among citizens of a society in which they promise the state or government not to commit offenses against other citizens (to follow the rules of a society), and in turn, they gain protection from being violated by other citizens; violators will be punished.

Social Darwinism: The belief that only the beneficial (or fittest) societal institutions or groups of people survive or thrive in society.

Social disorganization: See *Chicago School of criminology*.

Social dynamics: A concept proposed by Comte that describes aspects of social life that alter how societies are structured and that pattern the development of societal institutions.

Socialist feminism: Feminist theories that moved away from economic structure (e.g., Marxism) as the primary detriment for females and placed a focus on control of reproductive systems. This model believes that women should take control of their own bodies and their reproductive functions via contraceptives.

Social sciences: A category of scientific disciplines or fields of study that focus on various aspects of human behavior, such as criminology, psychology, economics, sociology, or anthropology; they typically use the scientific method for gaining knowledge.

Social statics: A concept proposed by Comte to describe aspects of society that relate to stability and social order, which allow societies to continue and endure.

Soft determinism: The assumption that both determinism (the fundamental assumption of the Positive School of criminology) and free will or free choice (the fundamental assumption of the Classical School) play roles in offenders' decisions to engage in criminal behavior. This perspective can be seen as a type of compromise or middle-road concept.

Somatotyping: The area of study, primarily linked to William Sheldon, that links body type to risk for delinquent and criminal behavior. Also, as a methodology, it is a way of ranking body types based on three categories: *endomorphy, mesomorphy,* and *ectomorphy* (see related entries).

Somotonic: The type of temperament or personality associated with a mesomorphic (muscular) body type; these people tend to be risk taking and aggressive.

Specific deterrence: Punishments given to an individual are meant to prevent or deter that particular individual from committing crime in the future.

Spuriousness: When other factors (often referred to as Z factors) are actually causing two variables (X and Y) to occur at the same time; it may appear as if X causes Y, when in fact they are both being caused by other Z factor(s). To account for spuriousness, which is required for determining causality, researchers must ensure that no other factors are causing the observed correlation between X and Y. An example is when ice cream sales (X) are related to crime rates (Y); the Z variable is warm weather, which increases the opportunity for crime because more people and offenders are interacting.

Stake in conformity: A significant portion of Toby's control theory, which applies to virtually all control theories and refers to the extent to which individuals have investments in conventional society. It is believed, and supported by empirical studies, that the higher the stake in conformity an individual has, the less likely he or she will engage in criminal offending.

Stigmata: The physical manifestations of atavism (biological inferiority), according to Lombroso; he claimed that if a person had more than five, he or she was a born criminal, meaning that a person (or feature of an individual) is a throwback to an earlier stage of evolutionary development and inevitably would be a chronic offender. Examples are very large ears or very small ears.

Strain theory: A category of theories of criminal behavior in which the emphasis is placed on a sense of frustration (e.g., economy) in crime causation, hence, the name *strain* theories.

Subterranean values: Those norms that individuals have been socialized to accept (e.g., violence) in certain contexts in a given society; an example would be the popularity of boxing or Ultimate Fighting Championship events in American society, even though violence is generally viewed negatively. Another example is the romanticized nature and popularity of crime movies, such as *The Godfather* and *Pulp Fiction*.

Superego: A subconscious domain of the psyche, according to Freud; it is not part of our nature but must be developed through early social attachments; it is responsible for our morality and conscience; it battles the subconscious drives of the id.

Swiftness of punishment: Swiftness of punishment is one of the key elements of deterrence; the assumption is that the faster punishment occurs after a crime is committed, the more an individual will be deterred in the future.

Tabula rasa: The assumption that when people are born, they have a blank slate regarding morality and that every portion of their ethical or moral beliefs is determined by the interactions that occur in the way they are raised and socialized. This is a key assumption of virtually all learning theories.

Techniques of neutralization: See *neutralization theory.*

Telescoping: The human tendency in which events are perceived to occur much more recently in the past than they actually did, causing estimates of crime events to be overreported; in the NCVS measure asking respondents to estimate their victimization over the last 6 months, respondents often report victimization that happened before the 6 month cutoff date.

Temporal lobes: The region of the brain located above our ears, which is responsible for a variety of functions and is located right above many primary limbic structures that govern our emotional and memory functions.

Temporal ordering: The criterion for determining causality; requires that the predictor variable (X) precedes the explanatory variable (Y) in time.

Testability: Refers to the extent that a theoretical model can be empirically or scientifically tested through observation and empirical research.

Theoretical elaboration: A form of theoretical integration that uses a traditional theory as the framework for the theoretical model but also adds concepts or propositions from other theories.

Theoretical reduction: See *up-and-down integration*.

Theory: A set of concepts linked together by a series of propositions in an organized way to explain a phenomenon.

Thornberry's interactional model: This integrated model of crime was the first major perspective to emphasize reciprocal, or feedback, effects in the causal modeling of the theoretical framework.

Trajectory: A path that someone takes in life, often due to life transitions (see *transitions*).

Transitions: Events that are important in altering trajectories toward or against crime, such as marriage or employment.

Twins separated at birth studies: Studies that examine the similarities between identical twins who are separated in infancy; research indicates that such twins are often extremely similar even though they grew up in completely different environments.

Twin studies: Studies that examine the relative concordance rates for monozygotic versus dizygotic twins, with virtually every study showing that identical twins (monozygotic) tend to be far more concordant for criminality than fraternal (dizygotic) twins.

Uniform Crime Report (UCR): An annual report published by the FBI in the DOJ, which is meant to estimate most of the major street crimes in the United States. It is based on police reports and arrests throughout the nation and started in the 1930s.

Up-and-down integration: A type of theoretical integration that is generally considered the classic form of theoretical integration because it has been done relatively often in the history of criminological theory development. This often involves increasing the level of abstraction of a single theory so that postulates seem to follow from a conceptually broader theory, such as differential reinforcement theory assuming virtually all of the concepts and assumptions of differential association theory.

Utilitarianism: A philosophical concept that is often applied to social policies of the Classical School of criminology, which relates to the idea of the greatest good for the greatest number.

Viscerotonic: The type of temperament or personality associated with an endomorphic (obese) body type; these people tend to be jolly, lazy, and happy-go-lucky.

Zone in transition: In the Chicago School or social disorganization theory, this zone (labeled Zone II) was once residential but is becoming more industrial because it is being invaded by the factories; this area of a city tends to have the highest crime rates due to the chaotic effect that the invasion of factories has had on the area.

Credits and Sources

Chapter 1

Image 1.1: © 2009 Jupiterimages Corporation.

Image 1.2: © 2009 Jupiterimages Corporation.

Image 1.3: © James P. Blair/Getty Images.

Image 1.4: © Getty Images.

Figure 1.1: National Center for Health Statistics (2002). *Vital Statistics.* Washington, DC: U.S. Department of Health and Human Services.

Chapter 2

Image 2.1: © Getty Images.

Image 2.2: © Getty Images.

Image 2.3: © 2009 Jupiterimages Corporation.

Chapter 3

Figure 3.1: Created by Stephen G. Tibbetts.

Chapter 4

Image 4.1: © 2009 Jupiterimages Corporation.

Image 4.2: © Getty Images.

Image 4.3: © 2009 Jupiterimages Corporation.

Images 4.4a, b, c: Created by Sandy Sauvajot.

Chapter 5

Image 5.1: © 2009 Jupiterimages Corporation.

Figure 5.1: Based on figure presented by Anthony Walsh in "Genetic and Cytogenetic Intersex Anomalies," *International Journal of Offender Therapy and Comparative Criminology,* 39 (1995), 162.

Image 5.2: © AP Photo/M. Spencer Green.

Chapter 6

Figure 6.1: Created by Stephen G. Tibbetts.

Figure 6.2: Created by Stephen G. Tibbetts.

Chapter 7

Figure 7.1: Created by Stephen G. Tibbetts.

Image 7.1: Clifford Shaw and Henry McKay, *Juvenile Delinquency in Urban Areas* (Chicago: University of Chicago Press, 1972), 69. Copyright © The University of Chicago Press.

Chapter 8

Image 8.1: Used with permission of the American Sociological Association. www.asanet.org

Figure 8.1: Based on figure presented by Richard Tremblay and D. LeMarquand in "Individual Risk and Protective Factors," in *Child Delinquents: Development, Intervention, and Service Needs,* eds. Rolf Loeber and David Farrington (London: Sage, 2001), 154.

Figure 8.2: Created by Stephen G. Tibbetts.

Figure 8.3: Created by Stephen G. Tibbetts.

Figure 8.4: Created by Stephen G. Tibbetts.

Figure 8.5: Created by Stephen G. Tibbetts.

Figure 8.6: Created by Stephen G. Tibbetts.

Chapter 9

Image 9.1: © 2009 Jupiterimages Corporation.

Image 9.2: © 2009 Jupiterimages Corporation.

Chapter 10

Figure 10.1: Created by Stephen G. Tibbetts.

Image 10.1: © 2009 Jupiterimages Corporation.

Chapter 11

Table 11.1: From Frank S. Pearson and Neil Alan Weiner, "Toward an Integration of Criminological Theories," *Journal of Criminal Law and Criminology 76* (1985): 116–50 (Table, 130). © 2004–2008 *The Journal of Criminal Law and Criminology*—Northwestern University School of Law.

Figure 11.1: Delbert Elliott, David Huizinga, and Suzanne Ageton, *Explaining Delinquency and Drug Use* (Thousand Oaks, CA: Sage, 1985), 66.

Chapter 12

Image 12.1: © AP/Worldwide Photos.

Index

About the Author

Stephen G. Tibbetts is a professor in the Department of Criminal Justice at California State University, San Bernardino. He earned his undergraduate degree in criminology and law (with high honors) from the University of Florida and his masters and doctorate degrees from the University of Maryland, College Park. For more than a decade, he worked as an officer of the court (juvenile) in both Washington County, Tennessee, and San Bernardino County, California, providing recommendations for disposing numerous juvenile court cases. He has published more than 40 scholarly publications in scientific journals (including *Criminology, Justice Quarterly, Journal of Research in Crime and Delinquency, Journal of Criminal Justice* and *Criminal Justice & Behavior*), as well as seven books, all examining various topics regarding criminal offending and policies to reduce such behavior. One of these books, *American Youth Gangs at the Millennium*, was given a Choice award by the American Library Association as an Outstanding Academic Title. He received the Outstanding Professional Development Award from the College of Social and Behavioral Sciences at California State University, San Bernardino, in 2009. One of his most recent books, *Criminals in the Making: Criminality Across the Life Course* (Sage, 2008), was recently lauded by *The Chronicle of Higher Education* as one of the key scholarly publications in advancing the study of biosocial criminology.